PHYSICS
LABORATORY MANUAL

PHYSICS LABORATORY MANUAL

THIRD EDITION

Clifford N. Wall, *Ph. D.*

Professor of Physics, Emeritus, University of Minnesota

Raphael B. Levine, *Ph. D.*

Director, Metropolitan Atlanta Council for Health

Fritjof E. Christensen, *M. A.*

Associate Professor of Physics, St. Olaf College

PRENTICE-HALL, INC.

Englewood Cliffs, New Jersey

10 9 8 7 6 5 4 3 2 1

ISBN: 0-13-674101-0

Printed in the United States of America.

PRENTICE-HALL INTERNATIONAL, INC., *London*
PRENTICE-HALL OF AUSTRALIA, PTY. LTD., *Sydney*
PRENTICE-HALL OF CANADA, LTD., *Toronto*
PRENTICE-HALL OF INDIA PRIVATE LIMITED, *New Delhi*
PRENTICE-HALL OF JAPAN, INC., *Tokyo*

Contents

vi CONTENTS

EXPERIMENT

APPENDIX III

TABLES

Preface

The intent and form of the original manual is retained in this revision. But the content and treatment of subject matter have been changed to keep pace with some of the noteworthy developments in physics in the past decade. Material that now seems peripheral or outmoded has been replaced by matter of more significance. New and better laboratory equipment has become available as well as modern computer facilities. These all require some modification of laboratory procedure and method. The choice of what is to go, what is to remain, and what is to be added is a matter of judgment and experience. In this respect the authors (now three instead of two) prefer to take a moderate rather than an extreme position.

The major changes in this revision are as follows:

1. The treatment of errors has been enlarged to include both the normal and the Poisson distribution functions, as well as the method of least squares.

2. Substantial changes have been made in five mechanics experiments. Experiments 1 and 2 in the revision are new ones on random fluctuations; the old ones on density and equilibrium have been deleted. Experiment 3 on two-body collisions has been extended to include the linear air track. Experiment 4 on the acceleration of a falling body now includes a section on the method of least squares in the analysis of data. A complete computer program in Fortran IV for this section is given in Appendix I. Experiment 6 on centripetal force in uniform circular motion now includes a second part (B) on the stability of the motion.

3. The experiments in heat remain the same except for the deletion of the experiment on linear expansion and a significant change in experiment 21 on the gas thermometer. Here the new 1954 standard fixed point of the Kelvin temperature scale, i.e., the triple point of water, is introduced and used in defining the ideal-gas temperature.

4. Major changes have been made in many of the experiments on electricity and magnetism. The use of the oscilloscope has been greatly extended. New experiments on resonance, transistors, and diode power supplies have been introduced. The experiments on thermionic emission and the magnetron have been rewritten for the Ferranti GRD7 tube in place of the FP400 tube which is no longer available. The Child-Langmuir experiment has been deleted.

5. The experiments in light have been modified by replacing the photometer experiment with a new experiment on single and double slit diffraction using a gas laser (He-Ne) as a source. Also experiment 77 on the photoelectric tube has been completely revised for the measurement of h/e.

6. In atomic physics the Franck-Hertz experiment has been added.

7. Appendix I now contains a short treatise on the method of least squares including a suitable computer program for experiment 4. In the same appendix there is a derivation of Poisson's distribution function. Appendix II now includes a note on transistors. The physical tables in Appendix III have been updated and a short table of values of the normal distribution functions has been added.

The authors express their gratitude to the many users of the manual who have made helpful suggestions. The senior author is especially indebted to the laboratory coordinators at the University of Minnesota who have always acted as a clearing house for information about the manual.

PHYSICS
LABORATORY MANUAL

Introduction

PURPOSE OF LABORATORY WORK

Physics is both a body of knowledge and a method of inquiry. The knowledge rests upon a firm foundation of massive physical facts. The method of inquiry uses the tools of rational thought and experimental procedure. It is difficult to see how any student of general physics can gain real insight into the materials and methods of physics without some experience in the laboratory.

One of the chief aims of laboratory work is to give the student *direct experience* with some of the massive physical facts of nature upon which physical principles rest. Too often it is assumed that the student already possesses these facts because he has read about them in the textbook or heard about them in the lecture. This is definitely not what is meant by the phrase "direct experience." For example, a student gets direct experience with the acceleration caused by gravity when he goes into the laboratory and determines this acceleration with a free-fall apparatus. Reading or hearing about it is not sufficient. Some bed-rock information is essential for a proper understanding of physics. Without it, physics becomes authoritarian in character, and the student gets the impression that physics is largely an intellectual exercise in which particular conclusions are ground out of general principles, principles that from time to time have been handed down from heaven in some mysterious manner. Such an attitude toward physics is surely false and may become dangerous.

A second, equally important objective of laboratory work flows from the fact that physics is a method of inquiry that involves both the tools of rational thought and the apparatus of experimental procedure. It is the unique task of laboratory work to develop in the student some understanding of experimental procedure. This aim can best be achieved by giving him some experience in solving simple experimental problems. To do so, the student must become familiar with a large number of common laboratory instruments. He must deal with real equipment rather than with its ideal counterpart. He is exposed to the performance limitations of such equipment in range, accuracy, sensitivity, and so on. Errors caused by these limitations invariably accompany the use of the equipment. The student must take these errors into account before he can draw any conclusions from his experimental results. Hence, error analysis is an essential and difficult component of all good laboratory work. It is one of the worthwhile outcomes of laboratory work.

PURPOSE OF THE LABORATORY MANUAL

This physics laboratory manual is intended to help the beginning student make effective use of time spent in the laboratory. Explicit instructions are given concerning the operation of equipment and the method of procedure. These instructions should be regarded as a chart or map that will enable the inexperienced student to make his way through an experiment without getting completely lost on the way and without damaging the apparatus. Alternative methods of procedure exist and may be used by the student, provided that he assumes the responsibility of reaching the same goals under the existing conditions of space, time, and equipment.

The authors are well aware that a laboratory manual such as this one, which gives explicit descriptions and instructions, may be misused by some students to perform experiments in a purely mechanical fashion without thought or understanding. On the other hand, the manual can be of real service to the serious student who wishes to make a respectable amount of progress in the laboratory in the face of the almost always prevailing conditions of limited time, space, equipment, and instructional staff.

Our experience, which is borne out by that of other teachers, is that if a student wants an excuse for performing his work mechanically and unintelligently, such a book may indeed supply it, but to men who are beginners and want to make the best use of their opportunities it is a very real help, and if large classes are to be taken by any reasonable number of demonstrators, it is almost a necessity.[1]

This quotation taken from *A History of the Cavendish Laboratory* appears in Chapter 9 (Development of Teaching of Physics) and concerns the publication in 1896 of one of the first physics laboratory manuals. It forms a fitting conclusion to this examination of the purpose of a laboratory manual.

A. ERRORS AND UNCERTAINTY

In laboratory work it is necessary to transform a set of ideal mental operations into a corresponding set of real physical operations that can be performed in the laboratory. The correspondence is seldom, if ever, perfect. As a consequence, errors creep into the physical operations leading to results with some degree of uncertainty.

[1]L. R. Wilberforce in *A History of the Cavendish Laboratory 1871–1910* (London: Longmans, Green & Co. Ltd., 1910), 262.

Before such results can be reliably interpreted or used, some estimate of their uncertainty is necessary. In the case of a numerical result the uncertainty is usually expressed in fractional or percentage form, e.g., 1 part in 100 (1 %), 3 parts in 1000 (0.3 %), etc. Hence, one of the important things to be learned in the laboratory is how to make reasonable estimates of the errors involved in any physical measurements and how to handle the propagation of these errors, i.e., to know what these errors contribute to the uncertainty of the final result in an experiment.

In the following sections we shall discuss some of the important concepts and procedures in error analysis starting with the two general types of errors: systematic errors and random errors.

1. Systematic Errors. Systematic errors occur in an experiment because of faulty methods, defective measuring apparatus, or incomplete working equations. They are usually constant and therefore definite in sign and magnitude. They cannot be reduced by taking the average of a number of similar measurements, because the same error is included in each measurement.

These errors are often more significant than random errors discussed in the next section. They may be reduced by modifying the method, calibrating the measuring apparatus, or adjusting the working equations. Such procedures are generally described as corrections. For example, the reading of a micrometer caliper may not be zero when the jaws are closed. This so-called zero error must be corrected for in using the instrument; otherwise, all measurements made with the caliper will be in error by the same amount.

2. Random Errors. Random errors are present in almost all measurements. They arise because of uncontrollable conditions affecting the observer, the measuring device, and the quantity being measured. As a consequence of these errors, careful measurements of the same physical quantity made in the same way with the same equipment yield different values. Any quantity that has this property is called a random or statistical variable, since its use involves statistical and probability theory.

It is usually assumed that random errors have, at least approximately, a normal or Gaussian distribution represented graphically by the well-known bell-shaped curve. On this basis, these errors are as likely to be positive as negative and more likely to be small than large. Their effect may therefore be minimized by taking a number of measurements of the quantity to be determined and using the arithmetic mean of the measured values as the best estimate of the true value of the quantity. For example, if n measurements of the quantity are made, all equally reliable, and have the values $X_1, X_2, \ldots, X_n$, then the arithmetic mean $\bar{X}$ is defined by the relation

$$\bar{X} = \frac{1}{n}(X_1 + X_2 + \cdots + X_n) = \frac{1}{n} \sum_{i=1}^{n} X_i \qquad (I.1)$$

If any value X is added and then subtracted from the right side of Eq. (I.1), the equation becomes

$$\bar{X} = X + \frac{1}{n} \sum_{i=1}^{n} (X_i - X) \qquad (I.1a)$$

This is often a more convenient equation to use than the defining equation in computing $\bar{X}$.

In the previous paragraph we have spoken of the "true value" of a measured quantity. What is the meaning of this phrase?

For purely random errors, the true value of a measured quantity is a useful statistical parameter in the theory of measurement even though its exact determination is impossible. We define the true value as the mean value of an *infinite* set of measurements made under constant conditions, and represent it by the symbol μ (Greek letter mu, equivalent to m). As so defined, the true value μ is not a random variable but a constant whose value we are trying to estimate by statistical methods. Also we must realize that this true value depends not only on the quantity being measured but likewise on the method and apparatus of measurement. A change in method or apparatus will probably yield a different true value because of a change in the systematic errors.

Since we cannot make an infinite set of measurements to evaluate μ, we make a finite set and regard this finite set of measurements as a random sample of the infinite set. It then turns out, on the basis of the statistical theory of sampling, that the mean value $\bar{X}$ of our sample set of n measurements is the best estimate of the value of μ that we can obtain without making additional measurements. Obviously, the bigger our sample, the better is our estimate of the value of μ. The relation between μ and $\bar{X}$ is often expressed in the form $\mu \cong \bar{X}$, that is, μ is approximately equal to $\bar{X}$.

By use of the measured values X_i, the mean value $\bar{X}$, and the true value μ, we define the error E_i and the deviation D_i of any measured value X_i by the equations

$$E_i = X_i - \mu \tag{I.2}$$

and
$$D_i = X_i - \bar{X}, \qquad i = 1, 2, 3, \ldots, n \tag{I.3}$$

For an infinite set of measurements $\bar{X}$ is equal to μ (by definition) and there is no difference between errors and deviations. But for a finite set of measurements there is a difference, since in this case the value of $\bar{X}$ as given by Eq. (I.1) is only an approximation to the value of μ. For interesting and illuminating discussions on these matters see below.[2,3]

The *precision* with which a physical quantity is measured depends inversely upon the spread or dispersion of the set of measured values X_i about their mean value $\bar{X}$. If the values are widely dispersed, the precision is low (the imprecision is high) and conversely.

There are several different expressions for the dispersion of a set of measured values. The most useful one because of its direct connection with the normal error law is the *standard deviation*, usually represented by the symbol σ (Greek letter sigma, equivalent to s). The standard deviation σ is defined as the square root of the mean-square deviation or mean-square error for an *infinite* set of measurements. The square of the standard deviation is called the *variance*.

The standard deviation σ is the second useful statistical parameter in the theory

[2]D. C. Baird, *Experimentation: An Introduction to Measurement Theory and Experiment Design* (Englewood Cliffs, New Jersey: Prentice-Hall, Inc., 1962), Chap. 2.

[3]N. C. Barford, *Experimental Measurements: Precision, Error and Truth* (Reading, Massachusetts: Addison-Wesley Publishing Company, Inc., 1967), Chap 1.

of measurement. But just as in the case of the parameter μ, its exact determination is impossible and for the same reason, i.e., the requirement for an infinite set of measurements. The best that we can do is to make a finite set of measurements and, using it as a sample of the infinite set, compute the best estimate of σ, which we represent by the symbol s.

It can be shown that the best estimate of σ for a finite set of n measurements is given by the relation,

$$s = \sqrt{\frac{(X_1 - \bar{X})^2 + (X_2 - \bar{X})^2 + \cdots + (X_n - \bar{X})^2}{n - 1}}$$

$$= \sqrt{\frac{\sum_1^n (X_i - \bar{X})^2}{n - 1}} \tag{I.4}$$

where $s \cong \sigma$. Note the use of $n - 1$ rather than n in Eq. (I.4) for determining s. For large values of n the effect on the value of s is negligible and s may be regarded as the root-mean-square deviation. But for small values of n the effect is noticeable, and for $n = 1$ the value of s is indeterminate. It is only the excess of the number of measurements above one (the number of degrees of freedom) that is taken into account in computing s. Also note that it is always possible to substitute for the sum of the squares of the deviations in Eq. (I.4), i.e., $\sum_1^n (X_i - \bar{X})^2$, the equivalent expression,

$$\sum_1^n (X_i - X)^2 - n(X - \bar{X})^2 \tag{I.4a}$$

where X may have any value you choose.

The chief significance of the two parameters μ and σ lies in their connection with the normal law of errors. This law predicts that the probabilities of finding the value of a single measurement in the intervals $\mu - \sigma$ to $\mu + \sigma$, $\mu - 2\sigma$ to $\mu + 2\sigma$, $\mu - 3\sigma$ to $\mu + 3\sigma$, are respectively 0.683, 0.955, 0.997, to three places. This means that we would expect about 68% of our X_i values to lie in the interval $\bar{X} - s$ to $\bar{X} + s$, 95% in $\bar{X} - 2s$ to $\bar{X} + 2s$, and practically all of them to lie in $X - 3s$ to $X + 3s$ *provided* n is large enough to allow $\bar{X}$ to replace μ and s to replace σ. This proviso is troublesome for it means among other things that our computed values of $\bar{X}$ and s on the basis of a finite number of measurements are only estimates of the "true" values, μ and σ. In other words, if we repeat our n measurements of the physical quantity in question, we would get a different set of X_i and have different values of $\bar{X}$ and s. Therefore $\bar{X}$ and s are also random variables, each with their own mean value and dispersion. This would lead to an unending sequence of equations if it were not for a vital property of the normal law—its reproductive property. This property not only tells us that the different means and dispersions from different random samples of n measurements are normally distributed but also enables us to determine these dispersions by theory without actually repeating the set of n measurements.

The dispersion of the sample means about the true mean is called the *standard error* (dispersion of the means) and is given by the relation,

$$\sigma_m = \frac{\sigma}{\sqrt{n}} \cong \frac{s}{\sqrt{n}} \tag{I.5}$$

where σ_m is the standard error. The normal law predicts that the probabilities of finding one of the sample means, e.g., ours, in the intervals $\mu \pm \sigma_m$, $\mu \pm 2\sigma_m$, $\mu \pm 3\sigma_m$, are respectively 0.68, 0.95, 0.99+. Note that the standard error is inversely proportional to the square root of n, the number of measurements, and goes rather slowly to zero as n increases indefinitely.

Just as there is a dispersion of $\bar{X}$ values about the true mean μ, there is a corresponding dispersion of s values about the true standard deviation σ. If σ_s represents this latter dispersion, then

$$\sigma_s = \frac{\sigma}{\sqrt{2(n-1)}} \cong \frac{s}{\sqrt{2(n-1)}} \tag{I.6}$$

which also vanishes as n goes to infinity.

The value of σ_s given by Eq. (I.6) is a statistical measure of the absolute error in the value of s due to a finite (rather than an infinite) number of measurements. This error is written as $|s - \sigma|$ and, by the normal law, is more than likely to be less than σ_s (68% chance), is much more than likely to be less than $2\sigma_s$ (95% chance), and is almost certain to be less than $3\sigma_s$ (99% chance). If we take the more than likely case, then the error in s for ten measurements may be as much as 25% of s. And since s is used to compute the standard error, then σ_m may also be in error by as much as 25% of σ_m. Even for 50 measurements the error may still be as much as 10% of the computed value. For this reason the values of s, σ, and σ_m should be written with only one or two figures, and the number of measurements should always be stated.

We conclude this section on random errors with an example. The diameter of a metal cylinder is carefully measured 10 times with a micrometer caliper having negligible zero correction. By use of these data determine the best estimate of the value of the diameter and the standard error in the value.

The measured values and their deviations are shown in Table I.1.

Table I.1

i	X_i cm	$(X_i - X)$ cm	$(X_i - X)^2$ cm^2
1	4767×10^{-4}	4×10^{-4}	16×10^{-8}
2	4765	2	4
3	4761	−2	4
4	4758	−5	25
5	4763 (= X)	0	0
6	4769	6	36
7	4762	−1	1
8	4762	−1	1
9	4763	0	0
10	4762	−1	1
Sum	$10\bar{X}$	2×10^{-4}	88×10^{-8}

Note that we have chosen $X = X_5 = 4763 \times 10^{-4}$ because it appears to be close to the mean value (a guess). We now compute $\bar{X}$ and s in order to estimate the values of μ, σ, and σ_m.

By Eq. (I.1a):

$$\bar{X} = 4763 \times 10^{-4} + \tfrac{1}{10}(2 \times 10^{-4}) = 4763.2 \times 10^{-4} \text{ cm.}$$

By expression (I.4a) we compute the sum of the squares of the deviations from the mean, i.e.,

$$\sum_1^n (X_i - X)^2 - n(X - \bar{X})^2 = 88 \times 10^{-8} - 10(4763 - 4763.2)^2 \times 10^{-8}$$
$$= (88 - 0.4) \times 10^{-8}$$
$$= 87.6 \times 10^{-8} \text{ cm}^2$$

By Eq. (I.4):

$$s = \sqrt{\frac{87.6 \times 10^{-8}}{10 - 1}} = 3.1 \times 10^{-4} \text{ cm}$$

By Eq. (I.6):

$$\frac{\sigma_s}{s} = \frac{1}{\sqrt{2(10 - 1)}} = 0.24$$

which indicates a possible error in s by as much as 24% of s.

The estimates of μ, σ, and σ_m for ten measurements are:

$$\mu \cong \bar{X} = 4763 \times 10^{-4} \text{ cm}$$
$$\sigma \cong s = 3.1 \times 10^{-4} \text{ cm}$$
$$\sigma_m \cong \frac{s}{\sqrt{n}} = 1.0 \times 10^{-4} \text{ cm}$$

Therefore: cylinder diameter $= 4763 \times 10^{-4}$ cm with a standard error of 1.0×10^{-4} based on ten measurements.

Several things should be noted about this example.

1. The work is done systematically in order to avoid mistakes.

2. Only significant figures, i.e., all figures up to and including the first doubtful figure, are included in the results. There are four figures in the value of the diameter with the last figure 3 being doubtful. There are two figures in the standard error with the second figure 0 being very doubtful.

3. The approximate distribution of the values of the ten measurements about the mean is in rather good agreement with that given by the normal law of errors. Seven measurements of the ten have values falling in the interval $\bar{X} \pm s$, and all ten fall within the interval $\bar{X} \pm 2s$. The normal law figures are 68% in the first interval and 95% in the second.

In this example we have neglected to take into account any systematic error. We stated in the problem that the zero-point error (reading of micrometer when jaws are closed) was zero. Suppose the reading had been 2×10^{-4} cm. It is not necessary to go back to the original values and decrease each one by 2×10^{-4} cm. It is only necessary to decrease the value of $\bar{X}$ by that amount. There would be no effect whatever on the values of the dispersion parameters σ, σ_m, and σ_s.

3. Normal Distribution Function. In most physical measurements we find, after correcting for systematic errors, that the remaining random errors have a characteristic distribution about the mean. This distribution is called a *normal* or *Gaussian distribution.* The primary justification for use of this distribution is based on experience, but there is also theoretical support. It can be shown by the central limit theorem in mathematics that any observable error which is the sum of a large number of small independent errors is normally distributed to a high degree of approximation.

Also any error which is a linear function of other errors normally distributed is itself normally distributed. Thus the normal distribution is in a sense the great grandfather of many other distributions.

In order to describe in some detail the normal distribution function we return to our numerical example in Sec. 2 where we tabulated the data for ten equally reliable measurements of the diameter of a cylinder. If instead of ten measurements we had made a thousand measurements, this sort of tabulation would have been very inconvenient, since i would run from 1 to 1000 and Table I.1 would be 100 times as large.

There is an alternative procedure that is much simpler. We note that the measured values X_i are sometimes the same and that they lie in a restricted range. We also note that the numeric part of the value in each case consists of four digits, the number of significant figures. This number is determined by the measuring device used; e.g., the smallest division on the micrometer is 10^{-3} cm and we can only estimate to $\frac{1}{10}$ of this division. Therefore the smallest unit that this measuring device can measure is 10^{-4} cm, and when our measurement is expressed in such units, the result is always an integer. Thus the possible values that we can get in our measurements form a discrete set of integral values. We can therefore tabulate our measuring data in somewhat different form than that in Table I.1. This new form is shown in Table I.2.

Table I.2

x_i = integral scale values in units of 10^{-4} cm.*
n_i = absolute frequency of occurrence of x_i in the measurements.
N_i = accumulated absolute frequency.

i	x_i	n_i	N_i
1	4757	0	0
2	58	1	1
3	59	0	1
4	60	0	1
5	61	1	2
6	62	3	5
7	63	2	7
8	64	0	7
9	65	1	8
10	66	0	8
11	67	1	9
12	68	0	9
13	69	1	10
14	70	0	10

*The lowercase x_i in this table should not be confused with the uppercase X_i in Table I.1.

Table I.2 is often called a *frequency table* and has the distinct advantage over Table I.1 of permitting a change in the total number of observations without any material change in the form or size of the table. In column 2 of Table I.2 we list in integral order all integral scale values of x_i in the extended range of the measured values regardless of their occurrence in the observations. In column 3 we indicate the absolute frequency n_i (an integer) with which each of the x_i values occurs in our measurements including 0 for no occurrence. In column 4 we list the accumulated frequency N_i for each x_i, i.e., the sum of all preceding frequencies plus that at x_i.

With the information from Table I.2, it is now possible to plot either one or both of two closely related graphs called *histograms* that represent our observed distribution; i.e., we can plot n_i versus x_i or N_i versus x_i. Either one of these may be compared with the theoretical-normal or Gaussian distribution.

In general it is preferable to plot the accumulated frequencies N_i versus x_i, especially when the total number of observation is small, i.e., of order 10 as in our case. One important reason for this choice of procedure is that the relative fluctuations in the values of n_i are more extreme than those in the values of N_i, since the summation process in obtaining the N_i tends to nullify the fluctuations. In addition, the comparison between the observed distribution and the normal distribution is mathematically more straightforward.

However, if we use the graph of the accumulated frequencies N_i versus x_i, then the corresponding theoretical normal curve is not the familiar bell-shaped differential normal curve but rather the less familiar S-shaped integral normal curve. Both of these curves are shown in Fig. I.1. Curve A (the familiar one) is the so-called differential normal curve while curve B is the corresponding integral normal curve. The

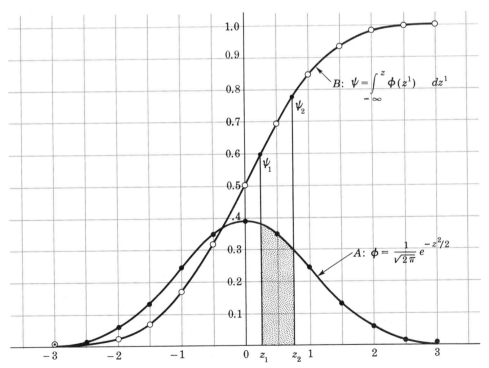

FIGURE I.1   Differential and integral normal distribution curves.

relation between the two curves is as follows: Any ordinate of curve B equals the scale area under curve A from the extreme left to the z position of that ordinate, and any ordinate of curve A equals the corresponding scale slope of curve B.

The independent variable z is just the deviation from the mean measured in units of the standard deviation; i.e.,

$$z = \frac{x - \mu}{\sigma} \cong \frac{x - \bar{x}}{s}$$

Thus before we can compare our observed distribution with the appropriate normal curve, we must first compute $\bar{x}$ and s in order to estimate the values of the two fundamental parameters, μ and σ. Both curves A and B are normalized so that the area under curve A is 1 and curve B monotonically increases from 0 to 1. Tabulated values of $\phi(z)$ and $\psi(z)$ are given in Appendix III, Table S.

Curves A and B are essentially probability curves. For example, suppose a set of measurements such as ours has a normal distribution with estimated values of μ and σ. Then the probability that the value of z for a single measurement value x will lie in any interval z_1 to z_2, as shown in Fig. I.1, is just equal to the shaded area under curve A based on the interval, or to the difference in the ordinates of curve B at the ends of the interval, i.e., $\psi_2 - \psi_1$, as shown in Fig. I.1.

We return now to the task of comparing our observed distribution of measured values with the predicted normal distribution. To do this we plot the values of the accumulated frequencies N_i (ordinates) as a function of the values of x_i (abscissae) and connect the plotted points by an ascending curve, as shown in Fig. I.2(b). For comparison we show in the same figure the corresponding integral normal curve, exactly equivalent to curve B in Fig. I.1 except for a multiplier of 10. We could have avoided using a multiplier by plotting relative values of the accumulated frequencies, i.e., N_i/n, where n is the total number of observations.

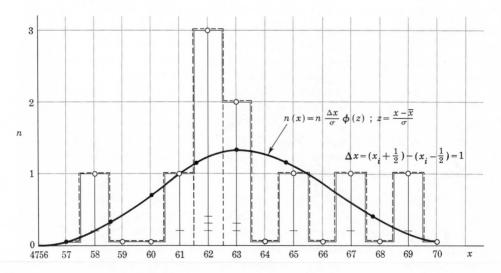

FIGURE I.2(a)   Frequency histogram.

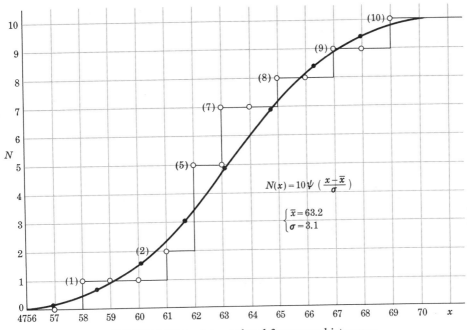

FIGURE I.2(b) Accumulated frequency histogram.

In any event the agreement between our staircase curve and the integral normal curve seems to be quite satisfactory. To better the agreement we would have to increase the total number of observations by a substantial amount. It can be shown that as n increases indefinitely, the value of N_i/n, for fixed i, approaches a definite limit lying somewhere in the interval 0 to 1. This limiting value is precisely the probability that any one measurement will have a value less than or equal to x_i. In the same way it can be shown that the limiting value of n_i/n as n increases indefinitely is just the probability that a measurement will have the value x_i. These statements are presumably true for any type of distribution, normal or not.

In Fig. I.2(a) we have plotted n_i as a function of x_i where the n_i are represented by rectangles of height n_i with a unit base centered at the point x_i. The scale area of the ith rectangle just equals n_i so that the scale area of all the rectangles is $n = 10$. The corresponding theoretical curve A is shown for comparison. Fluctuations in the values of n_i make the comparison difficult. Undoubtedly curves I.2(b) are better for comparison purposes.

4. Propagation of Errors. The experimental determination of the value of a physical quantity such as velocity or pressure is seldom obtained by direct measurement. Usually the quantity to be determined is a known function of one or more independent fundamental physical quantities such as length, time, and mass. The normal procedure is to measure the independent quantities and then to compute the value of the dependent quantity by using the measured values in the known relation. But since the measured values contain errors that are propagated into the computed value, this computed value will also be in error. The purpose of this section is to show how to evaluate the error in the computed value of the dependent quantity in terms of the errors in the measured values of the independent quantities. Since the laws of propagation are somewhat different for the two general types of errors, systematic and random, we first consider systematic errors, then random errors, and finally mixed errors.

a. *Systematic errors.* Let us suppose that the dependent quantity R to be determined is related to two (for simplicity) independent measurable quantities X and Y by the equation,

$$R = f(X, Y) \tag{I.7}$$

where f is a known function of the independent variables X and Y, and R is the dependent variable. Equation (I.7) is valid for both measured and "true" values of X and Y. In order to distinguish between the two kinds of values, let X_0, Y_0, and R_0 designate true values and X, Y, and R, measured values. The Eq. (I.7) refers to measured values, and the equation

$$R_0 = f(X_0, Y_0) \tag{I.8}$$

refers to true values.

In order to get an expression for errors, we subtract Eq. (I.8) from Eq. (I.7) and get

$$R - R_0 = f(X, Y) - f(X_0, Y_0) \tag{I.9}$$

The right side of Eq. (I.9), for *sufficiently small values* of $(X - X_0)$ and $(Y - Y_0)$, may be approximated by a linear function of these small quantities (Taylor's theorem). We assume this to be a valid approximation for errors and write

$$(R - R_0) = \frac{\partial f}{\partial X}(X - X_0) + \frac{\partial f}{\partial Y}(Y - Y_0) \tag{I.10}$$

The partial derivatives $\partial f/\partial X$ and $\partial f/\partial Y$ represent, respectively, the rate of change of R with respect to X and to Y. They are usually evaluated at X_0, Y_0, but for small intervals they may be evaluated anywhere in the interval, provided the function varies slowly. Now, by definition,[4] $(X - X_0)$, $(Y - Y_0)$, and $(R - R_0)$ are the respective errors in X, Y, and R. Thus Eq. (I.10) tells us how to compute the error in R for *given* errors in X and Y. We shall call Eq. (I.10) the *determinate-error equation* for Eq. (I.7), which tells us how to compute R for *given* values of X and Y.

It is customary to write Eq. (I.10) in the differential form,

$$\Delta R = \frac{\partial f}{\partial X}\Delta X + \frac{\partial f}{\partial Y}\Delta Y \tag{I.11}$$

where $\Delta R = R - R_0$, $\Delta X = X - X_0$, and $\Delta Y = Y - Y_0$. In this form we see that the error equation (I.11) is just the differential of its parent, Eq. (I.7). Often it is more convenient to work with relative or fractional errors, especially when the func-

---

[4] The error is the quantity that must be added to the true value to obtain the measured value. The error with its sign reversed is called the correction.

tion $f(X, Y)$ is a product of powers of X and Y. In this case Eq. (I.11) is divided by Eq. (I.7) to produce the relative error equation. This procedure is exactly equivalent to taking the logarithmic differential of Eq. (I.7). Equations (I.7) and (I.11) are sufficient to handle the propagation of any small systematic errors in X and Y. Since a systematic error is definite in sign and magnitude, a known systematic error in X can be completely represented by ΔX, and that in Y by ΔY. Then by Eq. (I.11) we may compute the corresponding systematic error in R, that is, ΔR. The true value of R will then be $R_0 = R - \Delta R$, where R is the indirectly measured value. An alternative procedure is to compute $X_0 = X - \Delta X$, and $Y_0 = Y - \Delta Y$ first, and then use Eq. (I.8) to compute R_0. Either procedure gives the same value of R_0.

b. *Random errors.* Let us assume that we have corrected our measured values of X and Y for any significant systematic errors and have now to deal only with random errors. In contrast to systematic errors, random errors are indefinite in sign and magnitude and for this reason are indeterminate. We can only speak of the probability of the distribution of their signs and magnitudes. To assess these errors we must make repeated measurements of each independent quantity in order to get best estimates of the values of the two fundamental parameters, μ and σ, as explained in Sec. 2.

Random errors in the directly measured quantities X and Y will produce random errors in the computed quantity R. This means that the computed values of R will have a distribution with its own characteristic mean value and standard deviation. Two important questions arise. First, what is the relation between the mean value of R and the mean values of X and Y? And second, what is the relation between the standard deviation of R values and those for the X and Y values?

The first question about mean values may be answered by showing that Eq. (I.8) is valid for mean values as well as for true values. To prove this let X_i be any measured value of X, and Y_j any measured value of Y, and R_{ij} the corresponding computed value of R. Then Eq. (I.10) may be written in the form,

$$R_{ij} - \bar{R} = \frac{\partial f}{\partial X}(X_i - \bar{X}) + \frac{\partial f}{\partial Y}(Y_j - \bar{Y}) \tag{I.12}$$

where $\bar{X}$ and $\bar{Y}$ are mean values of X_i and Y_j values, but where $\bar{R}$ is simply $f(X, Y)$ evaluated at $X = \bar{X}$ and $Y = \bar{Y}$. If we now sum Eq. (I.12) over all i and j, i.e., over all measured values of X and Y, the right side of the summed equation must vanish, since $\bar{X}$ and $\bar{Y}$ are mean values. Therefore $\sum_{ij} (R_{ij} - \bar{R})$ vanishes, and hence $\bar{R}$ must be the mean of the computed R_{ij} values.

In order to answer the second question about standard deviations, we square both sides of Eq. (I.12) obtaining,

$$(R_{ij} - \bar{R})^2 = \left(\frac{\partial f}{\partial X}\right)^2 (X_i - \bar{X})^2 + \left(\frac{\partial f}{\partial Y}\right)^2 (Y_j - \bar{Y})^2$$
$$+ 2\left(\frac{\partial f}{\partial X}\right)\left(\frac{\partial f}{\partial Y}\right)(X_i - \bar{X})(Y_j - \bar{Y}) \tag{I.13}$$

When we sum both sides of this equation over all i and j, the sum of the cross-product terms will vanish. The remaining sums are sums of squares of deviations from the mean for R, X, and Y. They may be replaced by the squares of the standard deviations (variances) of R, X, and Y, since the common multiplier cancels out.[5] Therefore we get

$$\sigma_R^2 = \left(\frac{\partial f}{\partial X}\right)^2 \sigma_X^2 + \left(\frac{\partial f}{\partial Y}\right)^2 \sigma_Y^2 \tag{I.14}$$

This important variance relation enables us to obtain σ_R from the values of σ_X and σ_Y coming from direct measurements of X and Y.

It may be shown that Eq. (I.14) is still valid if we substitute standard errors for standard deviations. Finally we should note that this variance relation is valid even though the distribution is not normal.

We have put the distinguishing subscripts R, X, and Y on the sigmas in Eq. (I.14) for proper identification. This is a satisfactory notation as long as no other letter subscripts are needed. However Eq. (I.14) is also valid for standard errors designated by σ_m so that, in this case, double subscripts would be required, e.g., σ_{mR}. An alternative notation[6] often used is $\sigma\{R\}$ for σ_R, and $\sigma\{\bar{R}\}$ for σ_{mR}. The curly brackets around R (or $\bar{R}$) indicates that σ depends upon all values of R (or $\bar{R}$).

We can easily extend the results in this section to any number of independent variables, say n, by a slight change in notation. Instead of representing the independent variables by X, Y, ..., we represent them by Z_1, Z_2, ..., Z_n and write $R = f(Z_1, Z_2, \ldots, Z_n)$ instead of $R = f(X, Y)$.

In a great many problems R is a linear function of the variables and may be written in the form

$$R = a_1 Z_1 + a_2 Z_2 + \cdots + a_n Z_n \tag{I.7a}$$

with known coefficients a_1, a_2, ..., a_n. In this case the variance relation becomes

$$\sigma^2\{R\} = a_1^2 \sigma^2\{Z_1\} + \cdots + a_n^2 \sigma^2\{Z_n\} \tag{I.14a}$$

a natural extension of Eq. (I.14).

So far we have assumed that the variables X and Y originally used, or the new ones just introduced, were in fact different random variables, representing different physical quantities. It turns out that this is not necessary. All that is required is that the measured values of X and Y, or Z_1, Z_2, ..., Z_n, be uncorrelated, i.e., independent of one another.[7] Thus it is possible, for example, that the mean measured values of Z_1, Z_2, ..., Z_n represent n measurements of the same quantity represented by a single random variable. With this interpretation, we can prove the relation between standard deviation and standard error as expressed by Eq. (I.5).

Consider the defining equation for the mean value of a set of n equally reliable measurements of the same physical quantity represented by the random variable Z. This equation is

$$\bar{Z} = \frac{Z_1}{n} + \frac{Z_2}{n} + \cdots + \frac{Z_n}{n} \tag{I.7b}$$

---

[5]If we take n measurements of X and m measurements of Y, then the common multiplier is nm provided n and m are sufficiently large to allow us to replace $n - 1$ by n and $m - 1$ by m.

[6]Niels Arley and K. Rander Buch, *Probability and Statistics* (New York: John Wiley & Sons, Inc., 1950), pp. 55–56, 75–76. This book gives an excellent treatment of errors with numerous examples in Chap. 11.

[7]Ibid.

which has the same form as Eq. (I.7a). We may regard $\bar{Z}$ as a linear function of the measured values of $Z_1, Z_2, \ldots, Z_n$ in the sense that a repetition of these measurements would give a slightly different set of values leading to a slightly different value of $\bar{Z}$. Suppose we imagine taking n sets of the n measurements. Then there would be n values of Z_1, n values of $Z_2, \ldots, n$ values of Z_n, and n values of $\bar{Z}$. From these values we could compute not only the mean values of $Z_1, Z_2, \ldots, Z_n$, and $\bar{Z}$, but also their standard deviations. These standard deviations for sufficiently large values of n must satisfy the variance relation corresponding to Eq. (I.14a), that is,

$$\sigma^2\{\bar{Z}\} = \frac{\sigma^2\{Z_1\}}{n^2} + \frac{\sigma^2\{Z_2\}}{n^2} + \cdots + \frac{\sigma^2\{Z_n\}}{n^2} \tag{I.14b}$$

where R has been replaced by $\bar{Z}$, a_1^2 by $1/n^2$, etc. But now all of the variances on the right side of Eq. (I.14b) have practically the same value $\sigma^2\{Z\}$, that is, the variance of the first set of n measurements that we actually did make. When we substitute this variance for each of the others in Eq. (I.14b), the equation reduces to

$$\sigma\{\bar{Z}\} = \frac{\sigma\{Z\}}{\sqrt{n}} \tag{I.14c}$$

the exact equivalent of Eq. (I.5) giving the standard error in terms of the standard deviation and the number of measurements. A similar but more general relation between standard error and standard deviation exists for the method of least squares. See Appendix I, Note A.

The following numerical example illustrates the procedures involved in treating the propagation of random errors in indirect measurements. Suppose we wish to determine the volume V of a metal cylinder by direct measurements of its diameter D and length L where only random errors are considered. Let us further suppose that ten measurements of the diameter give a mean value of $D = 0.4760$ cm and an approximate standard deviation of $\sigma_D \cong 3.1 \times 10^{-4}$ cm, while ten measurements of the length give $\bar{L} = 8.99$ cm and an approximate standard deviation of $\sigma_L \cong 0.010$ cm.

We first calculate the volume by writing Eq. (I.7) in the form,

$$V = f(D, L) = \frac{\pi D^2 L}{4} \tag{I.15}$$

and setting $D = \bar{D}$ and $L = \bar{L}$. Thus

$$\bar{V} = \frac{\pi(0.4760 \text{ cm})^2(8.99 \text{ cm})}{4} = 1600 \text{ cm}^3$$

Next we compute the standard deviation and standard error in V in terms of those in D and L by use of the variance equation (I.14). However, because V is a product function of D and L, it is much simpler to use relative values of σ rather than absolute values. We get the equation for relative values by simply dividing the variance equation by V^2. When we do this, the relative variance equation may be written as

$$\left(\frac{\sigma_V}{V}\right)^2 = \left(\frac{2\sigma_D}{D}\right)^2 + \left(\frac{\sigma_L}{L}\right)^2 \tag{I.16}$$

In this example $\sigma_D/D \cong 3.1/4760 = 0.00065$ and $\sigma_L/L \cong 1.0/899 = 0.0011$. When we substitute these values in Eq. (I.16), we get

$$\left(\frac{\sigma_V}{V}\right)^2 \cong (0.0013)^2 + (0.0011)^2 = 2.9 \times 10^{-6}$$

Consequently $\sigma_V/V \cong 1.7 \times 10^{-3}$, $\sigma_V \cong 2.7$ cm^3, and $\sigma_{mV} = \sigma\{\bar{V}\} = \sigma_V/\sqrt{n} \cong 0.9$ cm^2.

The fact that the estimated values of σ_D and σ_L are based on ten measurements apiece means that each has a fair chance (1 in 3) of being in error by 25% or more. It is quite reasonable, therefore, to round off the values of σ_V and σ_{mV} to 3 cm³ and 1 cm³ respectively.

The final results should be reported as follows: $\bar{V} = 1600$ cm³ with a standard error of 1 cm³ based on ten measurements each of D and of L. Shorthand expressions,[8] such as $\bar{V} = (1600 \pm 1)$ cm³, should be avoided, even if it is clear from the context that 1 cm³ is the standard error, because this expression leaves the false impression that the true value of the volume of the cylinder in cm³ lies somewhere in the interval 1599 to 1601. However, if we are able to establish reasonable upper bounds on the magnitudes of the random errors and constant errors in the result, then the *overall upper bound—the uncertainty—*may be used in the shorthand expression. But we should state how we arrived at this uncertainty. For example, we could take *three times* the *standard error* as a reasonable upper bound on the random errors, and estimate a reasonable upper bound on the constant errors by using some fraction of the smallest scale divisions on our measuring instruments as a starting point. In this example the former uncertainty is ± 3 cm³. If the latter uncertainty also turned out to be ± 3 cm³, then the overall uncertainty in the value of the volume would be ± 6 cm³. The result could then be written (1600 ± 6) cm³, where 6 cm³ represents an *overall* estimated upper bound on the magnitudes of the random and constant errors.

c. *Mixed errors: error intervals or uncertainties.* It is very seldom that the significant errors in the measured quantities are either purely systematic or purely random. They are most often a mixture of both types of errors. Frequently a systematic error is present for which we know only an upper bound on its magnitude, e.g., a voltmeter with an accuracy of $\pm 2\%$ of full scale reading. Or, in a dynamic experiment, we may be able to make only a single measurement, e.g., time of flight. In such cases a satisfactory error analysis is difficult, if not impossible, without extensive supporting work such as meter calibration, repetition of the experiment, etc.

Also the propagation equation for systematic errors is linear (I.11), while that for random errors is quadratic (I.14). Thus we must make separate computations for these two types of errors.

Up to this point we have assumed that systematic errors were corrected before we treated random errors. This is not a necessary assumption. It can be shown that for mixed errors Eq. (I.11) is still valid provided $\Delta R = \bar{R} - R_0$, $\Delta X = \bar{X} - X_0$, and $\Delta Y = \bar{Y} - Y_0$, and Eq. (I.14) is valid even though constant errors are present. Therefore we can compute by use of Eq. (I.11) the systematic error ΔR in R (or at least an upper bound) just as though there were no random errors; and we can compute by Eq. (I.14) the variance σ_R^2 in R just as though there were no systematic errors.

However the amount of work required to carry out the foregoing procedures is so considerable that it is rarely done except in advanced laboratory practice. In the general physics laboratory we most often use a simpler but less exact method of error (or uncertainty) intervals.[9]

[8]Churchill Eisenhart, "Expression of the Uncertainties of Final Results," *Science* **160** (June 14, 1968): 1201–4.

[9]Haym Kruglak and John T. Moore, *Basic Mathematics for the Physical Sciences* (New York: McGraw-Hill Book Company, 1963). See Chap. 6 for a very good elementary discussion of measurements and errors with many examples.

In the error interval method we indicate the amount of uncertainty in our "measured" value of X, for example, by means of a definite interval that encloses or brackets this value and indicates its accuracy. The size of the interval is estimated by the observer on the basis of the type of equipment and operations used in making the measurement. Experience, of course, plays a large part in making a suitable estimate of the error interval. But it is just this experience that constitutes one of the worthwhile outcomes of laboratory work.

It is difficult to establish any general rules for estimating the size of an error interval. Perhaps, the most that can be said about the error interval in any direct measurement is that we should choose it large enough to be *quite* confident (95% level) that the true value of the measured quantity would fall within the interval, but at the same time, small enough to reflect the accuracy of our actual measurement. For example, if the significant error in a measured quantity is its systematic error with only an upper bound on its magnitude known, e.g., $\pm 2\%$ of full scale deflection, then the size of the corresponding error interval should be at least as large as this upper bound. On the other hand, if the significant errors were purely normal random errors, then the size of the error interval should be at least three times the standard error. The frequent practice of using the smallest scale division on the measuring device as the error interval should be used with caution and only when successive measurements give practically the same result. In this case the measurement is scale-limited.

For the case of purely random errors in a directly measured quantity, the size of the error interval may also be expressed in terms of the standard deviation σ provided the number of observations n is taken into account. For $n \geq 10$, the size of the error interval may be taken as σ, but for $n < 10$, it is best to use 2σ to be on the safe side. Or, even better, we can simply use the mean deviation disregarding signs. The mean deviation is always larger than the standard deviation (about 1.25σ) and is easier to compute.

We express the results of a direct measurement of a physical quantity in the form $X \pm \Delta X$ where X is the assigned value (usually the mean of several measured values) and $|\Delta X|$ is the size of the corresponding error interval.[10] For convenience we shall call ΔX the *indeterminate error* in X, since its sign is unknown and its magnitude is an estimated upper bound on the most likely errors. In a similar manner $Y + \Delta Y$ gives the measured value of Y and its error interval.

Once we have obtained the values of $X \pm \Delta X$ and $Y \pm \Delta Y$, we can compute not only the value of R but also its error interval. Furthermore we can be sure that the true value of R will lie somewhere in this interval provided the true values of X and of Y lie somewhere in their respective intervals. The converse is not necessarily true.

There are two different methods by which ΔR may be computed. The obvious method is to compute the values of R by use of Eq. (I.7) for the four sets of arguments: $(X + \Delta X, \ Y + \Delta Y)$, $(X + \Delta X, \ Y - \Delta Y)$, $(X - \Delta X, \ Y + \Delta Y)$, and $(X$

---

[10]The symbol $|\Delta X|$ means the absolute value of ΔX, i.e., its magnitude regardless of its sign, and represents the size of the error interval. The total width of the interval is $2|\Delta X|$.

$- \Delta X, Y - \Delta Y$). The difference between the largest and smallest values of R obtained in this computation is twice the size of the error interval for R. This method while always valid is often tedious and cumbersome. The computations must be made with great care since the size of the error interval is one-half the difference between two numbers that are very nearly equal. Mistakes in computation are likely to prove fatal.

A more direct method is to compute the size of the error interval for R directly from Eq. (I.11). In order to do this we let ΔX and ΔY be the indeterminate errors in X and Y respectively and give them the signs (either $+$ or $-$) that will make ΔR as large as possible. This maximum value is always given by the equation

$$|\Delta R| = \left| \frac{\partial f}{\partial X} \Delta X \right| + \left| \frac{\partial f}{\partial Y} \Delta Y \right| \tag{I.17}$$

where $|\Delta R|$ is the size of the error interval for R.

Equation (I.17) is the basis for several different rules concerning the propagation of indeterminate errors by use of error intervals. The following three simple rules based on this equation suffice to handle a large proportion of the error problems encountered in the physics laboratory.

Rule I. If the result R is the sum or difference of two measured quantities X and Y, the indeterminate error in R is the sum of the errors in X and Y.

EXAMPLE I(a):

> Mass of bulb with air: 66.928 $\pm$ 0.001 gm
>
> Mass of bulb empty: 66.682 $\pm$ 0.001 gm
>
> Mass of air: 0.246 $\pm$ 0.002 gm

Note that, although the mass of the bulb is reliable to about 1 part in 67,000, the mass of the air is reliable to only 1 part in 123 or 0.8%. Also notice that the errors are added even though the masses are subtracted.

Rule II. If the result R is the product or quotient of two measured quantities X and Y, the *percentage error* in R is the *sum* of the *percentage errors* in X and Y.

EXAMPLE I(b):

> Mass of object: $M = 345.1 \pm 0.1$ gm
>
> Volume of object: $V = 41.55 \pm 0.05$ cm³
>
> Density of object: $D = \dfrac{M}{V} = \dfrac{345.1}{41.55} = 8.306$ gm/cm³
>
> Percentage error in M: $\dfrac{\Delta M}{M} \times 100 = 0.03$
>
> Percentage error in V: $\dfrac{\Delta V}{V} \times 100 = 0.12$

Percentage error in D: $\dfrac{\Delta D}{D} \times 100 = 0.15$

Error in D: $\Delta D = 0.012$ gm/cm^3

Density of object: $D = 8.31 \pm 0.01$ gm/cm^3

Note that in this case the error in the result affects the third place in the density. Hence, only three figures in D need be retained.

Rule III. If the result R is some power n of the measured quantity X, then the *percentage error in R is n times the percentage error in X.*

EXAMPLE I(c):

Diameter of a sphere: $d = 7.65 \pm 0.03$ cm

Volume of sphere: $V = \frac{1}{6}\pi d^3 = 234$ cm^3

Percentage error in d: $\dfrac{\Delta d}{d} \times 100 = 0.4$

Percentage error in V: $\dfrac{\Delta V}{V} \times 100 = 3 \times 0.4 = 1.2$

Error in V: $\Delta V = 3$ cm^3

Volume of sphere: $V = 234 \pm 3$ cm^3

5. Propagation of Determinate Errors. The rules for the propagation of determinate errors are based upon the same analysis as those for indeterminate errors. However, in this case, the errors have a *definite* sign that must be taken into consideration in combining errors.

6. General Example. Suppose we wish to compute the density D of a metal cylinder from measurements of its mass m, its length l, and its diameter d. At the same time we wish to compute the error in D resulting from errors in the measured quantities m, l, and d. We know that the density (mass per unit volume) is given by the equation

$$D = \frac{4m}{\pi d^2 l} \tag{I.18}$$

In order to obtain the corresponding error equation we take the differential of the logarithm of Eq. (I.18) and get

$$\frac{\Delta D}{D} = \frac{\Delta m}{m} - 2\frac{\Delta d}{d} - \frac{\Delta l}{l} \tag{I.19}$$

This Eq. (I.19) shows us exactly how the errors Δm, Δd, and Δl combine to give the error ΔD. If the errors are determinate (have a definite sign), then Eq. (I.19) is used as it stands. In this case it is perfectly possible that the errors on the right side of Eq. (I.19) might balance out leaving $\Delta D = 0$.

If, however, the errors are indeterminate ($\pm$), then it is clear that the signs in Eq. (I.19) should be chosen in such a manner as to give the largest value of ΔD. This value may be achieved by simply adding the various error terms on the right side of Eq. (I.19), regardless of signs. Hence, for indeterminate errors, the error equation may be written

$$\frac{\Delta D}{D} = \frac{\Delta m}{m} + 2\frac{\Delta d}{d} + \frac{\Delta l}{l} \qquad (I.20)$$

This is the general procedure for obtaining the indeterminate-error equation from the determinate-error equation. The determinate-error equation, in turn, can always be derived from the working equation by the method we have used above.

7. Significant Figures. In recording data and results it is customary to keep only figures that are trustworthy and have some significance. They are called significant figures, and are always determined by the amount of error in the value they express. An illustration of the idea of significant figures is shown in Example I(a). The mass of the bulb with air is given as 66.928 $\pm$ 0.001 gm. There are five significant figures in this value; the last one, 8, is in doubt by one unit as indicated by the amount of error, $\pm$0.001. On the other hand, the mass of the air written as 0.246 $\pm$ 0.002 gm has only three significant figures; the last one, 6, is in doubt by two units. The first 0 in this value is not counted as a significant figure because it is put in simply to emphasize the position of the decimal point. In Example I(b) the mass of the object has four significant figures, its volume also has four, but its density only has three, because the error in the density affects the third digit and makes it doubtful. Notice that the density was computed to be 8.306 but because of the error it was "rounded off" to be 8.31. In casting off nonsignificant figures, if the value of the rejected figures is one-half or greater than one-half unit in the last place retained, increase the last digit retained by 1; if it is less than half, leave this digit unchanged.

It is clear that the amount of error in any measured or computed quantity determines the number of significant figures in the value of that quantity. Therefore, in recording the value of any quantity, all figures up to and including the first affected by the error should be retained. For example, if the error in the value of a quantity is 1 part in 100 (1%), it is fairly evident that the number of significant figures in that value will never be more than three, although it may sometimes be only two. Consider a 1% error in the values 5.024, 1.135, and 9.807. These errors (to one place) are respectively 0.05, 0.01, and 0.1. Thus the values with their errors may be written 5.02 $\pm$ 0.05, 1.14 $\pm$ 0.01, and 9.8 $\pm$ 0.1. Notice that the first two values have three significant figures, although the last one has only two figures.

In computing with logarithms it is advisable to use a five-place table when the errors are approximately 0.01%, a four-place table for errors of about 0.1%, and a slide rule for errors of about 1%. Where angles are involved, errors of 0.01, 0.1, and 1% call for angles expressed to the nearest 1, 6, and 30 min, respectively.

In writing a very small or a very large number it is customary to express it as a power of 10. The number is written as the product of two factors. The first factor contains as many digits as there are significant figures, the decimal point usually appearing to the right of the first digit. The second factor is a power of 10. Thus the speed of light is written 2.99776×10^{10} cm/sec and implies that this speed has been determined to six significant figures.

8. Vector Errors. Many physical quantities such as forces or velocities are more conveniently expressed in vector form (directed line segments) than in scalar form (rectangular components). In the foregoing treatment of errors we have considered only scalar quantities. That analysis is sufficient to handle vector quantities and vector equations, since these may always be replaced by their scalar components. For example, the rectangular components of a vector of magnitude r which makes an angle θ with the x axis in the xy plane has the rectangular components,

$$\left. \begin{array}{l} x = r \cos \theta \\ y = r \sin \theta \end{array} \right\} \tag{I.21}$$

If r and θ are the measured quantities with errors respectively of Δr and $\Delta \theta$ (expressed in radians), it may be shown with the use of calculus that the errors Δx and Δy in the rectangular components are given by the equations

$$\left. \begin{array}{l} \Delta x = \Delta r \cos \theta - y \Delta \theta \\ \Delta y = \Delta r \sin \theta + x \Delta \theta \end{array} \right\} \tag{I.22}$$

Once the errors in the rectangular components of a vector are obtained, the errors in any sum or difference of rectangular components may be computed by the methods already given.

However, it is often advantageous to deal directly with vector quantities and their errors. The following error analysis does so for coplanar vectors that are to be added or subtracted.

Suppose we measure the magnitude and direction of some vector **F**, such as a force vector.[11] "Let us designate the magnitude of this vector by f and the angle which this vector makes with the x axis in the xy plane by θ. Since f and θ are measured quantities there will be an error associated with each of these measurements. Let these errors be $\pm\Delta f$ and $\pm\Delta \theta$. If we draw the vector **F** in the xy plane as shown in Fig. I.3, starting it at the origin O, we are uncertain about the position of the end

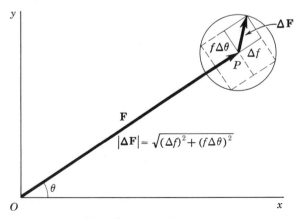

FIGURE I.3 Vector error.

[11]Letter symbols representing vectors are in boldface type.

point P because of the errors Δf and $\Delta \theta$. To indicate this uncertainty we draw the vector $\mathbf{F}$ as though f and θ were exact and then construct a region of uncertainty about the end point P. This region is essentially a small rectangle of half-length Δf and of half-width $f\Delta\theta$ with point P at its center. We can then be confident that the vector $\mathbf{F}$, considering errors, will have its end point lying somewhere in this rectangle. It is usually convenient to substitute for this rectangle its circumscribed circle, which has a radius equal to

$$\sqrt{(\Delta f)^2 + (f\Delta\theta)^2}$$

This substitution increases somewhat the region of uncertainty, but it has the distinct advantage of enabling us to represent the region of uncertainty by a single quantity—the radius of the circle. We then regard any radius of this small circle as a possible *vector error* $\Delta \mathbf{F}$ in the original vector $\mathbf{F}$. This statement implies that we know nothing about the direction of the real error in $\mathbf{F}$, but are certain that its magnitude is at most equal to the radius of the circle. Obviously this overstates the case for the error in $\mathbf{F}$, but in the treatment of errors it is always wise to be on the safe side.

Now, in the vector addition (or subtraction) of two or more measured coplanar vectors, it may be shown that their vector errors must always be added vectorially to obtain the vector error in the resultant. The student should convince himself of this fact by working out a simple example. Thus for the vector equation

$$\mathbf{R} = \mathbf{F}_1 + \mathbf{F}_2 \qquad \text{(vector)} \qquad \text{(I.23)}$$

there corresponds the vector error equation

$$\Delta\mathbf{R} = \Delta\mathbf{F}_1 + \Delta\mathbf{F}_2 \qquad \text{(vector)} \qquad \text{(I.24)}$$

Note the similarity of these vector equations to those in which scalar quantities are added (or subtracted). However, the error Eq. (I.24) is a vector, not a scalar equation; i.e., $\Delta\mathbf{R}$, $\Delta\mathbf{F}_1$, and $\Delta\mathbf{F}_2$ are all small vectors. In general, we cannot carry out the operation indicated in the equation because we know only the magnitudes of $\Delta\mathbf{F}_1$ and $\Delta\mathbf{F}_2$, not their directions. We cannot therefore determine the vector $\Delta\mathbf{R}$ by means of this equation. The best we can do is to determine the maximum magnitude that $\Delta\mathbf{R}$ can have. Obviously, this magnitude designated by $|\Delta\mathbf{R}|$ must be less than or at most equal to the sum of the magnitudes of $\Delta\mathbf{F}_1$ and $\Delta\mathbf{F}_2$. To be on the safe side we use the equality sign and write

$$|\Delta\mathbf{R}| = |\Delta\mathbf{F}_1| + |\Delta\mathbf{F}_2| \qquad \text{(I.25)}$$

where

$$\left. \begin{array}{l} |\Delta\mathbf{F}_1| = \sqrt{(\Delta f_1)^2 + (f_1\Delta\theta_1)^2} \\ |\Delta\mathbf{F}_2| = \sqrt{(\Delta f_2)^2 + (f_2\Delta\theta_2)^2} \end{array} \right\} \qquad \text{(I.26)}$$

In using Eq. (I.26) $\Delta\theta_1$ and $\Delta\theta_2$ must be expressed in radians.

In Fig. I.4 we represent graphically the propagation of errors in the addition (or subtraction) of measured vectors. The vectors $\mathbf{F}_1$ and $\mathbf{F}_2$ are combined by the parallelogram method to determine their resultant $\mathbf{R}$. Around the end point of each of the vectors $\mathbf{F}_1$ and $\mathbf{F}_2$ we draw a small circle whose radius, given by Eq. (I.26), represents the error in that vector. Around the end point of the resultant vector $\mathbf{R}$ we

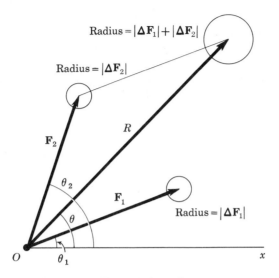

FIGURE I.4   Propagation of vector errors.

construct a circle whose radius is the sum of the other two radii. This circle represents the error in $\mathbf{R}$ in the sense that the end point of vector $\mathbf{R}$ will certainly lie somewhere in this circle, provided the end points of vectors $\mathbf{F}_1$ and $\mathbf{F}_2$ lie somewhere in their respective circles.

A special case of considerable importance arises when three coplanar force vectors are in equilibrium. Their resultant in this case is zero. But an actual experiment involving such force vectors seldom, if ever, produces this result. The reason for the failure is obvious, since each of the measured force vectors has its attendant vector error. The magnitude of the computed resultant force in this case cannot be expected to be exactly zero. However, its magnitude should never be greater than the sum of the magnitudes of the vector errors in the three force vectors.

B. GRAPHICAL METHODS

Frequently, the relation between two varying quantities may be clearly shown by means of a graph or curve. The independent variable is usually plotted along the x axis (abscissa) and the dependent variable along the y axis (ordinate).

The choice of scales is arbitrary and should be made on the basis of convenience and completeness of representation. In general, each scale unit should be chosen in such a manner that one-tenth of the smallest division on the coordinate paper represents a unit of the last significant figure of the measurement.

It is not necessary that the intersection of the two axes correspond to the zero point of each scale. The scale values assigned to this intersection point should be such that the curve occupies as much of the coordinate paper as possible and still satisfies the provisions of the previous paragraph.

The experimental values obtained are represented on the coordinate paper by means of sharp dots with small circles drawn about them.

A *smooth* curve drawn as nearly as possible through these points, so that very few points are far from the curve and so that there are as many points on one side of it as on the other, will graphically represent the observations. The exact form of the curve is a matter of judgment. An error in an observation may be indicated by the erratic location of a point.

A title for the curve should always accompany the curve. Also, the coordinates along each axis should be labeled with a statement of the quantity plotted and the units in which it is expressed.

A typical graph is shown in Fig. 44.2.

C. METRIC SYSTEM

In general, all measurements in the laboratory are to be made in the metric system of units, unless distinctly required otherwise. The fundamental units of length, mass, and time in the metric system are either (1), the centimeter, the gram, and the second (cgs units), or (2), the meter, the kilogram, and the second (mks units).

Upon these three fundamental units are founded the many derived units in physics, such as the dyne, erg, joule, watt, and so on. There are also related units such as the calorie, degree centigrade, ampere, and others. These units are generally described in connection with the experiment in which they occur.

A table of metric and English equivalents may be found in your general physics textbook.

D. GENERAL INSTRUCTIONS

1. Laboratory Materials. In addition to the laboratory manual you will need a notebook.

2. Rules of Conduct:

(a) Much of the apparatus is delicate and must be handled with care. The student should *never* try to operate the apparatus until the signal to proceed has been given.

(b) Breakage of any part of the apparatus always means an interruption in the experiment and a handicap to the group working with it. The student should examine the apparatus carefully before using it and should take every possible precaution to prevent injury to it.

(c) If any apparatus is found in unworkable condition or any breakdown occurs during the experiment, it should be reported at once to the instructor so that repairs can be made as soon as possible.

(d) Calculations and scratch work are to be done in the laboratory notebook provided by the student, not on the table tops.

(e) Each student in a group is individually responsible for the condition in which its location is left at the end of the period.

(f) Apparatus is not to be moved from one location to another without the instructor's permission.

E. CONCERNING THE EXPERIMENTS

The experiments in this manual cover a wide range of general physics topics and make use of a large assortment of standard laboratory equipment. The order of the experiments follows the conventional pattern beginning with those in mechanics (Part I) and ending with those in atomic physics (Part VI). The experiments are numbered consecutively in each part. There are gaps in the numbering system between parts in order to permit the introduction of additional experiments, should the need arise, without revision of the numbering system. This system is merely a matter of convenience. The experiments need not be done in the order given in the manual. By proper selection the instructor can modify the order to fit the needs of almost any general physics course. For example, the laboratory course could start with Exp. 12 on specific gravity, or with Exp. 70 on reflection and refraction, or even with Exp. 81 on the cloud chamber.

The level of difficulty of the experiments ranges from the very easy to the very hard with the bulk of the experiments lying somewhere between these extremes. As an aid to the instructor in the selection of experiments, we have separated the experiments into three groups with respect to their level of difficulty. Group I contains the relatively easy experiments; Group II, those of intermediate difficulty; and Group III, those that require considerable previous laboratory experience. The approximate distribution of the experiments among the three groups is as follows:

Group I. Exps. 1, 2, 3A, 4 (Part I), 5 (Part I), 6A, 7, 8, 9, 12, 13, 20, 23, 32, 33, 34, 35, 39, 60, 61, 62, 70, 71, 76, 81.

Group II. Exps. 3B, 4 (Part II), 5, 6B, 10, 11, 14, 21, 30, 31, 36, 37, 40, 41, 42, 43, 44, 45, 46, 72, 73, 74, 75, 77, 82.

Group III. Exps. 22, 38, 47, 48, 49, 80.

Note that three of the experiments (3, 6, and 44) are double experiments consisting of two independent experiments, A and B, covering the same subject matter. Each A or B division should be regarded as a distinct experiment.

In general, the experiments in Group I can be used without change in a laboratory course with two-hour laboratory periods. There will be ample time for the student to perform the experiment and take the data during the period. But computations and report writing normally have to be done outside the laboratory.

The experiments in Group II are more difficult than those in Group I and frequently contain more experimental material than can be covered in a two-hour laboratory period. However, any one of these experiments can be adapted to the two-hour period by a proper selection of the material given in the manual. Who should make this selection? The authors believe that the person in charge of the laboratory with his knowledge of the nature of the course, the type of students involved, and the conditions in the laboratory is best qualified to make this selection wherever and whenever it seems advisable.

The experiments in Group III are the most difficult experiments in the manual and require considerable laboratory sophistication on the part of the students. Note that these experiments occur throughout the manual and are distributed according to subject matter. In our opinion none of these can be satisfactorily performed in a single two-hour period. If these experiments are assigned, arrangements should be made to extend the laboratory period to three or four hours.

MECHANICS

Random Fluctuations:
Scatter of Shots

Object: To study the scatter pattern of a metal ball projectile fired repeatedly from a spring gun under the same conditions.

Apparatus: Spring gun with a metal ball projectile mounted on ballistic pendulum apparatus (Cenco-Blackwood or Beck Ball Pendulum), carbon paper, $\frac{1}{4}$ in. cross-section paper, adhesive tape, and a wooden target box. As shown in Fig. 5.2 in Exp. 5, the apparatus containing the gun is mounted on a table and the target box is taped to the floor in an appropriate position (several feet away from the gun) to receive the metal ball fired from the gun. The bottom of the target box is lined with carbon paper attached to the cross-section paper. When the ball strikes this lining, it produces a carbon spot on the cross-section paper at the impact point thus recording the striking position of this shot. For repeated firings, the transverse and longitudinal scatter pattern of the set of shots is clearly revealed and recorded.

Theory: The student should study Secs. A.2 and A.4 in the Introduction in preparing for this experiment since the basic theory is given there.

The scatter or dispersion of the shots striking a target when fired from the same gun under the same conditions is an excellent example of random fluctuations due to a large number of small disturbances affecting the system. In this experiment we shall make a quantitative study of the *longitudinal* dispersion of ten shots fired under the same conditions.

We first determine the mean longitudinal position $\bar{x}$ of the set of shots recorded on the target paper with respect to an arbitrary transverse base line drawn on this paper. We then compute the standard deviation σ and standard error σ_m of the set. Actually the computed values of these distribution parameters are only estimates of the true values, since the set of ten shots is only a small sample of the infinite set that is assumed to give true values.

The next step in the study is to compare the observed distribution of the set with the normal distribution law. To do this we proceed by the methods given in Sec. A.4 of the Introduction. An approximate check on the normality of the distribution is to compare the number of shots falling in the intervals $\bar{x} \pm \sigma$, $\bar{x} \pm 2\sigma$, and $\bar{x} \pm 3\sigma$ with those predicted by the normal law. A more complete check is to construct the staircase curve of accumulated frequencies for the distribution and to compare it with the corresponding integral normal curve as in Fig. I.2(b) in the Introduction.

Method: Clamp the mounting of the spring gun to the table in order to fix its position. Swing the pendulum (not used in this experiment) up onto the rack so that it will not interfere with the free flight of the projectile (metal ball). Cock the gun by placing the ball on the end of the firing rod and pushing it back, compressing the spring until the trigger is engaged. Use a low spring tension for the first set of shots.

Fire the gun and note the point where the ball strikes the floor. At this point center the wooden target box with its paper recorder so that the open end of the box faces the gun. Tape the box to the floor and the paper recorder to the box so that neither will move when struck by the ball.

Fire the metal ball from the gun ten times using the same spring tension and the same paper recorder. It is essential that conditions remain constant during this set of shots.

Carefully remove and label the cross-section paper upon which is recorded the striking point of each of the ten shots. Generally there will be ten distinct spots on the paper corresponding to the ten shots. Occasionally, however, two spots will be partially or wholly superimposed forming a double spot. A record paper with one or more double spots is quite satisfactory for analysis provided you can distinguish between double and single spots; otherwise a second run of ten shots must be made with new target paper.

Repeat the experiment using a definitely higher spring tension in the gun. For this second case the target box must be moved farther away from the gun since the range will be increased.

The analysis for the longitudinal dispersion of the ten shots in each case is made as follows. Draw a set of rectangular axes on the target paper with the x axis in the longitudinal direction and the y axis in the transverse direction. It is convenient to draw the axes so that all of the target spots lie in the first quadrant of the xy plane. With this arrangement the x coordinates of the target spots represent the longitudinal positions of the spots relative to an arbitrary base line, i.e., the y axis.

On the target paper observe and tabulate the values of x_i, $i = 1, 2, \ldots, 10$, for the ten target spots using $\frac{1}{4}$ in. (cross-section unit) as a unit length. Estimate positions to one-half this scale division. Compute and record the mean position $\bar{x}$, the standard deviation σ, and the standard error σ_m, in these same units. Indicate by a red line at $x = \bar{x}$, the mean longitudinal position of the target spots. Also draw lines at $\bar{x} \pm \sigma$ and $\bar{x} \pm 2\sigma$. Count and record the number of spots lying in the interval $\bar{x} - \sigma$ to $x + \sigma$ and the number lying in the interval $\bar{x} - 2\sigma$ to $\bar{x} + 2\sigma$. Compare these numbers with those predicted by the normal distribution law.

On the target paper draw the accumulated frequency histogram (staircase curve) showing the longitudinal distribution of target spots, i.e., plot $N(x) = y$ versus x where $N(x)$ is the number of spots lying to the left of x. On this graph indicate by small circles the theoretical values of $N(x)$ given by the normal law for $x = \bar{x} - \sigma$, $x = \bar{x}$, and $x = \bar{x} + \sigma$. Use Table S in Appendix III to obtain these theoretical values.

Record: The data and results of this experiment are completely reported on the two target papers, one for low spring tension and the other for high spring tension in the gun. See Fig.1.1.

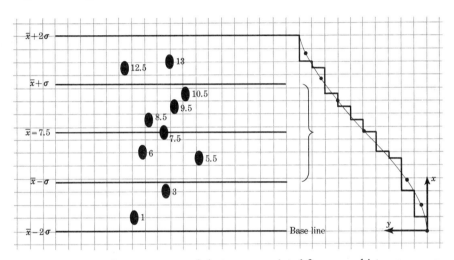

FIGURE 1.1 Scatter pattern of shots; accumulated frequency histogram.

QUESTIONS

1. Does the value of σ depend on the position of the base line for measuring positions of the target spots? Explain.

2. What fraction of the ten shots fall in the interval $\bar{x} \pm \sigma$ for case I, case II? Are these values compatible with those predicted by the normal law considering that the observed values of $\bar{x}$ and σ are only estimates of the true values?

3. Better values of $\bar{x}$ and σ could be obtained by increasing the number of shots in a set from ten to twenty. Would any unfavorable circumstances arise in this experiment as a result of this increase? Explain.

OPTIONAL EXPERIMENT

1. Instead of running two trials with different spring tensions in the gun, run two trials with the same spring tension. Under these circumstances the values of $\bar{x}$, σ, and σ_m for the two trials, I and II, may differ considerably. By use of your data and results and some additional statistical theory you should attempt to show that the difference in $\bar{x}$ values is not significant; i.e., the difference is most likely due to statistical fluctuation. Hint: Consider $\bar{x}(\mathrm{I}) - \bar{x}(\mathrm{II})$ and determine the standard error of this difference in terms of $\sigma_m(\mathrm{I})$ and $\sigma_m(\mathrm{II})$. Compare the magnitude of the difference of the means with that of its standard error. If this difference is less than three times the standard error of the difference, we usually assume that the difference is not significant.

Random Fluctuations:
Radioactive Decay

Object: To study the random fluctuations in the disintegration rate of a radioactive sample.

Apparatus: Scaler (Nuclear-Chicago #8770), detector (D-34 end window) plus mount, radioactive source (Sr^{90}, Cs^{137} or similar), and a timer. A description of the counter tube and scaler is given in Exp. 80 where this same equipment is used for another purpose.

Theory: Radioactive decay is the spontaneous and explosive disintegration of the nuclei of radioactive substances such as uranium or radium. In this explosive process the original nucleus usually emits an alpha particle (helium nucleus) or a beta particle (high speed electron) thus transforming itself into a different nuclear specie which may itself be radioactive. These microscopic nuclear explosions can be detected and counted by a Geiger counter tube, since the tube responds to any ionizing particle that enters the tube. In this experiment it is mostly β particles that are counted. Repeated counts over the same time interval of the β particles emitted by the source or its decay products show random fluctuations regardless of the precision of the detecting apparatus. These fluctuations stem from the statistical nature of nuclear decay which appears to be a purely random process.

It does not follow, however, that the randomness exhibited in nuclear decay obeys the normal distribution law. Rather the distribution is an asymmetrical (skew) type known as *Poisson's distribution*. This distribution is of great importance not only in particle physics but also in all fields where one is interested in the statistics of the number of like events occurring in some fixed time interval. Illustrations are: the number of calls at a telephone switchboard, the number of cars passing a given intersection, the number of suicides in a certain region, the number of corpuscles appearing in the field of a microscope, etc., each in some fixed interval of time.

Poisson's distribution is given by the relation,

$$P(n, v) = \frac{e^{-v}v^n}{n!} \tag{2.1}$$

where n is zero or any positive integer, v is a positive parameter, and $P(n, v)$ is the probability of observing the number n of like events in a given time interval. A detailed account of this distribution is given in Appendix I, Note B. There it is shown that the parameter v is equal to the weighted mean value of n, i.e., $v = \sum_{n=0}^{\infty} nP(n, v)$; and that the standard deviation σ is equal to $\sqrt{v}$.

Obviously the Poisson distribution differs from the normal or Gaussian distribution in several respects. It is discontinuous and asymmetrical rather than continuous and symmetrical. Also, it depends on a single parameter v rather than the two needed in the normal distribution. Nevertheless, for sufficiently large values of v, $v > 20$, the Poisson distribution becomes practically indistinguishable from a normal distribution of the random variable n with a mean value of v and a standard deviation of $\sqrt{v}$. In this experiment the value of v in counts per minute is considerably larger than 20 and hence the distribution may be treated as normal.

Method: Since high voltages are involved in this experiment, it is dangerous for beginning students to make the necessary electrical connections. These will be made in advance.

In order to operate the counter and scaler, and to obtain the data and results, take the following steps:

1. Turn the voltage control knob on the scaler to its lowest position (counterclockwise) and turn the count switch ON.

2. Turn the power switch ON and wait a few minutes for warm-up. Then slowly turn the voltage control knob clockwise, noting the corresponding increase in voltage reading, until the scaler begins to indicate pulses from the counter tube. These pulses should occur at about 800 v.

3. Observe and record the number of counts indicated on the scaler for a time interval of 60 sec using the timer, count switch, and reset button. The number should be larger than 100. If it is not, call the instructor before proceeding. Make nine more runs under the same conditions so that you have a set of ten counts each over a time interval of 60 sec.

4. Repeat step 3 obtaining a second set of data.

5. For each set of observations, compute and record the mean value of the counts and their standard deviation and standard error. In each case compare the computed standard deviation with that predicted by the Poisson distribution law, $\sigma = \sqrt{v}$.

6. Compare the mean values of the counts obtained from the two sets of data and examine the significance of any difference in the means. See Optional Experiment 1.1 for a possible test of significance.

Record: (*Partial Sample*)
Data and Results: Set I.
Apparatus Numbers:
  Timer # _____
  Scaling unit # _____
  Shield and counter tube # _____
  Time interval = 60 sec.

Trial	Count	Deviation	Square Deviation
i	n_i	$n_i - \bar{n}_i$	$(n_i - \bar{n}_i)^2$
1	231	19	361
2	203	−9	81
3	—	—	—
.	.	.	.
.	.	.	.
.	.	.	.
10	—	—	—
Sum	2125	0	1635

$$v \simeq \bar{n}_i = 212 \text{ counts} \qquad \sigma \simeq \sqrt{\frac{1635}{10-1}} = 13.5 \text{ counts}$$

$$\sigma_{\text{theor}} = \sqrt{v} \simeq 14.5 \text{ counts} \qquad \sigma_m \simeq \frac{13.5}{\sqrt{10}} = 4.3 \text{ counts}$$

QUESTIONS

1. What fraction of your ten counts, for each set of data, lies in the interval $\bar{n}_i \pm \sigma$? Compare these with the value given by the normal law.

2. Draw the accumulated frequency histogram (staircase curve) for either set of data and compare it with the integral normal curve.

Two-Body Collision:
Mass, Momentum,
and Energy

A. AIR TRACK

Object: To study energy and momentum principles as they relate to inelastic and elastic collisions of two gliders on a linear air track.

Apparatus: Air track, spark tape, 2 gliders, 2 spark timers, and a compressed air source. There are essentially two types of air tracks: the trough and the inverted trough. In each type the gliders move along the track on a cushion of air.

Theory: In this experiment we shall study the behavior of two colliding bodies on a linear air track (see Fig. 3.1). Because the linear air track is used, the motion is one dimensional, thereby simplifying the analysis of the motion. For the various collisions we may write the following equation:

$$M_1 U_1 + M_2 U_2 = M_1 V_1 + M_2 V_2 \qquad (3.1)$$
$$\underset{\text{(before)}}{} \qquad \underset{\text{(after)}}{}$$

Eq. (3.1) states that the momentum is conserved and, interestingly, is true for both elastic and inelastic collisions. If, however, one were to investigate the energy rela-

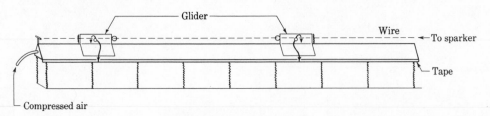

FIGURE 3.1   Air track.

tionship under the two situations, one learns that the mechanical energy is conserved for the elastic case only, that is;

$$\tfrac{1}{2}M_1 U_1^2 + \tfrac{1}{2}M_2 U_2^2 = \tfrac{1}{2}M_1 V_1^2 + \tfrac{1}{2}M_2 V_2^2 \tag{3.2}$$

For the inelastic case, the initial kinetic energy of the system is greater than the final.

Method: Each glider is complete with both spring and Velcro-tape bumpers. The spring loop is designed to produce collisions that approximate perfect elasticity. The Velcro-tape bumper is one of several different types that may be used for the inelastic collision. Depending on the type of collision the front bumpers will be either spring or Velcro-tape and, to achieve the opposite collision, one merely rotates each glider 180°.

In order to measure the velocities of the gliders both before and after the collision it is necessary to record the positions of the gliders at appropriate times. This is accomplished by using spark timers and spark-sensitive tape (see Fig. 3.2). Usually the track is designed to accommodate the spark tape and connectors.

Fig. 3.3 shows a typical set of points from which velocity is to be determined. Several methods may be used to compute the velocity of the glider, including the least squares method. However, to simplify the computation it is suggested that you consider the following method:[1]

1. Choose a displacement interval that consists of an odd number of dots as shown in Fig. 3.3.

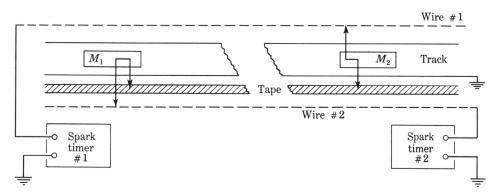

FIGURE 3.2   Glider connections with two timers.

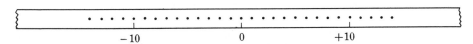

FIGURE 3.3   Spark tape; recorded glider positions.

[1]This method is based on the simple relation

$$R_j - R_{-j} = 2v\tau j$$

where R_j is defined in Step 4
   v is the magnitude of the average velocity over the distance of $R_j - R_{-j}$.
and τ is the time interval of the sparker

2. Mark the central dot as shown.

3. Place the edge of a metric rule against the series of dots with its zero point to the left of the interval.

4. Read off the scale positions of the dots from left to right. Record as R_{-10}, $R_{-9}, \ldots, R_{10}$.

5. Because the readings can be estimated to within 0.5 mm, each reading should be listed as 6.50, 7.55, etc.

6. Sum up the readings on each side of the central dot separately, e.g.,

$$\sum_{j=1}^{j=m} R_j \quad \text{and} \quad \sum_{j=1}^{j=m} R_{-j}$$

7. The difference between the sums may then be written as

$$\sum_{1}^{m} (R_j - R_{-j}) = 2v\tau \sum_{1}^{m} j$$

8. The above equation my be written

$$v\tau = \frac{\sum_{1}^{m} (R_j - R_{-j})}{2 \sum_{1}^{m} j} \tag{3.3}$$

Hence

$$v\tau = \frac{\sum_{1}^{10} R_j - \sum_{-1}^{-10} R_j}{110}$$

where τ is the time interval and $m = 10$.

Best results are obtained when relatively low velocities are used. An analysis showed that relative velocities between 20 and 40 cm/sec gave results which were within 1% error for momentum in both cases and for energy in the elastic collision. However, a change in velocity to 60 cm/sec for the same situations caused errors of approximately 10%.

If one could be sure that the interaction forces at the time of the collision took place along a line connecting the center of masses of the gliders, then good experimental results would be expected. When this condition exists no appreciable torques will be acting on the gliders.

Caution: Check the level of the track. This is most easily accomplished by observing the motion of a glider placed on the track (compressed air applied). The track is horizontal when the glider remains practically stationary.

Part I. Elastic collision. Measure the masses of the gliders equipped with bumpers. For the first step it might be desirable to use gliders of equal masses. Give to one glider a velocity U_1 and allow it to strike a second glider whose initial velocity is zero ($U_2 = 0$).

Compute and compare:

(a) the momentum of the system before collision and the momentum after the collision, and

(b) the kinetic energies of the system before and after the collision.

Part II. Inelastic collision. Repeat the steps in Part I using the special inelastic bumpers.

Part III. Optional steps.

1. Use gliders of unequal masses and repeat the steps in Parts I and II.
2. Use equal-mass gliders and initial velocities different from zero for both gliders. Repeat the steps in Parts I and II.
3. Consider unequal masses and initial velocities different from zero for both gliders.

B. AIR TABLE

Object: To make an experimental study of the changes of motion that occur in the collision of two metal disks. Hence: (1) to determine the ratio of the inertial masses of the disks; (2) to show that the linear momentum of a system is always conserved in collision phenomena; and (3) to show that the linear kinetic energy of a system is not necessarily conserved in collision phenomena.

Apparatus: Collision apparatus (M. I. T. design),[2] source of compressed air, spark timer, conducting paper, metric ruler, and protractor. The collision apparatus consists essentially of two brass disks floating on compressed air just above the surface of conducting paper placed in the bottom of a rectangular tray. Thus the disks are free to move in a horizontal plane and may be made to collide with each other. A space-time record of motion of the centers of the disks is obtained on the conducting paper by use of a spark timer. The method of supporting the disks virtually eliminates sliding friction; therefore, the velocities of the disks before and after collision may be determined directly.

Theory: Consider the horizontal motion of a single disk that is free of all external forces except its weight and the upward thrust of the compressed air on which it "floats" in a state of equilibrium. Since the vertical forces balance out and there are no horizontal forces (sliding friction has been virtually eliminated), the disk will move in a horizontal plane in accordance with Newton's first law of motion for any extended rigid body; i.e., the center of mass of the disk will move with constant velocity. The disk may also rotate with constant angular velocity about a vertical axis through its center of mass. These two motions are the only ones the disk may have under the prescribed conditions. Furthermore, these motions are completely independent of one another; each depends on how the disk was originally set into motion. The disk at rest is simply a special case of the above motions.

Center of Mass. At this point in our discussion it is well to digress in order to consider the concept of the center of mass of a system of particles. This concept is extremely important in the study of mechanics, because in many, but not all problems, the entire mass of the system may be considered as concentrated at the center of mass of the system.

[2]Robert G. Marcley, *Am. J. Phys.* **28** (1960): 670.

The center of mass of any system of particles is defined in the following way. Consider a system of n particles, whose masses are $m_1, m_2, \ldots, m_n$, and whose positions in space with respect to a rectangular frame of reference are given by the position vectors, $\mathbf{R}_1, \mathbf{R}_2, \ldots, \mathbf{R}_n$. These are a set of n vectors drawn from the origin of the reference system to the position of each particle (see Fig. 3.4). The position vector $\mathbf{R}$ of the center of mass of this system of particles, with respect to the frame of reference, is defined by the vector equation

$$\mathbf{R} = \frac{m_1\mathbf{R}_1 + m_2\mathbf{R}_2 + \cdots + m_n\mathbf{R}_n}{m_1 + m_2 + \cdots + m_n} \qquad \text{(vector)} \qquad (3.4)$$

It is clear from this defining equation that the position vector $\mathbf{R}$ of the center of mass is a "weighted" average of the position vectors of all the particles, the "weight factor" in each case being the mass of the particle. Note well that the words "weighted" and "weight factor" used in this connection have nothing to do with the gravitational weights of the particles, but instead indicate the relative importance of the terms being averaged.

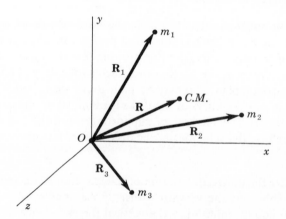

FIGURE 3.4   Diagram for $n = 3$.

Equation (3.4), which is a vector equation, can always be replaced by three scalar equations in the rectangular coordinates x, y, and z. This is left as an exercise for the student.

It can be shown by use of Eq. (3.4) and the calculus that the center of mass of any extended body of constant density and of regular geometrical form is at the geometrical center of the body, e.g., sphere, disk, cube, and so on. Furthermore, the center of mass of two extended bodies, such as two disks, can always be obtained by: (1) replacing each body with a single particle of mass equal to that of the body and located at the center of mass of the body; and (2) calculating the center of mass of the two resulting particles by use of Eq. (3.4).

Several important theorems in mechanics follow from Eq. (3.4) and the laws of motion of a single particle. A few of these theorems are stated below without proof. The proofs may be found in any textbook on theoretical mechanics.

1. In the statics of rigid bodies, the entire gravitational weight of a body may be considered to act at its center of mass; provided, of course, that the body is in a uniform gravitational field. In this case, the center of mass of a body is the same as its center of gravity.

2. In the kinematics of rigid bodies, the entire motion of the body is made up of a linear motion of its center of mass and an angular motion about its center of mass.

3. In the kinetics of rigid bodies, the total kinetic energy of the body is the sum of the kinetic energy of translation of its center of mass and the kinetic energy of rotation about an axis through its center of mass. The former is $\frac{1}{2}mv^2$, where m is the total mass of the body and v is the magnitude of the velocity of its center of mass. The latter is $\frac{1}{2}mk^2\omega^2$, where k is the radius of gyration of the body and ω is the magnitude of its angular velocity about the axis of rotation through the center of mass, provided the axis is fixed in direction.

4. In the dynamics of a rigid body, the center of mass of the body moves as though the entire mass of the body were concentrated there and all of the external forces were applied there. The linear or translational momentum of the body is $m\mathbf{v}$ (vector), where m is the total mass of the body and $\mathbf{v}$ is the velocity of its center of mass. Also, the body rotates under the action of the external forces about an axis through its center of mass. Its angular momentum about this axis is $mk^2\boldsymbol{\omega}$ (vector), provided the axis is fixed in direction.

The usefulness of the above theorems is apparent, since they enable us to treat the linear motion of an extended rigid body as though it were a single particle located at the center of mass.

Two-Body Collision: Momentum. We return now to the problem of the disks. The center of mass of each disk, including the apparatus permanently attached to it (lucite tube, electrode, and so on), will lie somewhere on the vertical axis of the disk since the attachments are all symmetrically placed with respect to this axis. Since the disks only move in a horizontal plane, their centers of mass will also move in that way. The space-time trace on the conducting paper will then represent the horizontal motion of the center of mass of each disk.

Suppose that the two disks, started by hand, move in such directions as to insure a collision. Before collision, and after the hand has been removed, the center of mass of each disk moves with constant velocity. Let these initial constant velocities be $\mathbf{u}_A$ and $\mathbf{u}_B$, the subscripts referring to the two different disks A and B. At a later time the disks will collide. During collision, the disks are in contact; the contact surface lying on the line of centers AB at the surfaces of the disks. During the time of contact, the disks exert forces on each other (action and reaction) applied at the surface of contact. Let $\mathbf{f}_A$ and $\mathbf{f}_B$ be the instantaneous forces exerted respectively on A and B. By Newton's third law of motion, these forces, $\mathbf{f}_A$ and $\mathbf{f}_B$, must, et every instant of time, be equal in magnitude, but opposite in sense of direction; i.e., $\mathbf{f}_A = -\mathbf{f}_B$. Note that there is no guarantee that these forces will lie along the line of centers AB, or that they will be constant in either magnitude or direction. In general, the small surface areas of the disks in contact during collision not only press against each other along the line of

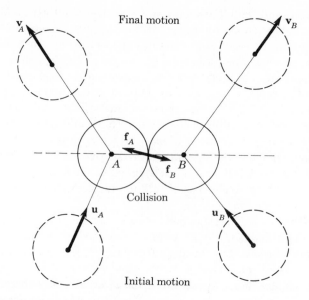

FIGURE 3.5 Two-body collision: reaction forces and velocity changes.

centers, but also tend to slide over each other. The former action produces central forces, whereas the latter produces tangential frictional forces (see Fig. 3.5).

Because of the highly variable and impulsive nature of the forces f_A and f_B during collision, it is customary in problems of this type to work with the impulses of these forces rather than with the forces themselves. The impulse of a force f during a time interval Δt, small enough so that the force is sensibly constant in this time interval, is $f\Delta t$ by definition. The effect of this impulse applied to any body is, by Newton's second law of motion, $m\Delta v$, where m is the mass of the body and Δv is the change in velocity of the center of mass of the body. Hence

$$f\Delta t = m\Delta v \qquad \text{(vector)} \qquad (3.5)$$

If a vector summation of Eq. (3.5) is made over the entire time of collision t, the left side of the resulting equation becomes the total impulse I exerted on the body during collision, and the right side becomes the total change of linear momentum of the body, i.e., the mass of the body m times the total change of velocity of its center of mass $(v - u)$. Thus, for the translational collision effect on the two disks A and B, we can write

$$\left. \begin{array}{l} I_A = m_A(v_A - u_A) \\ I_B = m_B(v_B - u_B) \\ I_A = -I_B \end{array} \right\} \qquad \text{(vector)} \qquad (3.6)$$

The first equation gives the translational effect of the collision on disk A, the second gives that effect on disk B, and the third relates I_A and I_B as action and reaction impulses between the disks. Obviously, $I_A = -I_B$ follows from the fact that $f_A = -f_B$ at every instant of time.

We have considered only the effects of the collision on the translational or linear motions of the disks. Rotational effects also exist, but we are not concerned with these in this experiment. Fortunately, the two effects are independent of one another; therefore, no error is introduced into the former by neglecting the latter.

By eliminating $\mathbf{I}_A$ and $\mathbf{I}_B$ in Eqs. (3.6), we obtain the single equation

$$m_A(\mathbf{v}_A - \mathbf{u}_A) = -m_B(\mathbf{v}_B - \mathbf{u}_B) \qquad \text{(vector)} \qquad (3.7)$$

Several conclusions of utmost importance in dynamics can be drawn from Eq. (3.7).

1. Since Eq. (3.7) is a vector equation, and since m_A and m_B are positive scalar quantities, this equation can only be satisfied providing that the vectors $(\mathbf{v}_A - \mathbf{u}_A)$ and $(\mathbf{v}_B - \mathbf{u}_B)$ are parallel, but opposite in sense of direction. Thus the *change* of velocity of the center of mass of A must be parallel, but opposite in direction, to that of B for any collision between A and B.

2. The ratio of the masses m_B/m_A must equal the inverse ratio of the magnitudes of the changes of velocity of A and B due to collision, i.e.,

$$\frac{m_B}{m_A} = \frac{|\mathbf{v}_A - \mathbf{u}_A|}{|\mathbf{v}_B - \mathbf{u}_B|} \qquad (3.8)$$

Equation (3.8) affords a means of determining the ratio of two masses in terms of purely kinematical quantities without making any use of the gravitational or weight property of mass, such as the use of a balance. The most sensitive tests fail to detect any difference between these two quite different methods of comparing masses.

3. The vector Eq. (3.7) may be rewritten in the form

$$m_A\mathbf{u}_A + m_B\mathbf{u}_B = m_A\mathbf{v}_A + m_B\mathbf{v}_B \qquad \text{(vector)} \qquad (3.9)$$

Each term in this equation is of the same form and represents a linear momentum, i.e., the mass of a body times the velocity of its center of mass. The left side of Eq. (3.9) is the initial momentum of the system before collision, whereas the right side is the final momentum of the system after collision. Thus Eq. (3.9) states that linear momentum is conserved in the collision process. This is a special case of the general principle of conservation of linear momentum, which holds not only for the collisions between disks, but for all bodies, large or small. Not only do metal disks obey the principle of the conservation of linear momentum, but so do galaxies, stars, men, atoms, nuclei, and electrons.

Two-Body Collision: Energy. Let us briefly consider the energy transactions that occur in the collision of the disks. The unwary student might assume that since linear momentum is conserved in the collision, mechanical energy would also be conserved. This assumption turns out to be invalid, in general, and is only true for what are called perfectly elastic collisions. Strictly speaking, the latter collisions are only possible in the sub-microscopic world of atoms and nuclei. In the macroscopic world of disks and billiard balls there is always some conversion of mechanical energy into heat energy during collision. Hence, the collision is inelastic; i.e., some mechanical energy is lost in the process. The disks possess both gravitational potential energy, owing to their positions, and kinetic energy, owing to their motions. Since they are constrained to move in a horizontal plane, their gravitational potential energy remains constant and does not change in collision. Thus any loss of mechanical energy due to collision must come at the expense of kinetic energy; i.e., the total kinetic energy of the disks after collision must be less than it was before collision. This statement does not mean, however, that the kinetic energy of translation of the system must necessarily decrease due to collision, since the system also possesses rotational kinetic energy that should be taken into account. In this experiment only the translational or linear kinetic energy of the system can be determined, not the rotational kinetic energy of the system. Fortunately, the linear kinetic energy usually dominates the rotational kinetic energy in this experiment, especially if some care is used in setting the disks into motion so that they initially have little or no rotational motion. Under these circumstances one finds that the total linear kinetic energy of the system after collision is usually less than that before collision. Thus the following inequality is nearly always satisfied

$$\tfrac{1}{2}m_A|\mathbf{v}_A|^2 + \tfrac{1}{2}m_B|\mathbf{v}_B|^2 < \tfrac{1}{2}m_A|\mathbf{u}_A|^2 + \tfrac{1}{2}m_B|\mathbf{u}_B|^2 \qquad (3.10)$$

The separate terms in this inequality represent linear kinetic energy, i.e., mass of body times the square of the magnitude of the velocity of its center of mass. Note that these terms are all scalar quantities. The left side of the inequality is the total linear kinetic energy after collision, whereas the right side is the kinetic energy before collision.

In this experiment we obtain as basic data a space-time record of the motion of the center of mass of each disk before and after collision. With these data we determine the velocities $\mathbf{u}_A$, $\mathbf{u}_B$, $\mathbf{v}_A$, and $\mathbf{v}_B$. From these initial and final velocities we obtain, either by graphical constructon or by trigonometrical analysis, the change of velocity of each disk $(\mathbf{v}_A - \mathbf{u}_A)$ and $(\mathbf{v}_B - \mathbf{u}_B)$ caused by the collision. The student must recognize that all of these quantities are vectors and must treat them accordingly. Once the vectors $(\mathbf{v}_A - \mathbf{u}_A)$ and $(\mathbf{v}_B - \mathbf{u}_B)$ are determined in magnitude and direction, the experimental validity of the conclusions drawn from Eq. (3.7) can be checked. If the masses of the disks A and B are not given, their ratio can be determined by using Eq. (3.8). In this case, Eq. (3.9), the conservation of linear momentum is automatically satisfied. If the masses are given or measured by use of a balance, then Eq. (3.9) can be checked experimentally. Finally, the validity of relation (3.10) can be checked.

It should be noted that in all of the foregoing relations involving either momentum or energy, the masses enter linearly. Therefore, only the ratio of masses is significant and the relations are valid no matter what the choice of unit mass may be, e.g., m_A may be chosen as a unit mass if it is desirable to do so.

Method: A schematic diagram of some of the essential features of the collision apparatus is shown in Fig. 3.6 (not drawn to scale). Each disk floats on a thin layer of compressed air issuing from a small hole centered in the bottom of the disk and supplied to the disk through a flexible hose. One end of the hose is attached to the lucite tube in the disk and the other end to a source of compressed air. The electrical lead from one terminal of the spark timer lies inside the hose. The hose to each disk is loosely supported several feet above the center of the tray in which the disks are placed.

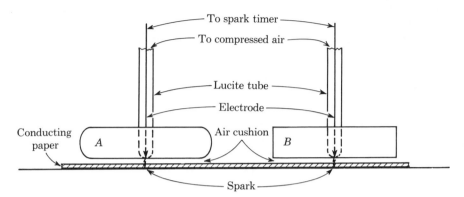

FIGURE 3.6 Collision apparatus.

This method of support tends to reduce external forces on the disks owing to hose connections. Small dots produced by sparks between the electrodes and the conducting paper record disk positions as a function of time on the conducting paper. The normal sparking rate is 60 per sec. Note that the shapes of the disks are not the same, A being rounded on the edge and B being flat. This is done to prevent any tipping of the disks during collision. It also enables one to distinguish between the disks without labelling them.

Examine the apparatus carefully, noting its construction and the manner in which air hose and electrical connections are made. Be very careful not to drop the disks in this operation. Connect the disks to the source of compressed air, but not to the spark timer as yet; then increase the air pressure slowly until the disks float freely on their respective air cushions. The disks will then start moving over the surface of the tray in response to any small, unbalanced horizontal forces. Level the tray and adjust the hose support so that the disks play over the central region of the tray and exhibit no tendency to slip to the sides of the tray.

Before attempting to collect any collision data by using the spark timer and conducting paper, make several preliminary trials without them. Gently set the disks in linear motion by hand, using the lucite tubes for this purpose, not the metal disks. This practice of handling only the lucite tubes of the disks, not the metal disks themselves, should be habitual in this experiment, since it is possible to get quite an electrical shock from the metal disks when the spark timer is ON. The initial velocities given to the disks should be about one foot per second, i.e., large enough to dominate any small erratic motions of the disks, but not large enough to produce appreciable permanent indentations in the disks when they collide. In starting a disk be careful not to give it any appreciable rotation and be sure to remove your hand well in advance of collision. Collision should occur somewhere near the center of the tray so that initial and final tracks are long enough to be analyzed. This can be accomplished by starting the disks either from adjacent or opposite corners of the tray. The change of velocity for each disk is more likely to lie along the line of centers in the former case than in the latter. Why? It is best not to try for a head-on-collision of the disks, for in this case it may be difficult, if not impossible, to distinguish between the initial and final tracks of each disk. It may be possible to get a collision in which, initially, one of the disks is at rest near the center of the tray. This rest condition is not an easy one to satisfy. Why?

After this preliminary work is done to your satisfaction, place a sheet of conducting paper (metallic surface down) flat on the bottom of the tray. Check to see that the air-borne disks move smoothly over it. Some releveling of the tray and readjustment of hose suspension may be necessary. Connect the electrical leads from the disks to the output terminals of the spark timer. Turn on the power source for the spark timer (110 v a-c), and allow the timer to warm up for a few minutes before attempting to use it. When everything is ready, start the disks in motion from appropriate starting points so that they will collide somewhere near the center of the tray. At the same time, depress the sparking switch on the spark timer, holding it down until the disks are just about to strike the edge of the tray after the rebound from collision. Release the sparking switch and stop the disks. A clear space-time trace of the track of each disk should appear on the conducting paper, somewhat similar to that shown by the dots in Fig. 3.7. For a good run the dots should lie on straight lines for several centimeters away from the points A and B and should be uniformly distributed in these intervals. Label the traces on the paper for this run. Be sure to show which disk, A or B, made which trace and indicate the direction of motion for each disk. Make one more trial of the experiment using a fresh paper. Vary the starting positions, initial velocities, and angles of attack, and repeat the above process. Each of the two students working on this experiment will then have two pairs of tracks but need analyze only one pair.

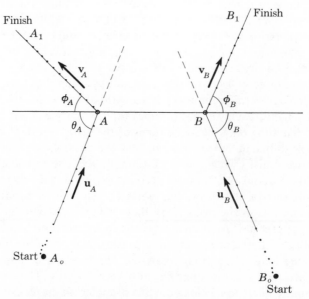

FIGURE 3.7   Space-time tracks of colliding disks.

Analysis of Data. An analysis of any one trial can be done as follows. On the conducting paper draw straight lines through the dots as in Fig. 3.7, thus obtaining the points A and B, i.e., the positions of the centers of the disks at collision. Do not use the starting portion of either track for this purpose, for there the disks are being accelerated by hand. Draw an extended straight line through the points AB and use this line as a reference line for determining directions. Determine the magnitude of each velocity $\mathbf{u}_A$, $\mathbf{v}_A$, $\mathbf{u}_B$, and $\mathbf{v}_B$. For this determination use the method described in Exp. 3A. Measure the angles ϕ_A, ϕ_B, θ_A, and θ_B indicated in Fig. 3.7 to within $\frac{1}{2}°$.

After having determined the magnitude and direction of the velocity vectors $\mathbf{u}_A$, $\mathbf{u}_B$, $\mathbf{v}_A$, and $\mathbf{v}_B$, proceed to determine the magnitude and direction of the change of velocity for each disk resulting from collision. There are three ways of doing so: (1) by graphical construction; (2) by trigonometrical analysis (sine and cosine laws); or (3) by the component method. We shall describe the graphical method and leave as possible exercises for the student the other two methods.

Draw the velocity vectors $-\mathbf{u}_A$, $\mathbf{v}_A$, $-\mathbf{u}_B$, and $\mathbf{v}_B$, using the extended line AB as a reference line. The velocity vectors $-\mathbf{u}_A$ and $-\mathbf{u}_B$ are just the vectors $\mathbf{u}_A$ and $\mathbf{u}_B$ reversed in direction. Using the parallelogram law, add vector $\mathbf{v}_A$ to the vector $-\mathbf{u}_A$, and vector $\mathbf{v}_B$ to the vector $-\mathbf{u}_B$. The resultant vectors $(\mathbf{v}_A - \mathbf{u}_A)$ and $(\mathbf{v}_B - \mathbf{u}_B)$ represent respectively the change of velocity of disk A and that of disk B resulting from collision. Measure and record the magnitudes of the vectors and the angles that each makes with the reference line (see Fig. 3.8 for an example).

The two velocity vectors $(\mathbf{v}_A - \mathbf{u}_A)$ and $(\mathbf{v}_B - \mathbf{u}_B)$ are of crucial importance in this experiment. First, according to theory, these vectors should be parallel to one another, but should point in opposite directions, since they give the directions of the

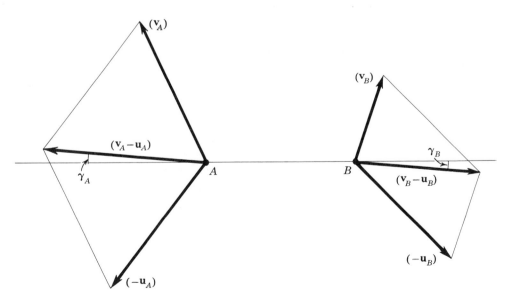

FIGURE 3.8   Graphical construction of velocity changes.

impulses of action and reaction on the two disks during collision. Therefore, γ_A should equal γ_B in Fig. 3.8 within the limits of experimental error. Measure and record these angles.

Second, the ratio of the magnitudes of these two velocity vectors should equal the inverse ratio of the masses of the two disks. By using Eq. (3.8), determine the ratio M_B/M_A. This ratio for any sequence of trials should be constant within the limits of experimental error. It is evident that determining M_B/M_A in this manner automatically satisfies Eq. (3.9), which expresses the principle of conservation of linear momentum for the system.

Finally, compute the initial and final linear kinetic energies of each disk, and check the validity of relation (3.10). In making this computation we can arbitrarily assign any value we choose for the mass M_A. It is most convenient to choose $M_A = 1$ mass unit.

Errors: Before we can draw any sure conclusions from the results of this experiment, it is necessary to make some estimate of the experimental errors in the data and their effect on the results. For example, it is not likely that the angle γ_A will be exactly equal to angle γ_B as determined in this experiment, although theory predicts their equality. In order to infer that they are equal within the limits of experimental errors we must know what these errors are.

Since we are dealing with vector quantities in this experiment, we must apply the corresponding error analysis. This type of analysis is discussed in Sec. 8 of the Introduction and, at this point, it would be wise for the student to review that section before proceeding further.

Make an error analysis similar to that given in the Introduction for your data and results, obtaining the errors in angles γ_A and γ_B, and those in the ratio M_B/M_A. Do not bother to make an error analysis of the kinetic energy relation unless your instructor asks for it.

Record: Tabulate your data and results as shown below. Be sure to include estimated errors in measured values and calculated errors in results.

TRIAL I

		A	B	
Initial:	displacement			
	time			
	$\lvert \mathbf{u} \rvert$			
	θ			
Final:	displacement			
	time			
	$\lvert \mathbf{v} \rvert$			
	ϕ			
Change:	$\lvert \mathbf{v} - \mathbf{u} \rvert$			
	γ			
Ratio:	$\dfrac{M_B}{M_A} = \dfrac{\lvert \mathbf{v}_A - \mathbf{u}_A \rvert}{\lvert \mathbf{v}_B - \mathbf{u}_B \rvert} = ($ $)$			

	A	B	$A + B$
Initial K.E. $(\frac{1}{2}M\lvert\mathbf{u}\rvert^2)$			
Final K.E. $(\frac{1}{2}M\lvert\mathbf{v}\rvert^2)$			

QUESTIONS

1. Under what circumstances would you expect the change of velocity of each disk due to collision between the disks to lie along the line AB?

2. Show by use of your data and results that the linear momentum of the system in any arbitrary direction is conserved in the collision process.

3. The uniform motion of the center of mass of the two disks is not changed during collision. Prove this fact by using your data and results for any one trial. HINT: determine the position of the center of mass of the two disks 0.1 sec before collision, at collision, and 0.1 sec after collision. The points so determined should lie on a straight line and be equally spaced on this line.

4. The coefficient of restitution (or resilience) of the two disks may be defined as the ratio of the relative velocity of recession of the two disks along their line of centers (after collision) to that of approach (before collision). Determine this coefficient, using your data and results. The value of this coefficient lies somewhere in the interval 0–1. It is 1 for a perfectly elastic collision and 0 for a completely inelastic collision.

OPTIONAL EXPERIMENTS

1. The analysis of the experimental data in this experiment may be made entirely in terms of component velocities before and after collision along the line AB and perpendicular to that line. This is perhaps a more systematic but less illuminating procedure than that given. With the instructor's permission you may subsitute this procedure for that given in the write-up.

2. The disks may be loaded with additional weights, thus changing the values of M_A and M_B. Try the experiment with loaded disks.

3. The interaction of the disks in collision may be changed by putting a large rubber band around each disk. Try this experiment.

Accelerated Motion:
Falling Body

LEAST SQUARES ANALYSIS

Object: To study the motion of a freely falling body. To show that its acceleration is constant. To determine this acceleration—the acceleration caused by gravity.

Apparatus: Free-fall apparatus, spark timer, waxed tape, plumb bob, steel rule, and vernier calipers. The apparatus consists of a supporting metal column with a solenoid at the top to hold and release the falling body (a metal cylinder). The column supports two vertical wires, one just in front and the other just in back of the falling cylinder. A long strip of waxed paper is held between the cylinder and the wire in back. Sparks produced by the spark timer jump from the wire in front to a tapered metal collar on the cylinder and then through the waxed paper to the grounded wire in back. Since these sparks are produced at a *constant* rate by the spark timer (usually 60 to 120 per sec), distance versus time for the freely falling body is recorded on the waxed paper (see Fig. 4.1 for the essential details).

Theory: In this experiment we study the motion of a freely falling body starting from rest and moving vertically downward along a straight line. The apparatus gives a record of the position of the body as a function of time. By an analysis of this record, we wish first to show that the motion is one of uniform or constant acceleration, and second, we wish to determine the value of this acceleration by different methods including the method of least squares.

In view of the above statement we have no right to use the equations of uniform acceleration in analyzing the data until we have shown that this is, in fact, motion of uniform acceleration.

The first step in the procedure is to get the relation between the distance of fall

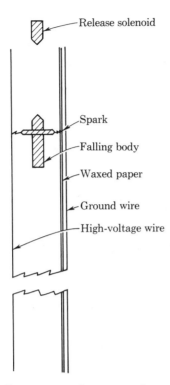

FIGURE 4.1   Free-fall apparatus (support column and spark timer not shown).

and the corresponding time of fall expressed in the form of a table of values. By using this table of values, we can plot a graph of distance versus time, obtaining a curve I similar to that shown in Fig. 4.2. The scale slope of this curve for any time t is the instantaneous velocity of the body at time t, since this slope represents the time rate of change of position of the body.

Using curve I, we can, in principle, construct a second curve II, giving the instantaneous velocity of the body as a function of the time, as in Fig. 4.3. The ordinates of curve II are the scale slopes of curve I plotted against the time. In actual practice this procedure is usually a very difficult one to carry out with any degree of accuracy. The important point here is the existence of curve II, since this curve has two properties that are very useful in analyzing straight line motion of any kind.

The first property of curve II is that its scale slope for any time t is the instantaneous acceleration of the body at that time. This follows from the definition of acceleration as the time rate of change of velocity. Therefore, the shape of curve II reveals the character of the acceleration. If, for example, curve II is a straight line, then we can be sure that the acceleration is a constant whose value is the scale slope of this straight line.

The second important property of curve II is that the scale area under any section

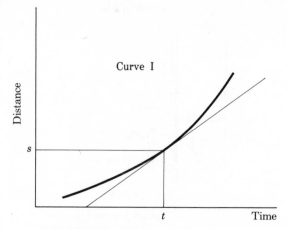

FIGURE 4.2 Distance vs time curve.

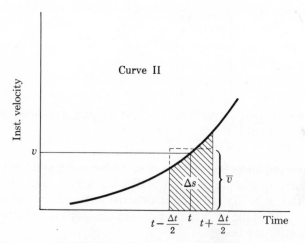

FIGURE 4.3 Velocity vs time curve.

of the curve (i.e., for any time interval) is equal to the distance traversed by the body in that time interval. For example, in Fig. 4.3, the scale area shown in the shaded section is equal to the distance traversed in the time from $t - \Delta t/2$ to $t + \Delta t/2$, where t is the time at the center of the interval, and the width of the time interval is Δt.

Now, the experimental record enables us to determine the distance of fall, i.e., the scale area under curve II, for a uniform sequence of time intervals since we can directly measure the distance of fall involved in each of the time intervals. Moreover, we can compute the average velocity of the body in any of these time intervals by dividing the distance of fall during the interval by the value of the interval. This average velocity is the average ordinate in the shaded section of Fig. 4.3; i.e., it is the height of a rectangle having the same area and the same base as the shaded area in the figure. Thus if $\bar{v}$ is the average velocity in this case and Δs is the distance traversed in the interval Δt, then $\Delta s = \bar{v}\Delta t$. It is clear from Fig. 4.3 that $\bar{v}$ equals the instantaneous velocity v at some point in the interval, but as a general rule, this point will neither lie in the center of the interval, nor will its position be independent of the size of the interval when the center of the interval is fixed. The one important exception to this general rule is the case when curve II is a straight line. In this case, and only in this case, both of the above conditions will exist; i.e., the average velocity $\bar{v}$ will equal the instantaneous velocity v at the center of the interval, and this average velocity will be independent of the size of the time interval as long as the center of the interval is fixed. Notice that, when curve II is a straight line, the shaded area in Fig. 4.3 becomes a trapezoid and the average height of this trapezoid $\bar{v}$ is just the velocity ordinate v at the time t, regardless of the width Δt of the trapezoid.

A test for uniform acceleration then amounts to computing the average velocity of the body for a sequence of time intervals about a fixed time-center of the intervals. If these average velocities turn out to be constant (within the limits of error of the experiment), then we can be sure that the acceleration of the body is constant. Furthermore, in this case, the average velocity in any time interval will equal the instantaneous velocity at the center of the time interval.

Even in the more general case of non-uniform acceleration, the above procedure enables one to estimate the instantaneous velocity at any time. Here it is found that the average velocity over the sequence of time intervals shows a definite trend as the time interval is shortened. It is then necessary to extrapolate to infinitely small intervals to obtain the instantaneous velocity at the center of the interval.

The preceding arguments concerning straight line motion of a body are not presumed to be rigorous. However, they can be made so using the calculus.

The space-time record of the falling body is registered on the tape as a linear sequence of dots showing the position of the body (edge of collar) as a function of time. Successive dots are separated in time by a constant time interval τ (usually $\frac{1}{60}$ sec), but the space intervals between successive dots increase because of the acceleration of the body.

In order to measure and tabulate the space and time coordinates of the dots, one of the dots is chosen as a reference or zero point for measuring distances and times. It therefore serves as an origin for both the space and the time coordinates of the dots. Actually any time and space origins may be chosen, but for reasons that will appear later, it is very advantageous to use the dot at the time-center of the sequence as a reference point. This means that the sequence of dots for analysis should consist of an odd number of dots. To satisfy this condition it may be necessary to cast out a few dots at the beginning or end of the tape. The choice of this central dot as a time-origin introduces a symmetry into the time coordinates of the other dots that simplifies both computations and results.[1,2] Having chosen the sequence, we label the reference dot with the index 0; the dots succeeding the reference dot we label in order of succession $+1, +2, \ldots, +m$ (the final dot), and the dots preceding the reference dot $-1, -2, \ldots, -m$ (the initial dot). Thus we can use the running index j to identify

[1] J. L. Bailey, *Ann. Math. Statistics* **2** (1931): 355.
[2] G. C. Cox and Margaret Matuschak, *Journal Phys. Chem.* **45** (1941): 362.

each of the dots by allowing j to take all integral values from $-m$ to $+m$. In this notation the total number of dots in the sequence is $2m + 1$, an odd number. The time coordinates of the dots will be $t_j = j\tau$ where j runs from $-m$ to $+m$. The corresponding position coordinates S_j of the dots are measured with respect to the central dot and given a positive sign for j positive, and a negative sign for j negative in accord with the usual sign convention. See Fig. 4.4.

With a table of j and S_j values available, we can at once apply the test for uniform acceleration by computing average velocities over a set of increasing time intervals centered on the reference dot $j = 0$. These average velocities by definition are:

$$\frac{S_1 - S_{-1}}{2\tau}, \frac{S_2 - S_{-2}}{4\tau}, \ldots, \frac{S_m - S_{-m}}{2m\tau} \tag{4.1}$$

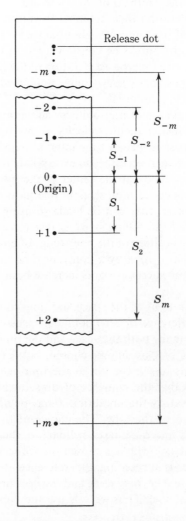

FIGURE 4.4   Dot pattern with position coordinates.

For uniform acceleration these ratios should all be the same within the limits of experimental error, and the common ratio should equal the instantaneous velocity v_0 that the falling body had at the reference dot $j = 0$.

If the test for uniform acceleration is satisfied, we can determine the instantaneous velocity v_j of the falling body at each of the dots (except for the initial and final dots) by computing the appropriate average velocity. Thus in general

$$v_j = \frac{S_{j+1} - S_{j-1}}{2\tau} \tag{4.2}$$

where j runs from $-m + 1$ to $m - 1$. We cannot determine v_{-m} and v_m by this method for obvious reasons.

By an exactly similar procedure we can determine the acceleration of the body at each of the dots (except for the first two and the last two in the sequence). The general equation for the acceleration a_j is

$$a_j = \frac{v_{j+1} - v_{j-1}}{2\tau} \tag{4.3}$$

where j runs from $-m + 2$ to $m - 2$. The values of a_j should be constant within the limits of experimental error.

The foregoing method of obtaining velocities from positions, and accelerations from velocities, is called the *method of successive differences*. The method is simple and direct but not accurate, since there is a large buildup of errors in the acceleration values and a disastrous nullification of original data if the acceleration values are averaged.

Another method of analysis is the *graphical method*. If we plot the values of v_j against those of $j\tau$, we should obtain a straight line whose scale slope is the acceleration of the falling body. This method clearly indicates the general features of rectilinear motion with constant acceleration, but it fails to give an accurate value for the acceleration. In this respect the graphical method and the method of successive differences have the same weakness in that the thing wanted (acceleration) is twice removed from the thing measured (position of dots).

The method of analysis that gives the best quantitative results in this experiment and in many others is called the *method of least squares*. See Appendix I, Note A, for a discussion of this method. Here we simply point out that the method of least squares is based on the assumption that the arithmetic mean of any number of equally reliable measurements of a physical quantity gives the most probable value of that quantity. The method makes the best possible direct use of the available data for computing results. But unfortunately the computations are often long and tedious and mistakes are not likely to balance out. In this situation a computer is a real asset. But even without computer facilities, the least squares computations in this experiment are sufficiently simple to be manageable.

In order to use the method of least squares in this experiment, we write the general equation of motion of the freely falling body in the usual form,

$$S = s_0 + v_0 t + \tfrac{1}{2}gt^2 \tag{4.4}$$

This equation gives the position S of the body at any time t in terms of its acceleration

g, its velocity v_0 at $t = 0$, and its position s_0 at $t = 0$. When we substitute the values of S_j and t_j from our data into Eq. (4.4), we get an equation for each dot in the sequence. The general form is

$$S_j \cong s_0 + v_0\tau j + \tfrac{1}{2}g\tau^2 j^2 \tag{4.5}$$

where j runs from $-m$ to $+m$. The dot equations (4.5) are only approximate as indicated by the symbol $\cong$, since they involve experimental values of S and t. However the value of t $(= j\tau)$ is usually known to a high degree of accuracy, since the spark timer normally uses the same source of electrical power as does the electric clock. Hence the primary source of errors in this experiment lies in the values of S_j. This is not only due to errors in measuring the positions of the dots but also is due to the fact that these positions may be in error because of small random fluctuations in the paths of the sparks. For example, if we stopped the falling body at some point in its path but kept the spark timer on, we would get on the tape a scatter pattern of dots very similar to the scatter pattern of shots obtained in Exp. 1. This scattering effect still persists when the body is in motion. Thus our method of analysis should attempt to nullify the effect of scattering by some averaging process. To do this we need a fairly large number of dots.

In order to simplify the dot equations (4.5) we let $s_0 = \epsilon$, $v_0\tau = b$, and $\tfrac{1}{2}g\tau^2 = c$ and regard ϵ, b, and c as our unknowns. These new unknowns then have the same units of length as the S_j. Equations (4.5) then become

$$S_j \cong \epsilon + jb + j^2 c, \qquad j = -m, \ldots, 0, \ldots, m \tag{4.6}$$

These equations when spread out are:

$$
\left.
\begin{aligned}
S_{-m} &\cong \epsilon - mb + m^2 c \\
&\ \vdots \\
S_{-2} &\cong \epsilon - 2b + 4c \\
&\ \vdots \\
S_0 &\cong \epsilon \\
&\ \vdots \\
S_{+2} &\cong \epsilon + 2b + 4c \\
&\ \vdots \\
S_{+m} &\cong \epsilon + mb + m^2 c
\end{aligned}
\right\} \tag{4.6a}
$$

Equations (4.6) or (4.6a) are called the *equations of condition* (or *observation*) in least squares nomenclature. We note that there is one equation for each dot, a total of $2m + 1$ equations. The equations are independent and equally reliable, since the S_j on the left are position coordinates of the dots all measured in the same way. One might argue that we have given undue weight to the central dot by choosing it as the zero point for measuring the other S_j. This would be true if we had assumed in Eq. (4.4) that s_0 ($\cong \epsilon$) was zero. But we did not do that. Instead we have included in the normal equations a *zero point correction* ϵ as one of the unknowns to be determined. But, although we have not assumed that ϵ is zero, we do assume that it will be small, since the true position of the falling body at $t = 0$ must be close to the position of the central dot.

The equations of conditions (4.6a) are incompatible whenever their number is greater than three (the number of unknowns). With three unknowns we need only three independent equations. In this experiment we may have nine times that number and there exists no single set of ϵ, b, c values that will satisfy all of the equations. The best that we can do is to seek a set of values that will most nearly satisfy all of the equations. The method of least squares tells us just how to treat the equations of condition in order to find this particular set of ϵ, b, c values. We shall first apply the method and then attempt to justify it.

For each unknown in the equations of condition (4.6 or 4.6a), the method of least squares prescribes a distinct operation on these equations that results in a single equation. Since in our case there are three unknowns, there will be three distinct operations giving three distinct equations called the *normal equations*. Once the normal equations have been obtained, they may be solved simultaneously for the three unknowns.

We start with the unknown ϵ. The prescribed operation is to multiply each equation of condition in (4.6a) by the coefficient of ϵ in that equation and then add all of the equations together. Since the coefficient of ϵ is 1 in all of the equations, this operation amounts to adding all of the equations of conditions. When we do this, we get the first normal equation

$$\sum_{-m}^{m} S_j = \sum_{-m}^{m} \epsilon + \sum_{-m}^{m} cj^2 \tag{4.7}$$

Note that the unknown b does not enter this equation because it has been balanced out in the summation process.

We repeat this same operation for the unknown b, but in this case the general coefficient of b is j. Therefore each equation of condition must be multiplied by its value of j before summing the equations. This operation gives us the second normal equation

$$\sum_{-m}^{m} jS_j = \sum_{-m}^{m} bj^2 \tag{4.8}$$

Here both ϵ and c have balanced out in the summation process and the value of b is given explicitly in terms of the measured values S_j.

Finally we repeat this operation for the unknown c. In this case the general multiplier of the equations is j^2 and in the summation b is balanced out. The third normal equation gives

$$\sum_{-m}^{m} j^2 S_j = \sum_{-m}^{m} j^2 \epsilon + \sum_{-m}^{m} cj^4 \tag{4.9}$$

The three normal equations are put in a more workable form by solving the first

(4.7) for c, the second (4.8) for b, and the third (4.9) for c. We get respectively:

(I)
$$c = \frac{\sum\limits_{-m}^{m}(S_j - \epsilon)}{\sum\limits_{-m}^{m} j^2} \tag{4.10}$$

(II)
$$b = \frac{\sum\limits_{-m}^{m} jS_j}{\sum\limits_{-m}^{m} j^2} \tag{4.11}$$

(III)
$$c = \frac{\sum\limits_{-m}^{m} j^2(S_j - \epsilon)}{\sum\limits_{-m}^{m} j^4} \tag{4.12}$$

The simplicity of the normal equations (4.10), (4.11), and (4.12) is due primarily to our choice of the central dot as a zero point for time measurements. Any other choice would have led to more complicated normal equations. The value of b is given directly by Eq. (4.11). The values of c and ϵ are given by the Eqs. (4.10) and (4.12). Since the central dot is also used as the zero point for space measurements, ϵ represents a small zero point correction in the observed position of the central dot.

We are primarily interested in the value of c $(= \frac{1}{2}g\tau^2)$, since it gives the acceleration of the falling body except for a constant multiplier. Thus for $\tau = \frac{1}{60}$ sec, $c = g/7200$. When we examine the two normal equations that involve c (4.10 and 4.12), we note that both have *small* corrections but that these corrections are not the same. It is easily shown that the correction in Eq. (4.12) is about one-half ($\frac{5}{9}$) that in Eq. (4.10). Therefore if we simply neglect these corrections by setting $\epsilon = 0$ in both equations, then Eq. (4.12) gives a better estimate of the value of c than does Eq. (4.10) unless ϵ is actually zero. In this latter case both equations give the same value of c. This allows us to test for the significance of the correction by computing and comparing the two approximate values of c, say c_1 and c_2, using the uncorrected equations

$$c \cong c_1 = \frac{\sum\limits_{-m}^{m} S_j}{\sum\limits_{-m}^{m} j^2} \tag{4.13}$$

and

$$c \cong c_2 = \frac{\sum\limits_{-m}^{m} j^2 S_j}{\sum\limits_{-m}^{m} j^4} \tag{4.14}$$

It can be shown to a high degree of accuracy that

$$c = c_2 + \tfrac{5}{4}(c_2 - c_1) \tag{4.15}$$

and that
$$c_2 - c_1 = -\frac{4\epsilon}{3m^2} \tag{4.16}$$

Thus a knowledge of the value of c_2 and of c_1 enables us to determine both c and ϵ.

In order to justify the method of least squares, we return to the incompatible equations of condition (4.6) and write them in the modified form

$$S_j - (\epsilon + bj + cj^2) = d_j, \qquad j = -m, \ldots, 0, \ldots, m \qquad (4.17)$$

where d_j, called *deviations*, are defined for any arbitrary values of ϵ, b, and c. The *principle of least squares* states that the most probable values of ϵ, b, and c are those that make the *sum* of the *squares* of the *deviations* an *absolute minimum*.

This accounts for the name "least squares". We assume that the values of ϵ, b, and c that satisfy this principle are indeed the values that "most nearly" satisfy the original equations of condition.

We show in Appendix I, Note A that the principle of least squares, when applied to our problem, imposes a number of restrictions (one for each unknown) on the deviations defined by Eqs. (4.17). In our case the three unknowns are ϵ, b, and c, and the corresponding restrictions on the deviations are that:

$$\sum_{-m}^{m} d_j = 0, \quad \sum_{-m}^{m} j d_j = 0, \quad \text{and} \quad \sum_{-m}^{m} j^2 d_j = 0$$

Now if we simply add all of the Eqs. (4.17) and set $\sum_{-m}^{m} d_j = 0$, we get the first normal equation (4.7) by a simple rearrangement of terms. In a similar manner we can get Eq. (4.8) and Eq. (4.9). Thus the method that we used in obtaining the normal equations is completely justified.

Method: It is necessary that the apparatus be carefully leveled before it is used. If the apparatus is of the wall type, i.e., fastened securely to the wall of the laboratory, it is likely to have been leveled when it was installed. If, however, the apparatus is of the movable type mounted on a tripod, then it will usually have to be leveled by use of a plumb bob or carpenter's level (see your instructor for the proper technique).

Inspect the apparatus, noting the release mechanism for the falling body, the way in which the spark timer is connected to the two wires (the high voltage terminal in the spark timer should be connected to the front wire), and the method of adjusting the waxed paper. Try a dry run without using the spark timer. Place the tip of the metal cylinder against the tip of the iron core of the release solenoid and close the appropriate switch. The cylinder should hang vertically downward without any swinging motion. Release the cylinder and observe its fall into the box below. *Be sure to keep your hands and other articles out of the path of the falling body*. The cylinder should fall smoothly between the front wire and waxed paper, without actually touching either.

Turn on the power supply for the spark timer and allow the tubes in the timer to warm up (about one minute). Replace the cylinder in its starting position (it must hang vertically without any swing). Press the spark switch on the spark timer. Then release the cylinder. Sparking at the collar of the cylinder should produce a series of visible dots on the waxed paper as the cylinder falls in front of the paper. The dots occur at equal time intervals, but not at equal distance intervals, because of the acceleration of the body.

Examine the dot pattern on the tape. If there are missing dots or double dots, the tape is unsatisfactory for analysis and the experiment should be repeated with a fresh tape.

When a satisfactory tape has been obtained, remove it from the apparatus, label it number 1, and replace it with a fresh tape. Repeat the experiment. Remove the second tape and label it number 2. Each student then has a different tape for analysis.

In order to measure the dot pattern on a tape, gently stretch it out full length on a table so that it lies flat and straight. The tension on the tape should be about the same as it was when the tape was suspended in the apparatus. Choose the sequence of dots to be analyzed. Start the sequence at a well defined dot a few millimeters below the release-dot as in Fig. 4.4 in order to avoid the confusion of dots at the starting point. The number of dots in the sequence should be odd and as large as the tape warrants. Label the central dot of the sequence 0 (the reference dot); label the dots preceding the reference dot in order $-1, -2, \ldots, -m$ and the succeeding dots $1, 2, \ldots, m$.

In order to determine the position coordinates S_j with respect to the reference dot it is not necessary to measure the S_j directly. It is better to use a double meter stick of 200 cm length for this purpose. Place the double meter stick on edge on the tape so that it is in alignment with the dots and reads in the direction of increasing j values of the dots. Then simply read off and record in centimeters the scale positions of the dots. Estimate scale positions to within $\frac{1}{2}$ mm. Let these scale positions of the dots be represented by R_j. Then the position coordinates of the dots with respect to the reference point are obtained by subtracting R_0 from each scale position value, i.e., $S_j = R_j - R_0$. Tabulate the values of j, R_j, and S_j as in Table 4.1 where some *sample* data and results are listed.

The check for uniform acceleration is now made by computing the value of each of the ratios:

$$\frac{S_1 - S_{-1}}{2\tau}, \frac{S_2 - S_{-2}}{4\tau}, \ldots, \frac{S_m - S_{-m}}{2m\tau} \tag{4.18}$$

It is convenient in these computations and in those that follow to use τ as our *unit* of *time*. For uniform acceleration the values should be the same within the limits of experimental errors and each ratio equals v_0 in cm/τ. There is no change in the results when R_j values are substituted for S_j values so that either may be used. Compute and record these ratios.

By use of the ratios (4.18) we can obtain a good value of v_0 by dividing the sum of the numerators of the ratios by the sum of the denominators. We get

$$v_0 = \frac{\sum_1^m (S_j - S_{-j})}{2 \sum_1^m j\tau} = \frac{\sum_1^m (S_j - S_{-j})}{m(m+1)\tau} \tag{4.19}$$

This method of averaging ratios is often called the *method of composition* and, in this case, gives a value of v_0 that is practically identical with the least squares value from Eq. (4.11). Therefore we shall use Eq. (4.19) rather than Eq. (4.11) for this purpose, since the computations are much easier.

Part I. Difference Method and Graphical Analysis. In this part of the experiment we shall use only the positive values of j and S_j starting with $j = 0$. These are sufficient

Table 4.1

Number of dots $= 2m + 1 = 29$, $\quad m = 14$
Unit time $= \tau(= \frac{1}{60} \text{ sec})$
Scale positions: R_j
Coordinate positions: S_j
$S_j = R_j - R_0$

j	R_j cm	S_j cm
-14	60.00	-41.10
-13	61.15	-39.95
.	.	.
.	.	.
.	.	.
-2	91.95	-9.15
-1	96.40	-4.70
0	101.10	0.00
1	106.05	4.95
2	111.35	10.25
3	116.85	15.75
4	122.60	21.50
.	.	.
.	.	.
.	.	.
14	195.50	94.40

$$\sum_{-14}^{14} S_j = 275.70 \text{ cm} \qquad \frac{S_1 - S_{-1}}{2\tau} = 4.82 \text{ cm}/\tau$$

$$\sum_{-14}^{14} |S_j| = 1016.5 \text{ cm} \qquad \frac{S_2 - S_{-2}}{4\tau} = 4.85 \text{ cm}/\tau$$

$$\vdots$$

$$\frac{S_{14} - S_{-14}}{28\tau} = 4.84 \text{ cm}/\tau$$

to reveal the general features of the motion of the falling body and furnish us with reasonably satisfactory results. For this purpose it is convenient to retabulate the positive values of j and S_j in a table with extra columns for computed values of velocity and acceleration.

Compute and record in Table 4.2 the instantaneous velocities v_j for $j = 1, 2, \ldots, m - 1$ by the method of differences suggested in the theory, Eq. (4.2). Repeat this difference process on the v_j values in order to obtain values of the accelerations a_j for $j = 2, 3, \ldots, m - 2$, Eq. (4.3).

Plot the S_j values from Table 4.2 as ordinates against the values of $j\tau$ as abscissas for $j = 0, 1, 2, \ldots, m$. Draw a smooth curve through the plotted points. The curve is a parabola. On the same graph and with the same abscissa plot the values of v_j as ordinates. These latter points should be on a straight line with a positive *scale slope* equal to the acceleration g in cm/τ^2 and with an intercept on the velocity axis equal to v_0 in cm/τ. Draw the best straight line you can through the $(j\tau, v_j)$ plotted points extending the line to get its intercept on the v axis. Record the value of this intercept and compare it with the value of v_0 given by Eq. (4.19). Determine the scale slope of this line in cm/τ^2 and compare it with the value of g given by Eq. (4.13).

Table 4.2

Unit time $= \tau$ ($= \frac{1}{60}$ sec)

$$v_j = \frac{S_{j+1} - S_{j-1}}{2\tau}$$

$$a_j = \frac{v_{j+1} - v_{j-1}}{2\tau}$$

j	S_j cm	v_j cm/τ	a_j cm/τ^2
0	0		
1	4.95	5.12	...
2	10.25	5.40	.25
3	15.75	5.62	
4	21.50		
...	...	...	...
14	94.40	...	...

Eq. (4.19): $v = \dfrac{\sum\limits_{1}^{14}(S_j - S_{-j})}{(14)(15)\tau} = \dfrac{1016.5}{210\tau} = 4.840$ cm/τ

Eq. (4.13): $g = \dfrac{2c_1}{\tau^2} = \dfrac{(2)(275.7)}{\tau^2(2030)} = 0.2720$ cm/τ^2

Part II. Least Squares. In order to make the computations for a least squares analysis, it is advantageous to modify Table 4.1 so that the two values of S with the same value of j except for sign appear on the same line in the table. This can be done by allowing j to run from 0 to $+m$ and assigning two values of the time and position coordinates to each value of j. The two time coordinates (not shown in the table) will be $j\tau$ and $-j\tau$. The corresponding position coordinates are S_j and S_{-j} shown in Table 4.3. This arrangement makes it easy to compute the values of $S_j + S_{-j}$ and $j^2(S_j + S_{-j})$ and their sums, which we need in the normal equations.

In Table 4.3 we list in the fourth column the computed values of $S_j + S_{-j}$, in the fifth column the values of j^2, and in the sixth column the computed values of $j^2(S_j + S_{-j})$. The sum of each column of figures is entered on the last line in the table. From these sums in the last line we compute the values of c_1, c_2, c, ϵ, and b as shown for the sample sums in Table 4.3.

Note that the value of c ($= \frac{1}{2}g\tau^2$) can be determined without knowing the values of ϵ and b. But we have included these latter two evaluations because they are necessary for the determination of the standard error in c. Also we have used Eq. (4.19) for computing the value of b rather than the more complex normal Eq. (4.11), since both give essentially the same result to four significant figures.

Errors: The method of least squares not only gives the most probable values of the unknowns ϵ, b, and c but it also gives the standard error in each of these quantities. But in order to determine the standard error in c, for example, it is first necessary to determine the minimum value of the $\sum\limits_{-m}^{m} d_j^2$. This evaluation although straightforward is exceedingly tedious. We substitute the least squares value of each of the unknowns into Eqs. (4.17), compute the value of each d_j, square this value, and then sum these squares over all values of j. This sum of squares will always be positive. The sum

Table 4.3

Position coordinates of dots with respect to central dot are S_j and S_{-j}.

j	S_j cm	S_{-j} cm	$(S_j + S_{-j})$ cm	j^2	$j^2(S_j + S_{-j})$ cm
0	0	0	0	0	0
1	4.95	−4.70	0.25	1	0.25
2	10.25	−9.15	1.10	4	4.40
.	.	.	.	.	.
.	.	.	.	.	.
.	.	.	.	.	.
14	94.40	−41.10	53.50	196	10,447.
Sum 105	646.10	−370.40	275.70	1015	34,703.

$$c_1 = \frac{\sum_1^{14}(S_j + S_{-j})}{2 \sum_1^{14} j^2} = \frac{275.7}{2030} = 0.1358 \text{ cm} \qquad \text{Eq. (4.13)}$$

$$c_2 = \frac{\sum_1^{14} j^2(S_j + S_{-j})}{2 \sum_1^{14} j^4} = \frac{34,703}{255,374} = 0.1359 \text{ cm} \qquad \text{Eq. (4.14)}$$

$$c = c_2 + 1.25(c_2 - c_1) = 0.1360 \text{ cm} \qquad \text{Eq. (4.15)}$$

$$\epsilon = -\tfrac{3}{4}m^2(c_2 - c_1) = -0.015 \text{ cm} \qquad \text{Eq. (4.16)}$$

$$b = \frac{\sum_1^{14}(S_j - S_{-j})}{2 \sum_1^{14} j} = \frac{1016.5}{210} = 4.840 \text{ cm} \qquad \text{Eq. (4.19)}$$

represents our best estimate of the sum of the squares of the deviations by which the observed dot-positions deviate from their least squares (or true) positions, i.e., from the quantities $\epsilon + bj + cj^2$.

We get the best estimate of the standard deviation of the measured S_j values by the familiar relation

$$\sigma \cong \sqrt{\frac{\sum_{-m}^{m} d_j^2}{2m - 2}} \qquad (4.20)$$

i.e., the root mean square deviation modified for three unknowns instead of just one. Once σ has been determined, it may be shown (see Appendix I, Note A) that the standard error in c may be calculated by means of the equation

$$\sigma^2\{c\} \cong \frac{(2m + 1)\sigma^2}{(2m + 1)\sum_{-m}^{m} j^4 - (\sum_{-m}^{m} j^2)^2} \qquad (4.21)$$

For $m = 14$, as in our sample data, Eq. (4.21) reduces to

$$\sigma^2\{c\} \cong 8.8 \times 10^{-6}\sigma^2 \qquad (4.22)$$

and then to

$$\sigma\{c\} \cong 3.0 \times 10^{-3}\sigma \qquad (4.23)$$

Thus the determination of $\sigma\{c\}$ by Eq. (4.21) depends directly upon the determination of σ by Eq. (4.20).

The complete data from which the sample data in Table 4.1 was taken gave (with the use of a computer) a value of σ equal to 0.028 cm. For normal random errors this means that the actual position of any dot on the tape has 68% chance of being within 0.028 cm of its true position, and that it is very unlikely that the error in its position would be more than 0.1 cm or 1 mm. Since the value of d_j for that dot represents the error in its position, we conclude that it is very unlikely that any of the d_j in Eqs. (4.17) would be more than 0.1 cm. This fact gives us a sensitive check on the accuracy of our data and results.

If computer facilities are available, proceed to Part III at this point. If not, carry out the following procedures.

From your data, calculate the least squares values of ϵ, b, and c by the method indicated following Table 4.3. Next, calculate a few values of d_j, say for $j = 0, \pm 1, \pm 10$, in order to see that you are in the right "ball park." If any of these values is greater in magnitude than 1 mm, this is practically prima facie evidence that you have made mistakes in your data or calculations.

Record the values of d_j for $j = 0, -1, 1, -10$, and 10. For the value of σ use the root-mean-square value of the deviations you have calculated or 0.03 cm, whichever is the larger. Then determine $\sigma\{c\}$. In these calculations it is not necessary to carry along more than two significant figures once the d_j for $j = 0, \pm 1, \pm 10$ have been computed. But the initial computation of the d_j by use of Eqs. (4.17) must be made with care.

For the remainder of the procedure, skip to Part IV.

Part III. Computer Determinations. Prepare a set of data for computer input in accordance with special instructions provided by your computer personnel. This will probably be done on a special form. In Appendix I, Note A, a computer program suitable for this experiment is reproduced. The data you will need to furnish includes the following:

1. A four-letter identification code (e.g., the first four letters of your last name),
2. The value of m (maximum value of j in Table 4.3),
3. Whether a table of residuals is desired ($1 = $ yes, $0 = $ no), and
4. A table of values of j, S_j, and S_{-j}, arranged in the same way as the first three columns of Table 4.3 (omit the zero values of j, S_j, and S_{-j}).

The computer program will make the calculations indicated following Table 4.3 and will print out results with your identification code at the top.

The first set of results includes the values of the constants from the normal equations. Examine these to see whether they are comparable to the sample results indicated below Table 4.3.

The second set of results consists of a table of values for d_j, the deviations in the individual measurements when using the least squares constants in Eq. (4.17). Examine these values. If any of them is greater than about 0.1 cm, you have probably made an error in measurement for the corresponding dot or in your calculations. Discuss any such cases with your instructor.

The third set of results contains error estimates in the computed values. The first of these, the "least sum squares deviations," is the sum of the squares of all the deviations, i.e., the value of the function $U(\epsilon, b, c)$. Moreover, it is the minimum value of this function, or the least squares value. The remaining values in the table are self-explanatory.

The final set of results is a table of the "Sum of squares of residuals." Each value in this table is comparable to the "least sum squares deviations" item above, except that they correspond not to the least squares values of c and ϵ substituted in Eqs. (4.17) but to a set of values *around* the least squares values. The entries in the table are therefore values of the function $U(\epsilon, b, c)$ in the neighborhood of the minimum, or least squares, value of U. Examine this table. Somewhere near the center of the table you will find a minimum value of U. Circle this value and note the corresponding values of c and ϵ. These should be very close to the least squares values printed in the first set of computer results.

It is possible from examination of the table to determine the standard errors in c and ϵ. By Eq. (4.20), we note that the value of σ^2 is given by $U_{min}/(2m-2)$. Divide your circled minimum value of U by the quantity $(2m-2)$ to determine the value of σ^2. This value should be in the vicinity of 0.001. Add this value to the value of U_{min} and reexamine the table of values of U. You will find that entries equal to the enlarged value of U surround the minimum value in the form of a tilted ellipse. Sketch in this "error eliipse". The significance of this ellipse is that 68% of measured values of U_{min} (from independent experiments on the same equipment) will fall within its boundaries.

Now draw a rectangle, with sides parallel to the sides of the table, just large enough to contain the error ellipse. The edges of this rectangle indicate maximum and minimum values of c and ϵ, corresponding to the standard errors in these quantities. From the table, determine these standard errors and record them.

Finally, it will be interesting to sketch in another ellipse, corresponding to 3σ for values of U. Multiply the previously determined value of σ^2 by 9, $(3\sigma)^2$ is $9\sigma^2$, and sketch in the indicated ellipse. This ellipse will contain approximately 99.5% of all values of U_{min} determined with the same equipment and is certainly a safe estimate of the upper and lower bounds of c and ϵ, when projected to the axes by means of the rectangle as before. Record these bounds.

Part IV. Final Calculations. Whether by hand calculations or by computer, you have determined values of σ and $\sigma\{c\}$. Computations from the complete example data (of which partial entries were shown in Tables 4.1, 4.2, and 4.3) give a value of $\sigma = 0.028$ cm. This, in turn, leads to a value of $\sigma\{c\} = 8.4 \times 10^{-5}$ cm. Since the example value of c is 0.1360 cm, the relative standard error in c is about 6 parts in 10,000. Therefore a reasonable upper bound in the error in c due to random errors would be about 3 times this or 2 parts in 1000. That is, the upper bound is taken to be about three times the relative standard error based on 29 observations. Thus the final value of c and g from the sample data could be written

$$c = [0.1360 \pm 0.00025] \text{ cm}$$

or

$$g = \frac{2c}{\tau^2} = [979 \pm 2] \text{ cm/sec}^2$$

Compute and record the value of g obtained from your data, its standard error, and the approximate upper bound on the errors. Compare your value of g with that given in Appendix III, Table P. The values of g given in Table P are not based upon free fall experiments but rather on pendulum experiments that are much more precise. The latter are likely to give slightly higher values of g. Is there any reason for this?

Record: Each student should analyze the data given on one of the tapes. Students should not use the same tape for analysis unless the instructor so directs.

QUESTIONS

1. An error of 0.01 % in the time intervals produced by the spark timer would produce what error in the value of the acceleration?

2. If S is a quadratic function of the time t, i.e., $S = \alpha + \beta t + \gamma t^2$, show (calculus) that v is a linear function of t and that the acceleration is constant.

OPTIONAL EXPERIMENT

1. There is another method of obtaining the value of g from this experiment once it has been established that the acceleration is constant. Since the body falls from *rest*, its instantaneous velocity v_0 at the central dot distance h_0 below its rest point is given by the equation

$$v_0^2 = 2gh_0 \tag{4.24}$$

The value of v_0 may be determined in the manner outlined in this experiment. The value of h_0 may be measured directly on the tape provided the rest position of the body is initially marked on the tape. Eq. (4.24) can then be used to calculate the value of g.

Velocity of
a Projectile

Object: To determine the initial velocity of a projectile (1) by measurements of range and fall, (2) by means of a ballistic pendulum.

Apparatus: Cenco-Blackwood ballistic pendulum, or Beck ball pendulum,[1,2] trip balance, metric steel scale or simple cathetometer, plumb bob, steel tape, and wooden target box with target paper (quadrille ruled paper under carbon paper under protective paper).

The Blackwood pendulum is a combination of a ballistic pendulum and a spring gun for propelling the projectile. The pendulum (Fig. 5.1) consists of a massive cylindrical bob *C*, hollowed out to receive the projectile, and suspended by a strong rod *K* whose upper end is pivoted at the top of a heavy support rod.

The projectile is a brass ball *B* that, when shot into the pendulum bob, is held there by the spring *S* in such position that its center of gravity lies in the axis of the suspension rod *K*. A brass indicator *I* is attached to the pendulum bob *C* in such a way that its tip indicates the height of the center of gravity of the loaded pendulum.

When the projectile is shot into the pendulum, it swings upward and is caught at its highest point by means of the pawl *P* that engages a tooth in the curved rack *R*.

The Beck ball pendulum is similar to the Blackwood pendulum except for some structural innovations. The pendulum truss is attached to a shaft that is permanently mounted at each end by a ball bearing. Because of this special mounting, the pendulum can be removed and replaced without disturbing the alignment and freedom of motion of the pendulum. The ball trapping mechanism is very effective. The spring gun is designed to fire at any one of three different spring tensions. A red dot

[1]Bernard O. Beck & Co.; P.O. Box 272, Arlington, Texas 76010.
[2]C. N. Wall, "Apparatus Review: The Beck Ball Pendulum," *Amer. J. Phys.* **36**, No. 12 (Dec. 1968): 1161.

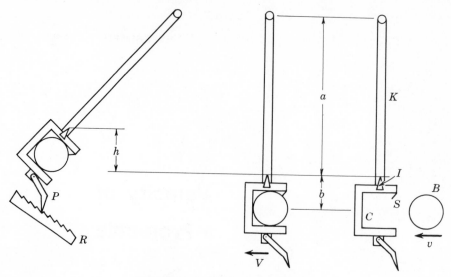

FIGURE 5.1 Ballistic pendulum.

on the pendulum just above the bob is the index point for measuring the vertical rise of the pendulum.

Theory: *Part I.* The initial velocity of a projectile shot *horizontally* from a gun and allowed to fall freely toward the earth may be determined in terms of the acceleration of gravity, the horizontal range of the projectile, and its vertical fall. In time t the projectile will fall vertically through a distance y given by the equation

$$y = \tfrac{1}{2}gt^2 \tag{5.1}$$

In the same time its horizontal displacement x will be given by

$$x = vt \tag{5.2}$$

where v is the initial horizontal velocity. See Fig. 5.2. If we eliminate t between these two equations and solve for v, we get

$$v = x\sqrt{\frac{g}{2y}} \tag{5.3}$$

By means of this equation we may compute v in terms of measured values of x and y and the value of g.

The determinate-error equation corresponding to Eq. (5.3) will be

$$\frac{\Delta v}{v} = \frac{\Delta x}{x} - \frac{1}{2}\frac{\Delta y}{y} \tag{5.4}$$

where the sign rules concerning determinate and indeterminate errors should be strictly observed.

Part II. Another method of determining the initial velocity of a projectile fired horizontally is the use of the ballistic pendulum.

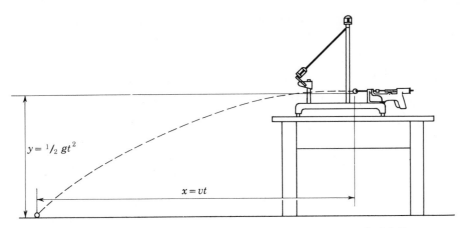

$y = \frac{1}{2} gt^2$

$x = vt$

FIGURE 5.2 Projectile motion: horizontal range and vertical fall.

Suppose the projectile of mass m and velocity v is fired into the pendulum initially hanging at rest in a vertical position. As a result of this collision, the pendulum bob with the projectile trapped inside it is given a velocity V. Since momentum is conserved even for an inelastic collision such as this, the following relation must be satisfied

$$mv = (m + M)V \tag{5.5}$$

The mass M in this equation is the *effective* mass of the pendulum rather than its real mass M_0 since the mass of the pendulum is actually distributed throughout the pendulum rather than being concentrated entirely in the bob of the pendulum. Only in this latter case would M and M_0 be the same, because only in this case would the entire mass of the pendulum have the same velocity V. Fortunately, in the apparatus used in this experiment most of the mass of the pendulum is concentrated in the bob so that the difference between M and M_0 is small. The relation between M and M_0 will be given later.

The velocity V given to the pendulum bob by the impact of the projectile causes it to swing up along a circular arc until the center of gravity of the loaded pendulum rises to such a vertical height h that its initial kinetic energy is entirely converted into potential energy, i.e.

$$\tfrac{1}{2}(M + m)V^2 = (M_0 + m)gh \tag{5.6}$$

Here again it is necessary to make the distinction between M and M_0.

By eliminating V between Eqs. (5.5) and (5.6), and then solving for v, we get

$$v = \frac{1}{m}\sqrt{2gh(M_0 + m)(M + m)} \tag{5.7}$$

The relation between M and M_0 is developed in the following section (fine print) and is shown to be

$$M + m = (M_0 + m)\frac{a}{a + b} \tag{5.8}$$

where a and b are the dimensions indicated in Fig. 5.1. Since b is small compared to a, the correction factor $a/(a + b)$ is only slightly less than unity.

In Eq. (5.7) we may substitute the value of $M + m$ from Eq. (5.8), obtaining

$$v = \frac{M_0 + m}{m}\sqrt{2gh\left(\frac{a}{a + b}\right)} \tag{5.9}$$

By the binomial theorem the quantity $\sqrt{a/(a + b)}$ is approximately equal to $1 - \frac{1}{2}b/a$, provided b is small compared to a. Hence Eq. (5.9) may be written as

$$v = \frac{M_0 + m}{m}\sqrt{2gh}\left(1 - \frac{b}{2a}\right) \tag{5.10}$$

The factor in parenthesis represents the correction due to the fact that not all of the mass of the pendulum is concentrated in the bob. If it were, this term would reduce to unity. Actually, this term has a value for the apparatus in this experiment of about 0.96 or 0.95 representing an error of 4 or 5%, if the term is not included.

The determinate-error equation for this part of the experiment may be obtained in the usual manner by taking the logarithmic derivative of Eq. (5.10). The factor in parenthesis is treated as a constant since small errors in M_0, m, b, and a will not appreciably change this factor. Thus we get

$$\frac{\Delta v}{v} = \frac{M_0}{M_0 + m}\left(\frac{\Delta M_0}{M_0} - \frac{\Delta m}{m}\right) + \frac{1}{2}\frac{\Delta h}{h} \tag{5.10a}$$

The relation between the effective mass $M + m$ of the loaded pendulum and its real mass $M_0 + m$ may be obtained from the formula for the kinetic energy of rotation of the loaded pendulum. As a result of the collision between projectile and pendulum, the loaded pendulum is given an initial kinetic energy of rotation of amount $\frac{1}{2}I\omega^2$, where I is the moment of inertia of the loaded pendulum about its axis of suspension, and ω is the angular velocity of the pendulum just after the collision. Now I in this formula may be replaced by $(M_0 + m)K^2$, where K is the radius of gyration of the loaded pendulum about its axis of suspension. Also ω may be replaced by $V/(a + b)$, where a is the distance between the axis of suspension and the center of gravity of the loaded pendulum, and where b is the distance between the center of gravity and the center of the bob. See Fig. 5.1. Thus

$$\frac{1}{2}I\omega^2 = \frac{1}{2}\frac{(M_0 + m)K^2}{(a + b)^2}V^2 \tag{5.11}$$

The right side of Eq. (5.11) may be written in the form

$$\frac{1}{2}(M + m)V^2$$

provided that $\qquad\qquad M + m = (M_0 + m)\dfrac{K^2}{(a + b)^2} \tag{5.12}$

Equation (5.12) gives the effective mass of the loaded pendulum in terms of its real mass and the constants K, a, and b. The constants a and b may be measured directly. The radius of gyration K could be determined by observing the period of oscillation of the loaded pendulum and applying the theory of the physical pendulum. This procedure is hardly necessary, however, because the pendulum is manufactured in such a way that its center of oscillation (or percussion) is approximately at the center of the bob at the distance $a + b$ from the axis of suspension. This means that this pendulum will oscillate like a simple pendulum of length $a + b$. Hence from the theory of the physical pendulum $K^2/a = a + b$. It follows then that the effective mass of the loaded pendulum is given by Eq. (5.8).

There is a second correction factor that may be significant and should be taken into consideration if necessary. In Eq. (5.6) we have assumed that in the swing of the pendulum no significant amount mechanical energy is dissipated due to friction at the bearings and of the pawl riding on the rack during a portion of the swing. This assumption may not be valid.

The friction effect of the pawl on the rack can be checked by repeating this portion of the experiment with the pawl held up from the rack by a rubber band. In this process, of course, the pendulum is not caught at its highest point by the pawl but swings back down where it must be stopped by hand before striking the gun. If this repetition leads to a significant increase in the amplitude of the swing of the pendulum and thus in the value of h, a correction may be necessary.

Let δ equal the small increase in the value of h when the friction of the pawl is eliminated. When this correction is put into Eq. (5.10), the resulting equation may be put in the form

$$v = \frac{M_0 + m}{m}\sqrt{2gh}\left(1 - \frac{b}{2a} + \frac{\delta}{2h}\right) \qquad (5.10a)$$

Note that the two corrections have opposite signs and thus may balance out. This actually seems to be the case in the Beck ball pendulum for the highest spring tension setting.

Method: *Part I. Initial Velocity from Measurements of Range and Fall.* Make sure that the apparatus is level and clamped in position on the table. In this part of the experiment the pendulum is not used and should be swung up onto the rack so that it will not interfere with the free flight of the ball.

Cock the gun by placing the ball on the end of the firing rod and pushing it back, compressing the spring until the trigger is engaged. Fire the gun and note the place where the ball strikes the floor. Place the wooden box with a carbon-paper recorder at this position. Shoot the ball five more times from the gun. Measure the horizontal range x of the ball along the floor from the point immediately below the projection point of the ball (use plumb bob) to the mean position of the points at which the ball strikes the bottom of the box on the floor. Estimate the average deviation of these points from the mean position and use this as the error in x.

At the same time measure the vertical fall y of the ball, making allowance for the box thickness and estimating the error in y.

By use of Eqs. (5.3) and (5.4) compute the initial velocity of the ball and the error in this velocity.

Part II. Initial Velocity by Use of Ballistic Pendulum. Release the pendulum from the rack and allow it to hang freely *without swinging. Without changing the spring tension* in the gun, load the gun and fire the ball into the pendulum bob. This will cause the pendulum with the ball inside it to swing up along the rack where it will be caught at its uppermost position. A scale along the outer edge of the rack provides a means for noting and recording the position of the pendulum on the rack. To remove the ball from the pendulum, push it out with the finger or with a rubber-tipped pencil, meanwhile holding up the spring catch.

Repeat this process four more times, recording each time the rest position of the pendulum on the rack. Determine the mean of these positions and set the pendulum at this position. Measure the height h_1 of the index point of the center of gravity above the base of the apparatus. Then release the pendulum, allowing it to hang in its lowermost position, and measure the height h_0 of the index point above the base. The difference between these readings gives the vertical height h through which the center of gravity of the loaded pendulum is raised as a result of the collision.

Weigh and record the masses of the pendulum and the ball. At the same time measure and record the lengths a and b of the pendulum. These distances need not be measured with extreme accuracy since they only appear in the correction term in Eq. (5.10).

From these data calculate the initial velocity v of the ball using Eq. (5.10). Also compute the *indeterminate* error in v. Compare the value of v obtained in Part II with that obtained in Part I.

Change the spring tension in the gun and repeat both Parts I and II.

Record:

App. No._____

Part I.

$g =$ Thickness of box $=$

Trial	x, cm	Height above floor, cm	y, cm	v_I, cm/sec
1				
2				

Part II.

$M_0 =$ $a =$
$m =$ $b =$

Trial	h_0, cm	h_1, cm	h, cm	v_{II}, cm/sec
1				
2				

QUESTIONS

1. In Part I of this experiment it is assumed that the floor is level. Suppose that this were not true but that the floor tipped down through a *small* angle θ in the direction of flight. Show that the true value of x would still correspond very closely to the value measured along the floor, but that the value of y as measured would have to be increased by $x\theta$ approximately. HINT: For small θ, $\cos \theta \doteq 1$; $\sin \theta \doteq \theta$.

2. Show by use of the binomial theorem that $\sqrt{a/(a + b)}$ reduces to $1 - (b/2a)$ for $b \ll a$.

OPTIONAL EXPERIMENT

1. Tilt the gun at various angles with the horizontal, keeping the spring tension constant. Modify Eqs. (5.1) and (5.2) to cover this case. Then determine v in terms of the measured values of x and y and the angle of tilt of the gun. Also determine experimentally the tilt angle of the gun for which x is a maximum. Is this tilt angle 45°? Explain.

Circular Motion:
Centripetal and Centrifugal Force
Stable and Unstable Motion

A. STABLE MOTION

Object: To determine the centripetal force on a body in uniform circular motion.

Apparatus: Centripetal force apparatus, motor-driven rotator, weight holder and weights, extender, template, divider, steel ruler, timer, and stroboscope.

The centripetal force apparatus (Cenco) as shown in Fig. 6.1 consists of a cylindrical bob B which slides freely on guide rods between two mechanical stops. A spring S of adjustable tension exerts the centripetal force upon the bob when the apparatus rotates about the vertical axis OV. This entire apparatus is mounted in a motor-driven rotator (not shown). By means of a variable-speed friction drive on the rotator, the rate of rotation of the apparatus can be adjusted so that the bob moves from its normal position to a predetermined position near the outer stop. There it actuates an indicator pointer P with its tip near the axis of rotation so that it is clearly visible at any angular speed.

Theory: When the apparatus shown in Fig. 6.1 is at rest, the spring S holds the bob B firmly against the inner stop. When we set the apparatus into rotation about its axis OV at a *gradually* increasing rate, we find that above a certain critical rate of rotation the bob will leave the inner stop and start moving out along the axis OH. If we now hold this rate constant, we find that the bob will do either one of two things: (1) it will come to rest relative to the frame of the apparatus at some point between the two stops and remain there; or (2) it will continue to move until it is brought to rest by the outer stop. In the former case the bob is said to be in *stable* circular motion, while in the latter case it is said to be *unstable*. It is the stable motion that we wish to use in Exp. 6A. We can obtain the stable condition by adjusting the initial tension in the spring so that its relative value is small rather than large.

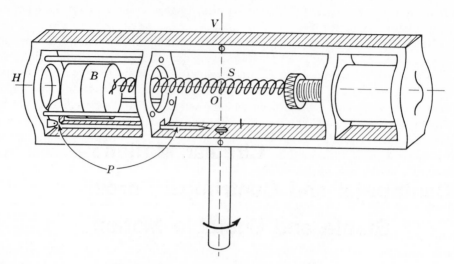

FIGURE 6.1 Centripetal force apparatus.

In stable motion the bob of mass m will move with constant speed v in a horizontal circle of radius r. The centripetal acceleration of the bob will be v^2/r directed toward the center of the circle (point O in Fig. 6.1). The spring force on the bob is the centripetal force F, the only unbalanced force on the bob except for the frictional force at the guide rods which is negligible. Therefore

$$F = -m\omega^2 r \tag{6.1}$$

where ω, the angular speed of rotation, is v/r and where the negative sign indicates an acceleration opposite in direction to that of increasing r values. Since $\omega = 2\pi n$, where $n = $ the number of revolutions per second, Eq. (6.1) can be put into the form

$$F = -m4\pi^2 n^2 r \tag{6.2}$$

The corresponding error equation is

$$\frac{\Delta F}{F} = \frac{\Delta m}{m} + 2\frac{\Delta n}{n} + \frac{\Delta r}{r} \tag{6.3}$$

By measuring m, r, and n, we can calculate the value of F by Eq. (6.2). But we can also measure F directly by loading the spring until it stretches the same amount as it does in the rotation experiment.

Method: Remove the centripetal-force apparatus from the rotator and examine it carefully to see how it works. Notice how the spring tension may be changed and also how the indicator level is actuated. Try to pull the cylindrical bob out to the end of the apparatus with your fingers. Note that a large force is required to do this.

Examine the rotator. Set the friction-drive wheel near the center of the driving disk and turn on the motor. Notice that the spindle of the friction-drive wheel turns very slowly, if at all. Try increasing the speed of the spindle by moving the friction disk away from the center of the driving disk. Engage and disengage the revolution counter. Stop the motor.

Reset the friction disk at the center of the driving disk, reclamp the apparatus on the rotator, and start the motor. Slowly increase the speed until the pointer rises and is even with the fixed index *I*. Unfortunately, it is impossible to keep the speed constant enough to hold the pointer in this position for any length of time without a continual readjustment of the position of the friction disk. One method of partially avoiding this difficulty is to run the apparatus at a speed *slightly* larger than the critical speed and then to apply a small amount of friction on the rotating spindle sufficient to reduce the speed to its critical value. The side of a pencil held against the spindle frequently works well in this respect.

Determine the value of *n* (revolutions per second) by counting the number of revolutions with the revolution counter over an interval of 2 min. To do this, record the initial reading of the counter, control the speed of the apparatus, engage the counter at the zero instant, disengage the counter at the end of 2 min, and record the final reading of the counter. To prevent the counter from spinning after disengagement, touch the counter gear lightly with the finger while it is being disengaged. Make two additional runs of 2 min each under the same conditions. Calculate the average *n* for the three runs. Use the mean deviation as Δn.

If a stroboscope is available, you may use it instead of the counter to determine the values of *n*. To do this illuminate the rotating apparatus with the flashing strobe light and bring it to "rest" by adjusting the frequency of the flashes. Be sure this latter frequency is equal to *n* and not some multiple or submultiple of *n*.

Remove the centripetal-force apparatus from the rotator and hang it from the support stand with bob down. Attach a weight holder to the bob and add weights until the bob reaches the same distance from the axis as it had when revolving. The total weight on the spring expressed in dynes (this includes the weight of the bob itself) is equal to the centripetal force *F*. The accuracy of the weights may be taken as $\pm 0.1\%$.

With the spring extended as in the previous paragraph, measure the radius *r* of rotation, i.e., the distance from the axis of rotation to the center line on the bob. If available, use the special extender and template for this measurement.

Calculate the centripetal force *F* by substituting the values of *n*, *r*, and *m* in Eq. (6.2). Compute the error in *F*.

Determine the percent difference between the observed and calculated values of *F*.

Change the spring tension and repeat the experiment.

Record:

App. No._____

Mass of bob:	$m =$
Radius of rotation:	$r =$
Time interval:	$t = 120$ sec

Case I.

No. of rev: $N = \begin{cases} \text{1st } (\quad) - (\quad) = (\quad) \\ \text{2d} \\ \text{3rd} \quad\rule{3cm}{0.4pt} \\ \text{Ave} \end{cases}$

No. rev per sec: $n =$

Centripetal force (calc): $F = (\qquad)$ dyne

Centripetal force (obs):

Bob	(	) gmf
Wt holder	(	)
Weights	(	)
Total	(	) gmf
F (obs) = (		) dyne

QUESTIONS

1. Show by dimensional argument that Eq. (6.2) gives the force in dynes provided m is in grams, r in centimeters, and n in revolutions per second.

2. Is one justified in using $\frac{22}{7}$ as the value of π in this experiment? Explain.

3. If, in Case I of this experiment, the apparatus were rotated with twice the angular speed needed for balance, with what force would the bob press against the end of the frame?

4. If this experiment were performed on Mars ($g_{\text{Mars}} = 0.4g_{\text{Earth}}$), which of your measured quantities would be different and by how much?

5. Compare the centripetal acceleration of the bob in one trial of this experiment with the acceleration of gravity.

6. If this experiment were performed in a freely falling elevator cage, which part of the experiment would succeed and which part would fail?

Digression: *Centrifugal and Other Inertial Forces.*

In the preceding experiment we said nothing about the existence of a centrifugal force pulling the bob away from the axis of rotation. The reason for this is quite simple: There was no such force acting upon the bob in the *frame of reference* we used to describe the motion of the bob. We used, without mention, a *laboratory frame of reference* for the purpose, i.e., one attached to the fixtures in the laboratory room and thus to the surface of the earth.

The laboratory frame of reference is, of course, normally used in most experiments without mention and is assumed to be an inertial reference system in which Newton's laws of motion are valid. Strictly speaking this is not true because of the rotation of the earth. But the effect of the earth's rotation on small scale phenomena, such as in this experiment, is quite negligible so we usually regard the laboratory frame of reference as an inertial frame.

For large scale phenomena on the earth such as weather, tides, ocean currents, etc., the rotation of the earth must be taken into account.

In order to introduce a *centrifugal* force into this experiment, we must change our frame of reference from that of the "stationary" laboratory to one fixed in the rotating apparatus i.e., a *rotating frame of reference*. In this new frame of reference we note that the bob is at rest. So by changing our reference system, we have converted the uniform circular motion of the bob into one of rest. This is a considerable simplification in the motion of the bob. But the big question that remains is whether we can modify Newton's second law so that it can still be used in the rotating frame of reference. The answer is yes. We have only to change the form of Eq. (6.1) without changing its substance, and then to interpret it in terms of the rotating frame of reference. To do this we write Eq. (6.1) in the form,

$$F + m\omega^2 r = 0 \tag{6.4}$$

The two terms on the left side of Eq. (6.4) are now regarded as forces on the bob acting along the longitudinal axis OH of the apparatus. There is no change in the interpretation of F, the centripetal force on the bob. But we now regard the term $m\omega^2 r$ as representing a *second force* on the bob that just balances out the centripetal force. This puts the bob into a state of equilibrium in the rotating frame of reference. We call this force the *centrifugal force* on the bob. The change from a laboratory frame of reference to a rotating frame of reference has transformed the centripetal acceleration in the laboratory system into centrifugal force in the rotating system.

As a result of the foregoing discussion on frames, we can describe the dynamics in Exp.6A in two different but equally valid ways depending upon the frame of reference we use.

(a) In the *laboratory frame of reference* the bob moves uniformly in a circle. Hence the bob is not in equilibrium but has a centripetal acceleration $(-\omega^2 r)$ directed toward the center of the circle. This acceleration, multiplied by the mass m of the bob, equals the unbalanced force of the spring acting upon the bob (the centripetal force) as in Eq. (6.1). There is no centrifugal force in this frame of reference.

(b) In the *rotating frame of reference* the bob is at rest. It has neither velocity nor acceleration. Hence the net force acting upon the bob is zero. There are two longitudinal forces acting on the bob: (1) the spring force pulling the bob toward the center of the apparatus; and (2) the centrifugal force pulling away from the center. These two forces balance out as in Eq. (6.4) leaving the bob at rest. There *is* a centrifugal force in this frame of reference but *no centripetal acceleration.*

Centrifugal force is a typical member of a general class of forces called *inertial forces.* One or more of these forces always arises whenever we use an accelerated frame of reference. It can be shown that the motion of a particle in an accelerated reference system is the same as though the system were an inertial reference system, *provided* the net force on the particle includes the inertial forces characteristic of the accelerated reference system. Centrifugal force, for example, is one of the inertial forces that always appears in a rotating frame of reference.

Two other inertial forces are worthy of comment. In a rotating frame of reference Coriolis[1] forces as well as centrifugal forces arise. A *Coriolis force* is a *transverse force* acting upon any particle in motion with respect to the rotating frame of reference. The magnitude of this force is $2m\omega u$, where u is the velocity of the particle with respect to the rotating reference system. The direction of this force is perpendicular to the velocity vector and to the axis of rotation. In Fig. (6.1) when the bob moves out along the guide rods, the Coriolis force on it is directed into the paper. A reversal of either u or ω reverses the direction of

[1]G. F. Coriolis (1792–1842), French civil engineer.

the Coriolis force. This force cannot affect the motion of the bob along the guide rods when there is no friction. Centrifugal and Coriolis forces are the two inertial forces that characterize a rotating reference system.

The other interesting inertial force is an antigravitational force in the frame of reference of a *freely* moving space ship. Here the gravitational forces of the neighboring astronomical bodies, earth, moon, sun, etc., accelerate the space ship in such a way that the characteristic inertial forces in this accelerated frame of reference balance out all of the external gravitational forces. Physical events inside the ship then go on as though there were no external gravitational forces, and the dynamics of these events become especially simple. Such a frame of reference is called a *Lorentz Inertial Frame*.

Inertial forces in accelerated frames of reference are similar to the gravitational weight force (*mg*) on a particle near the surface of the earth. For instance, any inertial force on a particle is always directly proportional to the mass of the particle. Furthermore, inertial forces, even though they obey Newton's second law of motion, do not obey Newton's third law. We cannot detect the reaction forces of the inertial forces, nor can we detect the reaction force to weight in a laboratory frame of reference. We need an inertial frame of reference for this purpose.

Inertial forces are often called fictitious forces. They are fictitious only in the sense that we do not "feel" these forces when we are not in the accelerated frame of reference. When we are in the accelerated reference system, such as a plane or a ship in rough weather, we have no difficulty whatever "feeling" the inertial forces.[2,3,4]

B. STABILITY CONDITIONS

Object: To study the stability conditions of uniform circular motion.

Apparatus: Same as for Part A. Not all models of the centripetal force apparatus (Cenco) have springs with sufficient adjustment to show instability. Only those with a scale at the adjusting collar seem to work.

Theory: As pointed out in Part A of this experiment, we can obtain either stable or unstable circular motion by appropriate adjustment of the initial tension in the spring of the centripetal force apparatus. For high initial tension the motion is likely to be unstable, while for low tension the motion is stable. What are the reasons for this?

In order to answer the preceding question, we must examine the complete equation of motion of the bob, not just the equilibrium equation as in Part A. Since the motion of the bob is simple linear motion with respect to the rotating apparatus, we choose for our reference system the *rotating frame of reference*. In this frame of reference the equation of motion of the bob is

$$F_i - F_s = ma_r \qquad (6.5)$$

where $-F_s$ is the force on the bob exerted by the spring, $+F_i$ is the *inertial* (centrifu-

[2]R. W. Pohl, *Physical Principles of Mechanics and Acoustics* (Edinburgh: Blackie & Son Limited, 1932), Chap. 8.

[3]Edwin F. Taylor and John A. Wheeler, *Spacetime Physics* (San Francisco: W. H. Freeman & Co., 1966), Chap. 1.

[4]Walter C. Michels, Malcolm Correll, and A. L. Patterson, *Foundations of Physics* (Princeton, New Jersey: Van Nostrand Reinhold Company, 1968), p. 183.

gal) force on the bob, and a_r is the acceleration of the bob along the guide rods of the apparatus at any radial distance r from the axis of rotation.

In Eq. (6.5) we use the normal sign convention that r is positive and that the positive direction is in the direction of increasing r values. With this convention the spring force is negative and the inertial force is positive.

There are other forces acting upon the bob, i.e., weight of the bob, normal force of the guide rods, and the Coriolis force. These are all transverse forces that balance out and contribute nothing to the acceleration a_r. The only exception is the frictional force of the guide rods on the bob. We have neglected the friction force in Eq. (6.5) because it is small compared to F_i and F_s. But we must always keep in mind that this friction force becomes important when $(F_i - F_s)$ is small.

Both F_i and F_s are specific functions of r that we need to know. The inertial force F_i as a function of r is, from Eq. (6.4),

$$F_i = m\omega^2 r \qquad (6.6)$$

The magnitude of the spring force is F_s which equals the tension in the stretched spring. This tension obviously depends upon the value of r. Hooke's law of elasticity tells us that the tension in the spring can always be written as a linear function of r. This fact can easily be verified by plotting measured values of F_s versus r. The graph will be a straight line. We assume that this is true and write F_s in the form

$$F_s = f + kr \qquad (6.7)$$

where k is the spring constant (constant for a given spring) and f is a constant dependent upon the adjustment of the initial spring tension in the apparatus. In terms of the straight line graph, k is the slope of the line and f is the intercept on the F_s axis. Obviously (for a given spring) k is fixed, but f may be varied by adjusting the threaded collar to which one end of the spring is attached.

We now substitute the values of F_i and F_s into Eq. (6.5) and get

$$m\omega^2 r - f - kr = ma_r \qquad (6.8)$$

the equation of motion of the bob relative to the apparatus. The values of r are confined to the interval $r_1 \leqq r \leqq r_2$, where $r = r_1$ and $r = r_2$ are, respectively, the positions of the bob at the inner and outer stops. In this interval $f + kr$ is always a positive quantity, since the spring is always under tension.

Therefore we can always adjust ω so that a_r vanishes at some point r_0 in the interval $r_1 \leqq r_0 \leqq r_2$. At this point the bob is in equilibrium (by definition). Let $\omega = \omega_0$ at this point. Then Eq. (6.8) becomes

$$m\omega_0^2 r_0 - f - kr_0 = 0 \qquad (6.9)$$

the equilibrium equation.

The nature of the equilibrium at $r = r_0$ for fixed ω_0 can now be determined by examining the motion of the bob in the immediate neighborhood of this equilibrium position. To do this we write Eq. (6.8) in the form

$$m\omega_0^2(r_0 + x) - f - k(r_0 + x) = ma_r \qquad (6.10)$$

where x is any small positive displacement of the bob from the equilibrium position. By use of Eq. (6.9) we reduce Eq. (6.10) to

$$m\omega_0^2 x - kx = ma_r \qquad (6.11)$$

Equation (6.11) tells us at once the nature of the equilibrium. If $m\omega_0^2 > k$, then the acceleration a_r has the same sign (direction) as the displacement x.

Therefore the displacement has produced an unbalanced force on the bob directed away from its equilibrium position. The bob will therefore leave this position, and the equilibrium is unstable. If on the other hand $m\omega_0^2 < k$, then the direction of the acceleration is opposite to that of the displacement, the unbalanced force on the bob is directed toward the equilibrium position, and the equilibrium is stable. Finally if $m\omega_0^2 = k$, then the acceleration is zero for any displacement and the bob is in neutral equilibrium.

Thus the sign of the quantity $(m\omega_0^2 - k)$ determines the nature of the equilibrium. But by Eq. (6.9) the intrinsic sign of f is the same as that of $(m\omega_0^2 - k)$. Therefore the stability conditions are as shown in the following table.

$(m\omega_0^2 - k)$	f	Equilibrium
$+$	$+$	Unstable
$-$	$-$	Stable
0	0	Neutral

An interesting graphical confirmation of the preceding results can be obtained by plotting graphs of F_s versus r and F_i versus r in the same diagram as shown in Fig. 6.2. For the case of low initial spring tension, lines (1) and (2) represent respectively the graphs of F_s versus r and F_i versus r. These two lines intersect at point S where $F_i - F_s = 0$ at $r = r_0$, the condition of equilibrium. This is a point of stable equilibrium, since a small *positive* displacement makes $F_s > F_i$, thus producing a *negative* net force on the bob and vice versa.

On the other hand, for large initial spring tension, lines (3) and (4) represent

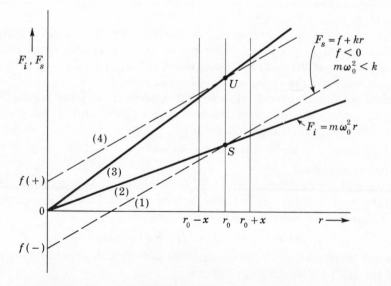

FIGURE 6.2   Graphical confirmation of stability conditions.

respectively F_i and F_s. These lines intersect at point U where again $F_i - F_s = 0$, at $r = r_0$. Again we have equilibrium at $r = r_0$, but now the equilibrium is unstable, since any small *positive* displacement makes $F_s < F_i$, thus producing a *positive* net force on the bob. Examine the case for a negative displacement.

Note that the F_s lines are all parallel having the same slope k but different intercepts f, whereas the F_i lines have different slopes $m\omega_0^2$ representing different values of ω_0 but the same zero intercept. Examine the case for neutral equilibrium.

Note also that an increase in the value of ω_0 makes r_0 larger for the stable condition but smaller for the unstable condition.

Method: In this experiment we first choose r_0, the equilibrium position of the bob, to be at the outer stop slightly beyond the indicator position. By making ω sufficiently large, we can always get the bob out to this position, where it is held by the outer stop. If we now *slowly* decrease ω until the bob just breaks contact with the outer stop as indicated by the fall of the pointer, then $r_0 = r_2$ and $\omega_0 = \omega_{02}$ represent equilibrium values.

Under the same spring condition we then choose r_0 to be at the inner stop, i.e., at $r_0 = r_1$. In this case we start with the bob held at the inner stop using a small value of ω. We now slowly increase the value of ω until the bob just breaks contact with the inner stop. At this point $r_0 = r_1$ and $\omega_0 = \omega_{01}$ represent equilibrium values.

By use of the stroboscope we can keep the rotating apparatus virtually "at rest" in the operations just described by adjusting the frequency of the light flashes to keep pace with the decreasing (or increasing) values of ω. At the onset of equilibrium we read the scale on the stroboscope (revolutions per minute) in order to determine ω_{02} and ω_{01}.

To test the stability of the motion we may have to reduce (or increase) the value of ω still more because of friction at the guide rods (neglected in the theory) and slight fluctuations in the value of ω. For stable motion the bob moves farther in (or out) but comes to rest and remains at some position between the stops. For unstable motion the bob continues to move until it strikes the other stop.

Determine ω_0 at each stop for maximum and minimum settings of the initial spring tension. For each setting make three observations of ω_0 at each stop. For each setting observe the stability of the equilibrium.

Determine the spring constant k by suspending the apparatus as in Exp. 6A and measuring the extension of the spring for an increase of 1 kg in the initial load. At the same time measure the values of r_1 and r_2.

Compute the values of ω_0, $m\omega_0^2$, $(m\omega_0^2 - k)r_0$ for each equilibrium point. Record the results as indicated in the following table.

		$m = (\quad)$ gm $k = (\quad)$ dyne/cm Stability		$r_1 = (\quad)$ cm $r_2 = (\quad)$ cm $m\omega_0^2$		
Initial tension	r_0 1, 2	U or S	ω_0 1, 2	$m\omega_0^2$ 1, 2	$(m\omega_0^2 - k)$ 1, 2	$(m\omega_0^2 - k)r_0$ 1, 2
Max	r_1 r_2					
Min	r_1 r_2					

Simple
Pendulum

Object: To investigate the relation between the period of a simple pendulum and its length; also to determine the acceleration of gravity.

Apparatus: Simple pendulum, timing device, meter stick, and vernier caliper. The simple pendulum consists of a small brass sphere B suspended from a rigid support S by means of a fine steel wire as shown in Fig. 7.1. The wire is fastened at its upper end to a small threaded rod fastened to a knife-edge holder K. The knife-edge rests in a groove on the support. A hole through the support under the middle of the knife-edge permits the pendulum to swing freely in a vertical plane about a line coincident with the knife-edge. The position of the nut N has been adjusted so that the effect of the supporting system on the period of the pendulum is negligible.

Theory: The vibration of a pendulum such as that shown in Fig. 7.1 is, for small amplitudes, an example of simple harmonic motion, the period of which is given approximately by the equation

$$T = 2\pi\sqrt{\frac{L}{g}} \tag{7.1}$$

where T = period, i.e., the time for a complete to-and-fro vibration, L = length from the point of suspension to the center of the bob, and g = acceleration of gravity.

Strictly speaking, Eq. (7.1) is only valid for infinitely small amplitudes and for a pendulum with all of its mass concentrated at the end of its suspension. However, the errors introduced by not being able to satisfy these conditions in the laboratory, are very small, provided that the amplitude of vibration does not exceed 2° and provided that the radius of the bob is small compared to the length of the pendulum.

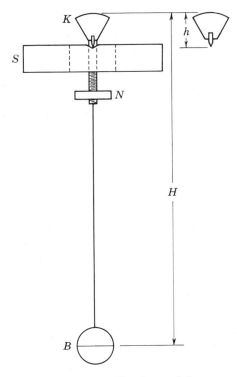

FIGURE 7.1   Simple pendulum.

Under these conditions Eq. (7.1) states that T is independent of the amplitude of vibration and is directly proportional to $\sqrt{L}$.

The vibration of a simple pendulum also provides a simple and accurate means of determining g, for both T and L can be determined in the laboratory, and hence g may be calculated by use of the equation

$$g = 4\pi^2 \frac{L}{T^2} \tag{7.2}$$

The determinate-error equation is

$$\frac{\Delta g}{g} = \frac{\Delta L}{L} - 2\frac{\Delta T}{T} \tag{7.3}$$

Method: Examine the suspension device for the pendulum, see that the knife-edge is properly placed in the V-shaped groove, and set the pendulum swinging in a small arc in a vertical plane perpendicular to the line of the knife-edge. It should swing freely and smoothly. The amplitude of the swing should not be more than about 4% of the length of the pendulum.

Determine the time for 100 complete vibrations. Make a second determination. The two times should not differ by more than a fraction of a second. If they do, an error in counting has been made, and a third count should be made. The counting should be started *after* the pendulum is swinging. The instantaneous position and direction of the pendulum bob at the zero count, i.e., when the time is zero, may be marked on the paper attached to the wall in back of the pendulum. This mark then serves as a reference point for counting vibrations.

After two satisfactory counts have been made, stop the pendulum, allowing it to hang vertically at rest. Measure the distance H from the top of the knife-edge holder to the center of the bob. Then measure the height h of the knife-edge holder itself with a vernier caliper. The difference between H and h will be the length of the pendulum L. Measure the diameter of the bob.

Make two additional runs of this experiment, using different lengths of suspension wire in each case.

Record all these data in tabulated form, including estimated errors. Calculate the value of g for each case by means of Eq. (7.2) and the error in g by Eq. (7.3). Compare these values with the accepted value for your geographical location. See Table P, Appendix III, to determine this value of g.

Record:

  App. No._____

		No. vib	Time, sec	H, cm	h, cm	T, sec	L, cm	g, cm/sec^2
1								
2								
3								

QUESTIONS

1. If the timing device in this experiment were in error by $\pm 0.1\%$, what approximate error in centimeters per second squared would be introduced into the value of g?

2. If an error of ± 1 vibration is made in counting the number of vibrations (assume to be approximately 100), what error would this introduce into the value of g?

3. It may be shown that practically no error in the period is introduced by the supporting system KN (Fig. 7.1) provided its natural period of oscillation is the same as that for the complete pendulum. What *physical* argument could be used to justify the above statement?

4. Show that the effective length of the laboratory pendulum is $L[1 + 0.1(d^2/L^2)]$ where d is the diameter of the bob. Calculate the error introduced in this experiment by neglecting the diameter of the bob.

OPTIONAL EXPERIMENT

1. The period of a simple pendulum depends upon its angular amplitude of vibration α. The theoretical correction factor for Eq. (7.1) is $[1 + \frac{1}{4}\sin^2(\alpha/2) + \frac{9}{64}\sin^4(\alpha/2) + \cdots]$. Check this correction experimentally.

Spiral
Spring

Object: (1) To determine the force constant of a spiral spring; (2) to determine its period of vibration with several different loads; (3) to compare the observed period with the calculated period.

Apparatus: Spiral spring and support, weight holder and weights, simple cathetometer for measuring displacements, and a timing device. The spiral spring in this experiment is a steel spring capable of supporting loads of several kilograms. It is suspended by a hook attached to a rigid framework of heavy metal rods. The cathetometer is a vertical metal rod in a tripod base with an engraved metric scale upon it. A vernier scale with a pointer is attached to a sleeve that can be moved along the main scale. Therefore, vertical displacements may be determined more accurately with this device than with an ordinary meter stick.

Theory: When a load is gradually applied to the free end of a spring suspended from a fixed support, the spring usually stretches until the tension in the spring just balances the weight of the load. Some springs, however, possess an initial strain and tension even without any apparent load. In this case it will be found that the coils of the spring are pressed tightly together. Thus the spring through this action furnishes its own load. If now a gradually increasing external load is applied to such a spring, the internal load is gradually relaxed without appreciable stretch of the spring until the coils of the spring are just pulled apart. Thereafter, there is only an external load on the spring, and the spring stretches in a normal manner. That is, within limits, the added load on the spring is directly proportional to the stretch of the spring and the spring obeys *Hooke's law*.

Under these conditions, the loaded spring, if set into vibration, will undergo harmonic motion with a period given by the equation

$$T = 2\pi \sqrt{\frac{M}{k}} \qquad (8.1)$$

where T = period of the motion, M = the effective mass of the vibrating system, and k = the spring constant, i.e., the ratio between the added force and the corresponding stretch of the spring. The effective mass M of the spring and its load is the mass of the load M_0 plus one-third the mass of the spring. Thus Eq. (8.1) may be written

$$T = 2\pi \sqrt{\frac{M_0 + (m/3)}{k}} \qquad (8.2)$$

The contribution of the mass of the spring to the effective mass of the vibrating system may be shown as follows. Consider the kinetic energy of a spring and its load undergoing harmonic motion. At the instant under consideration let the load M_0 be moving up with velocity v_0 as shown in Fig. 8.1. At this same instant an element of mass of the spring dm will also be moving up but with a velocity v which is smaller than v_0. It is fairly evident that the ratio between v and v_0 is just the ratio between y and y_0. Hence, $v = v_0(y/y_0)$. The kinetic energy of the spring alone will be

$$\int_0^{y_0} \tfrac{1}{2} v^2 \, dm$$

But dm may be written as $(m/y_0) \, dy$, where m is the mass of the spring. Thus the integral equals

$$\frac{1}{2} \left(\frac{m}{3} \right) v_0^2$$

The total kinetic energy of the system will then be

$$\frac{1}{2} \left(M_0 + \frac{m}{3} \right) v_0^2$$

and the effective mass of the system is therefore $M_0 + m/3$

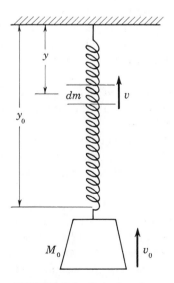

FIGURE 8.1   Spiral spring.

The determinate-error equation may be written approximately as

$$\frac{\Delta T}{T} = \frac{1}{2} \frac{\Delta M_0}{M_0} - \frac{1}{2} \frac{\Delta k}{k} \tag{8.3}$$

It is assumed that $m/3$ is small compared to M_0, therefore, an error in m will not appreciably alter the value of ΔT.

Method: *Part I. Determination of Force Constant.* Place an initial load on the spring (about 2 kg) to relax any initial set in the spring. Measure the spring stretch from this reference point; i.e., determine with the cathetometer the vertical position of some mark on the weight holder. Add 1 kg to the load and again determine the position of the same mark. Continue this process until the initial load has been *increased* by 4 kg. Then reduce the load in steps of 1 kg by taking position readings at each step until the load reaches its initial value.

Record these data in tabulated form. Plot the *added* load in grams (ordinate) against the stretch of spring in centimeters (abscissa). Note that the points lie on a straight line, indicating that the spring obeys Hooke's law. Draw the best possible straight line through the plotted points (see Introduction, Section B) and determine the slope of this line by choosing two points on the line, one near the origin with coordinates x_1 cm and y_1 gmf and the other near the upper end of the line with coordinates x_2 cm and y_2 gmf. The slope will be $(y_2 - y_1)/(x_2 - x_1)$ gmf/cm and the force constant will be this slope multiplied by 980.

It is rather difficult to compute the error in K. Try to estimate how much y_1, y_2 could be changed and still give a straight line that would fit the observed points. Also take into account the fact that the weights used in this experiment may be in error by $\pm 0.1\%$ and that the cathetometer readings are only good to 0.01 cm.

A better way to determine the value of k and its error once we know that the extension of the spring is a linear function of the load is to use the method of least squares. See Eq. A.6 in Appendix I, Note A. Let Y in that equation represent the position readings of the cathetometer and X equal the load on the spring. In this case B is essentially the inverse of the slope of your graph which gives k.

Part II. Determination of the Period of Vibration. Place a load of about 3500 gm on the spring, set the system into vertical vibration with an amplitude of about 5 cm, and determine the period of vibration.

Determine the time for 100 complete vibrations. Make a second determination. The two times should not differ by more than a fraction of a second.

Make two additional runs of this experiment using loads of about 5000 and 6000 gm.

Record in tabulated form the data from these runs and determine the period of the system in each run along with the error in the period.

Finally, calculate the periods for the three different loads using Eq. (8.2) and their errors using Eq. (8.3). Remember that M_0 is the mass of the *total load* on the end of the spring, including the mass of the weight holder.

Record: *Part I. Force Constant of Spring.*

Load, gm, Initial +	Readings		Ave stretch, cm
	↓ cm ↑		
0	31.75	31.77	0
1000	25.43	25.47	6.31
2000	...		...

From graph: $k = $ _____ ; $\Delta k = $ _____.

Part II. Period of Vibration.

Mass of spring: $= ($ $)$ gm
Mass of weight holder: $= ($ $)$ gm

	M_0, gm	No. vib		Time, sec	T (obs), sec	T (theo), sec	% diff
		1st	2nd				
1							
2							
3							

QUESTIONS

1. What constant fractional error would be introduced into the calculated period of the system in this experiment if the mass of the spring were neglected? Would this error be significant in this experiment?

2. It frequently happens in this experiment that the vibrating spring system, after a time, begins to swing to and fro in a vertical plane like a pendulum. This is an example of resonance; i.e., in both motions the system has about the same period. What change in the system could be made to prevent this resonance phenomenon?

3. If two springs with different spring constants k_1 and k_2 were hooked together in series so as to form a single spring for this experiment, what would be the spring constant of the combination?

OPTIONAL EXPERIMENT

1. Check experimentally your conclusions regarding the situation as outlined in Question 3.

Work and Power

Object: To determine the power output which a person can develop under certain conditions.

Apparatus: Prony brake, stop watch, scales.

Theory: Power is the rate of doing work. Its units are those of energy per second. Two special units are the horsepower, which is equal to 550 foot-pounds per second, and the watt, which is equal to 1 joule per second. A device used to measure the power output of a rotating machine is called the *Prony brake.* In its simplest form it consists of a friction band passed around a drum or wheel which is being rotated by the machine. Tension is kept on each end of the friction band by a spring balance. Rotating the wheel at a known rate, and knowing the tensions on the band and the diameter of the wheel, the power output can be calculated as follows.

When the wheel is rotating steadily as shown in Fig. 9.1(a) and the brake band and balances have assumed their equilibrium position, the new effect is the same as though at any instant the two cords were fastened to the rim and pulling with the forces F_a and F_b. See Fig. 9.1(b). (Note that these forces must be tangential ones since the force of friction can only be exerted in the direction of motion.) From this it may be seen that the resultant force tending to prevent turning of the wheel is $F_b - F_a$, acting at the rim. The work done against this force equals the product of the force and the distance traveled by the rim of the wheel. The average power exerted during the run is therefore

$$P = \frac{2\pi r N (F_b - F_a)}{t} \tag{9.1}$$

where $2\pi r =$ the circumference (the distance traveled in one turn), $N =$ the number of turns in the run, $t =$ the time of the run.

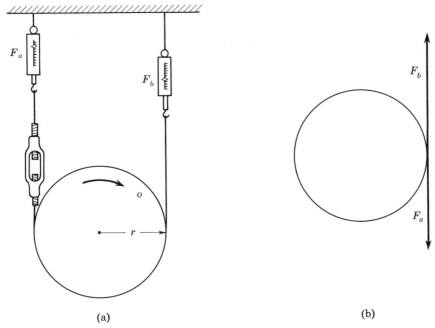

FIGURE 9.1 Prony brake.

Method: Adjust the turnbuckle until both balances read about 8 lb. Move the wheel *slightly* to make sure it is in its equilibrium position. Record the readings of the balances with the wheel at rest. Rotate the wheel at as fast a *steady rate* as possible for 30 sec. Count and record the number of revolutions made in this time, and the readings F_a and F_b while turning at the steady rate. Repeat for each partner.

Repeat the preceding paragraph of instructions for an equilibrium force of about 12 lb on each balance.

Again repeat, this time for 20 lb on each balance, and run for 10 sec. Measure the outside diameter of the wheel and the depth of the groove. To eliminate zero errors in the spring balance, find the difference between the readings at rest and in motion of *each* balance and add the absolute values of these differences. This sum is the net force acting on the rim $(F_b - F_a)$.

Run rapidly up a flight of stairs, from a running start, having your partner time the run with a stopwatch. Repeat for each student. Measure the heights of ten steps and record the average. Record the number of stairs and your weight.

Calculate the work and the power developed in each part of the experiment. Express the work in foot-pounds and the power in horsepower and in watts. Note that a person's power output depends mainly on the muscles involved.

Since this experiment is largely qualitative in nature it will not be necessary to make error calculations.

Record:

Trial	Student	Balance *A*			Balance *B*			Net force	No. of turns	No. of seconds	Work, ft-lb	Pwr, hp	Pwr, watts
		Rest	Turning	Diff	Rest	Turning	Diff						
I	1	8								30			
	2	8								30			
II	1	12								30			
	2	12								30			
III	1	20								10			
	2	20								10			
IV	1												
	2												

Diam wheel, over-all_____ }Radius to inside of groove_____

Depth of groove_____

Height (ave of 10) of a riser_____ }Height climbed_____

Number of steps_____

Weight, student 1_____

Weight, student 2_____

QUESTION

1. Transform the work done into ergs, joules.

Rotational
Motion

Object: To study the motion of a metal disk rotating about a fixed horizontal axle and driven by a falling weight. To show that its angular acceleration is constant and to determine experimentally the value of this acceleration. To compare the value of this acceleration with that predicted by Newton's laws of motion for the system.

Apparatus: Mounted metal disk free to rotate about a fixed horizontal axle, spark timer, waxed paper disk with scale, fish line, weights, and rubber cement. The essential details of the apparatus are shown in Fig. 10.1. The disk is driven by a falling weight attached to a fish line that is wrapped about the edge of the disk. A record of the angular position of the disk as a function of time is given by a series of dots produced on the waxed paper by equally timed sparks from the spark timer. The spark point is moved in toward the axle as the disk rotates in order to prevent overlap of the spark dots on the waxed paper.

Theory: The motion of a disk, free to rotate about a fixed horizontal axle and driven by a falling weight attached to one end of a cord that is wrapped about the disk, may be obtained by applying Newton's laws of motion for translation and for rotation. Let r = radius of disk, I = rotational inertia of disk about axis O, α = angular acceleration of disk due to the tension T in the cord, m = mass of falling weight, and a = acceleration of falling weight. There will be a small frictional torque acting upon the disk at the bearings. Let this frictional torque be ΔL (assumed to be constant). See Fig. 10.2. By Newton's laws we get

$$Tr - \Delta L = I\alpha \qquad \text{(for the disk)} \qquad (10.1)$$

and
$$mg - T = ma = mr\alpha \qquad \text{(for the weight)} \qquad (10.2)$$

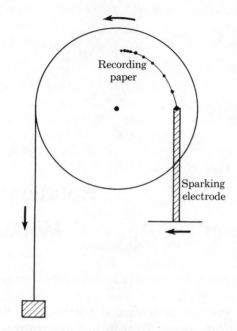

FIGURE 10.1 Rotational motion apparatus.

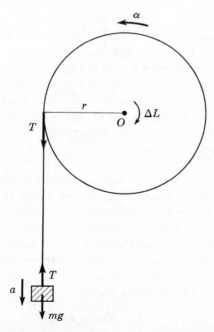

FIGURE 10.2 Accelerating forces on disk and falling weight.

We eliminate T from these equations and solve for α. Therefore

$$\alpha = \frac{mgr - \Delta L}{I + mr^2} \qquad (10.3)$$

The frictional torque ΔL may be eliminated from Eq. (10.3) by observing experimentally the mass Δm that must be attached to the cord to balance out the effect of the frictional torque. In this condition, the system once started with some initial velocity will continue to move with this velocity without acceleration. Any *additional* mass m_0 attached to the cord will produce acceleration. Thus the total mass m may be written as $m_0 + \Delta m$, where $(\Delta m)gr = \Delta L$. Thus Eq. (10.3) may be written in the form

$$\alpha = \frac{m_0 gr}{I + mr^2} \qquad (10.4)$$

Note that the total mass m still remains in the denominator of the equation. However, it is usually replaced by m_0, since Δm is generally much smaller than m_0 and, in addition, mr^2 is much smaller than I.

Equation (10.4) predicts that the angular acceleration of the disk will be constant and gives the value of α in terms of measurable quantities of the system and acceleration of gravity g. The value of I for the solid disk may be taken as $\frac{1}{2}Mr^2$, where M is the mass of the disk and r is its radius. The correction for the axle of the disk may be made, but it is usually negligible in this experiment.

By an analysis of the angular position-time record on the waxed paper disk, we can directly determine the angular acceleration of the disk for a given accelerating load m_0. Therefore it is possible to compare the predicted value of α with the measured value.

The analysis of the angular motion in this experiment could be made in the same manner as that of the linear motion in Exp. 4. But this experiment is much less precise and does not warrant the use of least squares. Hence we shall use a method equivalent to that suggested in the optional experiment at the end of Part II in Exp. 4. There Eq. (4.24) can be replaced by its angular equivalent

$$\omega_0^2 = 2\alpha\theta_0^2 \qquad (10.5)$$

where α is the constant angular acceleration of the disk, θ_0 is the total angular displacement of the disk from its rest position, and ω_0 is its angular velocity at $\theta = \theta_0$. The value of ω_0 may be determined from the angular equivalent of Eq. (4.19). The value of θ_0 may be obtained directly from the waxed paper disk. Equation (10.5) can then be used to calculate α.

In choosing the sequence of dots about the central dot at $\theta = \theta_0$ for the determination of ω_0, it may be advisable to use a time interval τ of $\frac{3}{60}$ sec instead of $\frac{1}{60}$ sec as in Exp. 4. This means that you count every third dot in the sequence. The total time interval of the sequence should be at least 1 sec.

Method: Cut a length of fish line that will reach from the disk support to the floor. Knot one end and secure it in the small notch on the circumference of the disk.

Attach a wire weight holder to the other end of the line. Turn the disk until the weight holder reaches the level of the support. Record the number of turns required.

In this experiment it is necessary to eliminate the effect of friction at the bearings of the disk. To do this, add small weights to the line until the system moves with *constant velocity* once it has been started. Note that the frictional effects are very small; thus the addition of one or two wire weight holders is usually sufficient. These weights should not be considered as part of the accelerating load since they are only used to balance out the frictional forces.

Carefully measure the diameter of the disk with a pair of calipers, estimating to tenths of a mm. Record this diameter along with your estimated error. Record the mass of the disk that is stamped on its surface.

Attach the waxed paper disk to the smooth face of the metal disk, using a little rubber cement applied near the edge of the disk. Center the paper on the disk as well as possible, having the angular scale recorded on the paper facing out. It may be that the scale reading on the paper disk is opposite to the direction that the disk turns in this experiment. If this is the case, it is better to re-label the scale on the paper than to try to use the original scale. Another way to avoid this scale difficulty would be to place the scale surface of the paper against the metal disk. Unfortunately, it will be found that the dots on the scale in this case are not as sharp as they are when the scale is facing out. Hence, it is better to use the first method, even though the scale is in the wrong direction.

Connect the high voltage terminal of the spark timer to the spark tip on the insulating rod and the ground terminal to the frame of the apparatus. If you are in doubt about these connections, consult the instructor. Connect the spark timer to the 110v a-c line and turn on the power supply in order to allow the tubes in the timer to warm up.

Attach 100 gm to the weight holder and turn the disk backward so that the cord is wrapped about the disk with no overlap and the weights are at the level of the support. Hold the system in this position until the suspended weights hang vertically with no swinging motion. Then release the disk, press the spark timer switch, and slowly move the spark tip toward the center of the disk in order to prevent overlap of the spark dots on the waxed paper. Turn off the spark as soon as the weights reach the floor. Stop the disk.

Remove the paper from the disk, being careful not to damage the spark tip or the record on the paper.

Repeat this experiment using a new disk of waxed paper.

Each student should individually analyze one of the records of this experiment. The record on the waxed disk consists of a locus of dots spiralling in towards the center of the paper disk. The dots are equally spaced in time, but not in angle, because of the angular acceleration.

Determine the experimental value of α by analysis of the position-time record on the waxed paper disk and the use of Eq. (10.5). Compare it with the theoretical value of α given by Eq. (10.4).

Record: Tabulate data and results.

QUESTIONS

1. What error in the theoretical value of α would be made if no correction were made for the frictional torque on the disk at the bearings? Is this error significant?

2. Estimate the error introduced into the value of I for the disk by failing to take into account the rotational inertia of the axle of the disk. Is this error significant?

Torsion Pendulum

Object: To determine (1) the moment of inertia of a torsion pendulum, (2) its constant of torsion, (3) the modulus of rigidity of the support rod of the torsion pendulum.

Apparatus: Torsion pendulum, heavy ring, two identical solid cylinders, caliper, micrometer caliper, steel scale, and a timer. The torsion pendulum consists of a solid disk suspended at its center by a metal rod. This rod is clamped in a rigid support at the top (see Fig. 11.1). Either the ring or the cylinders may be placed on the disk in order to change the moment of inertia of the pendulum.

Theory: If a body suspended by a rod or wire such as in Fig. 11.1 is given a small twist about the axis of suspension, it will oscillate with angular harmonic motion, the period of which is given by the equation

$$T = 2\pi \sqrt{\frac{I}{k}} \tag{11.1}$$

where T = period of oscillation,
   I = moment of inertia of the system about the axis of rotation, and
   k = torsion constant of the suspension, i.e., the constant ratio between the restoring torque and the angular displacement.

The torsion constant of the suspending rod is a function of the dimensions of the rod and the modulus of rigidity of its material. This modulus n is given by the equation

$$n = \frac{32lk}{\pi d^4} \tag{11.2}$$

where n = modulus of rigidity,
   l = length of rod,

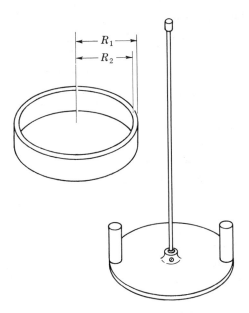

FIGURE 11.1   Torsion pendulum.

k = torsion constant, and

d = diameter of the rod.

In this experiment we wish to determine both I and k for the torsion pendulum. By observing the period of oscillation T of the pendulum, we can determine I/k by Eq. (11.1), but not I and k separately. A method of solving this problem is to *add* to the system a body of known moment of inertia I_0 and then to observe the new period of oscillation T_0. This gives the equation

$$T_0 = 2\pi \sqrt{\frac{I + I_0}{k}} \qquad (11.3)$$

By eliminating k between Eqs. (11.1) and (11.3) and solving for I, we get

$$I = q I_0 \qquad (11.4)$$

where

$$q = \frac{T^2}{T_0^2 - T^2}$$

In a similar manner we may eliminate I and solve for k. We get

$$k = 4\pi^2 p \frac{I_0}{T_0^2} \qquad (11.5)$$

where

$$p = \frac{T_0^2}{T_0^2 - T^2}$$

The error equations in this experiment are somewhat more complicated than those in preceding experiments. The simplest one is that for Eq. (11.2), which may be written for determinate errors as

$$\frac{\Delta n}{n} = \frac{\Delta l}{l} + \frac{\Delta k}{k} - 4\frac{\Delta d}{d} \qquad (11.6)$$

For Eq. (11.4) the corresponding determinate-error equation is

$$\frac{\Delta I}{I} = \frac{\Delta I_0}{I_0} + 2p\left(\frac{\Delta T}{T} - \frac{\Delta T_0}{T_0}\right) \qquad (11.7)$$

Notice that the fractional error in I may become very large if T_0 is not much larger than T, i.e., if p is large. In order to prevent this, I_0 should be made as large as possible.

Finally, the determinate-error equation for Eq. (11.5) is

$$\frac{\Delta k}{k} = \frac{\Delta I_0}{I_0} + 2q\frac{\Delta T}{T} - 2p\frac{\Delta T_0}{T_0} \qquad (11.8)$$

Here again we see the advantage of using a large value of I_0, for this makes both p and q small, thus keeping the fractional error in k small.

The "known" moment of inertia in this case, I_0, is that of a massive ring that has the same external diameter as the torsion disk. Actually its moment of inertia is not given, but it may easily be computed by means of the formula

$$I_0 = \tfrac{1}{2}M_0(R_1^2 + R_2^2) \qquad (11.9)$$

where I_0 = moment of inertia of the ring about its central axis,
      M_0 = mass of the ring,
      R_1 = external radius of the ring, and
      R_2 = internal radius of the ring.

The determinate-error equation for Eq. (11.9) is approximately

$$\frac{\Delta I_0}{I_0} = \frac{\Delta M_0}{M_0} + \frac{\Delta R_1}{R_1} + \frac{\Delta R_2}{R_2} \qquad (11.10)$$

since R_1 and R_2 are roughly equal to each other.

A second "known" moment of inertia is the combination of two solid metal cylinders, each of mass m_0 and radius r_0, standing upright at the edge of the torsion disk and on exactly opposite sides of the supporting rod. The moment of inertia of this combination may be computed by means of the theorem (Lagrange) that the moment of inertia of any body about any axis is its moment of inertia about a parallel axis through its center of gravity plus m_0h^2 where m_0 is the mass of the body and h is the perpendicular distance between the two axes.

To sum up, we may determine I and k for a torsion pendulum by finding its normal period T, and its new period T_0 when a known moment of inertia I_0 has been added.

Method: 1. Set the torsion pendulum (without the ring) into oscillation with an amplitude of about 10° to 20° after placing a chalk mark on its front edge. Determine its period by timing 50 complete oscillations. Repeat this procedure two more times. The times should not differ by more than a quarter of a period. Compute the period T and its error.

2. Place the ring on the disk, centering it carefully. Again set the system into oscillation and determine the new period T_0 in the same manner as before.

3. Compute the moment of inertia of the ring I_0 by measuring R_1 and R_2. The mass of the ring is stamped on its surface. Estimate the error in I_0.

4. By means of Eqs. (11.4) and (11.5) compute I and k for the torsion pendulum. Also determine the errors in I and in k by means of Eqs. (11.7) and (11.8).

5. Measure the diameter d and length l of the suspension rod. Take at least five measurements of d at various points along the rod. Compute the coefficient of rigidity n by means of Eq. (11.2); also compute the error in n by use of Eq. (11.6). Compare the value of n obtained with that given in Table M, Appendix III.

6. If time permits, repeat 2, 3, and 4, using the cylinders instead of the ring. Be sure that the two cylinders are placed on the disk in alignment with the index lines and with their surfaces just tangent to the circumference of the disk. Do not make error calculations for this part of the experiment.

Record:

App. No._____

Mass of ring: $M_0 =$

External diameter: $2R_1 =$

Internal diameter: $2R_2 =$

Length of rod: $l =$

Material of rod:

Diameter of rod: $d = \begin{cases} 1\underline{\hspace{2cm}} \\ 2\underline{\hspace{2cm}} \\ 3\underline{\hspace{2cm}} \\ 4\underline{\hspace{2cm}} \\ 5\underline{\hspace{2cm}} \\ \text{Ave}\underline{\hspace{2cm}} \end{cases}$

Trial	PERIOD T		PERIOD T_0	
	No. osc	Time	No. osc	Time
1				
2				
3				
Ave		$T =$		$T_0 =$

$I_0 =$ $\Delta I_0 =$

$q =$ $\Delta T =$

$p =$ $\Delta T_0 =$

$I =$ $\Delta I =$

$k =$ $\Delta k =$

$n =$ $\Delta n =$

QUESTIONS

1. Develop the determinate-error equations (11.7), (11.8), and (11.10) in this experiment.

2. Develop Eq. (11.9) by using the fact that I for a uniform solid cylinder around the same axis as the ring is $\frac{1}{2}Mr^2$.

3. Develop Eq. (11.2). Calculus problem.

OPTIONAL EXPERIMENTS

1. The value of k, and therefore n, for the suspension rod may be obtained more directly by applying known torques to one end of the rod (the other end is fixed) and noting the corresponding angular twists of the rod. Devise an apparatus for doing this and perform the experiment.

2. In the experiment we make use of the "known" moment of inertia of a massive ring as calculated by Eq. (11.9). Instead of doing this, we could experimentally determine its moment of inertia I_0 by treating it as a physical pendulum and allowing it to oscillate on a knife-edge suspension. Develop the necessary equations and perform the necessary experiment.

Specific Gravity:
Archimedes' Principle

Object: To determine the specific gravity of several different solids and of a liquid by the principle of Archimedes.

Apparatus: Balance with hydrostatic weighing device, solid specimens, liquid specimen, overflow vessel, sinkers, beaker, and hydrometer.

Theory: Specific gravity is defined as the ratio of the density of a substance to that of some standard substance—usually water. It may be obtained by comparing the weight of the substance to the weight of an equal volume of water. This may easily be done by using Archimedes' principle, which states that a body immersed in a fluid is buoyed up by a force equal to the weight of the displaced fluid. The volume of the displaced fluid, of course, will just equal the volume of the immersed portion of the body.

This principle may be used to determine the specific gravity of (1) solids more dense than water, (2) solids less dense than water, (3) liquids of any density.

1. *Solids more dense than water.* In this case the weight of the solid is obtained first in air (W_0), then in water (W_1). It is assumed, of course, that the solid will not dissolve in water. The specific gravity (S.G.) in this case is given by the formula

$$\text{S.G.} = \frac{W_0}{W_0 - W_1} \qquad (12.1)$$

2. *Solids less dense than water.* The solid is weighed in air (W_0). A sinker is then attached to it, and the system (solid plus sinker) is reweighed with the solid in air and the sinker in water (W_{01}). Finally the system is weighed again with both the solid and the sinker in water (W_{11}). The specific gravity of the solid is then given by the formula

$$\text{S.G.} = \frac{W_0}{W_{01} - W_{11}} \qquad (12.2)$$

3. *Liquids.* A solid body is weighed first in air (W_0), then in water (W_1), and finally in the liquid (W_2). The specific gravity of the liquid is then given by the formula

$$\text{S.G.} = \frac{W_0 - W_2}{W_0 - W_1} \tag{12.3}$$

Method: *Part I. Solids more dense than water.* Attach the solid specimen to the pan of the balance by means of a wire and determine its weight (W_0). Fill the overflow can with water, immerse the solid in the water, and catch the overflow in a beaker. Determine the weight of the solid immersed in the water (W_1). Calculate the S.G. of the solid by means of Eq. (12.1), and compute the approximate error in the result. Compare the loss of weight of the solid $(W_0 - W_1)$ with the weight of the water which overflowed into the beaker.

Repeat this experiment with a second solid specimen.

Part II. Solids less dense than water. Attach the solid specimen (block of wood) to the balance and determine its weight (W_0). Fasten a sinker (lead weight) to the block of wood, immerse it in water, then determine the weight of the system under these conditions (W_{01}). Finally immerse the entire system in water and determine the weight of the system (W_{11}). Calculate the S.G. of the wooden block by means of Eq. (12.2) and compute the approximate error.

Repeat this experiment with a second solid specimen (cork).

Part III. Liquids. Attach a small metal cylinder to the balance and determine its weight (W_0). Then immerse the cylinder in a beaker of water and redetermine its weight (W_1). Finally immerse the cylinder in a beaker of the liquid and determine its weight under this condition (W_2). Calculate the S.G. of the liquid and compute the approximate error.

Make a direct determination of the specific gravity of the liquid by means of a hydrometer and compare the two values.

Compare the values obtained with the accepted values in Table C, Appendix III.

Record: Tabulate data and results.

QUESTIONS

1. By using Archimedes' principle develop Eqs. (12.1), (12.2), and (12.3).

2. Bubbles of air are likely to attach themselves to the solid specimen when immersed in a liquid. What effect in general will this have on the calculated values of S.G.?

3. A Ping-Pong ball floats on water in a closed vessel partially filled with water. If the air pressure in the vessel is increased by pumping more air into the vessel, will the Ping-Pong ball rise or sink in the water? Explain.

4. Why can one neglect the buoyant effect of the air in this experiment?

5. A beaker partially filled with a liquid is weighed on a balance. When a stick is thrust into the liquid, there is an apparent increase in the weight of the beaker and its contents. Explain.

OPTIONAL EXPERIMENT

1. On the basis of Question 5, devise an experiment whereby the specific gravity of any liquid may be determined. Perform the experiment using several different liquids.

Surface
Tension

Object: To determine the surface tension of water (1) by a direct method, (2) by capillary action.

Apparatus: Micrometer microscope, wire frame, Jolly balance, capillary tube, and cathetometer.

Theory: Because of the mutual attraction of the molecules in a liquid, those in the interior experience forces that are nearly uniform in all directions, whereas those at the surface experience a net inward force of attraction. This means that work must be done to move a molecule from the interior of a liquid to its surface, i.e., the surface molecules possess more energy than those in the interior of the liquid. Thus the surface of a liquid possesses energy and therefore tends to act as if it were covered by a stretched film or membrane. This phenomenon is known as surface tension. The tension T of such a film is measured by the force F that it exerts per unit length along a line l in the surface, across which the measurement is made

$$T = \frac{F}{l} \qquad \text{(dyne/cm)} \tag{13.1}$$

The direct method of measuring surface tension is to lift a straight wire through the surface (Fig. 13.1). If this is done carefully, the film will be pulled up with the wire. If F is the force exerted by the film (two surfaces) on the wire, and L is the length of the wire, then

$$T = \frac{F}{2L} \tag{13.2}$$

The rise or fall of a liquid in a capillary tube may also be used to determine its surface tension. The surface tension in this case is given by the equation

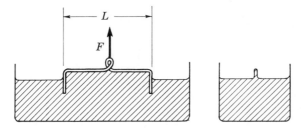

FIGURE 13.1   Direct measure of surface tension with wire frame.

$$T = \frac{dghr}{2 \cos \theta} \qquad (13.3)$$

where T = surface tension,
  d = density of the liquid,
  h = vertical rise of the liquid in the tube,
  r = internal radius of the tube, and
  θ = contact angle of the liquid with the tube wall.

See Fig. 13.2. Also see your general physics textbook for a development of Eq. (13.3).

Pure water in contact with clean glass has a contact angle $\theta = 0$. In this case the liquid (water) wets the glass.

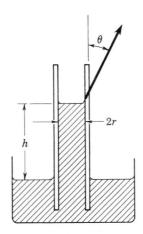

FIGURE 13.2   Measure of surface tension with capillary tube.

Method: *Part I. Direct Method.* Clean the wire frame by heating it to redness in an alcohol flame. *Do not handle it with the fingers, but use a pair of tweezers.* Attach it beneath the weight pan and indicator on the spring of the Jolly balance as shown in Fig. 13.3. Adjust the position of the spring until the center line of the indicator is even with the center line of the glass indicator tube. This should be accomplished fairly near the low end of the scale of the balance. Record the scale reading. Place

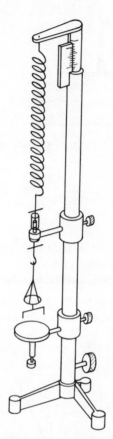

FIGURE 13.3   Jolly balance.

a 1-gm weight in the pan and again adjust the position of the spring until the indicator center lines coincide. Record the new scale reading. From these data the force constant of the spring may be obtained. Remove the 1-gm weight.

Place a beaker of *fresh* water on the platform, and raise the platform until the wire frame is immersed. (The tweezers may have to be used to accomplish this. NOTE: Do not put your fingers into the water. The slightest trace of oil will affect the results.) Lower the platform slowly until the frame begins to come through the surface of the water, drawing a film with it (Fig. 13.1). Continue to lower the beaker of water, but simultaneously adjust the spring tension so that the indicator center marks coincide at all times. As the wire frame draws the film farther out of the surface, a point will be reached where the wire continues to rise without a further lowering of the beaker. If the previous adjustments have been made slowly enough, this point will be easily recognized. As soon as it is reached, stop making further adjustments; the frame will gradually and then more rapidly continue to free itself. Record the setting of the scale. Repeat five times after uniform results are obtained, i.e., after successive scale readings agree to within 0.2 cm.

Measure carefully and record the outside length of the wire frame using a steel scale. From the known force constant of the spring determine the force exerted in breaking the film. Compute the surface tension and its error.

Part II. Capillary-tube Method. Select a piece of capillary tubing from the beaker of chromic acid solution and rinse thoroughly with water. Dip the tube deeply in a beaker of fresh water, and then withdraw it slowly until the liquid inside the tube begins to sink. Support the tube at this position vertically in the beaker, and measure, using the cathetometer, the height of the water in the tube above the surface of the water in the beaker. Repeat two more times.

Remove the tube and mount it under the microscope in such a way that its bottom end is above the tilted mirror of the microscope stand and its top end is in the focus of the microscope. Measure the inside diameter of the tube with the microscope (see Appendix II, Note C), taking readings when the cross hair is tangent to the bore of the tube and perpendicular to the direction of travel of the microscope. Turn the tube 90° and remeasure; use the average of the two measurements. Calculate the surface tension and its error and compare with the results from the direct method. Compare your values with that given in Table L, Appendix III.

Record:

Scale reading (no weight in pan)   _____

With 1 gm in pan   _____

Difference   _____

Force constant   _____ dynes/cm

Scale readings (wire breaks through)   _____

     _____

     _____

     _____

     _____

Average   _____

Net spring extension   _____

      to break through   _____

Corresponding force   _____

Length of wire frame   _____

Surface tension   _____ Error_____

	Trial 1	Trial 2	Trial 3
Height in tube	_____	_____	_____
Height in beaker	_____	_____	_____
Height h	_____	_____	_____
Average of h	_____		
Diameter of tube:		1st meas	90° meas
Left reading		_____	_____
Right reading		_____	_____
Difference		_____	_____
Average diameter	_____		
Surface tension	_____ Error_____		

QUESTIONS

1. Frequently the capillary-tube method gives a smaller value of T than the direct method. How would you explain this? HINT: What assumption was made in using Eq. (13.3)?

2. Surface tension may be defined as the amount of work required to form a unit area of surface of a liquid. Show that this definition also leads to Eqs. (13.1) and (13.2) in this experiment.

3. In Eq. (13.3) r refers to the internal radius of the tube at what point along the tube? Explain.

OPTIONAL EXPERIMENT

1. Determine the surface tension of ethyl alcohol.

Spherical Pendulum:
Angular Momentum

Object: To study the motion of the bob of a spherical pendulum by obtaining a space-time record of this motion. To analyze this space-time record and then show: (1) that angular momentum of the pendulum about its vertical axis is conserved and (2) that the orbit of the pendulum bob is an ellipse, provided that the maximum angular displacement of the pendulum is small ($<3°$).

Apparatus: A pendulum consisting of a long suspension wire (6–8 ft) attached to a spherical bob (1–2 lb), with a sparking electrode attached to the bottom of the bob; an adjustable plane table with leveling screws, recording paper (Teledeltos), spark timer, and connectors.

Theory: A simple pendulum is usually constrained to move in a fixed vertical plane. The path of the bob is, therefore, a circular arc whose radius is the length of the suspension. In this case the pendulum is said to have one degree of freedom, i.e., to describe its position requires the use of only one variable. If we remove the constraint of motion in a fixed vertical plane for the simple pendulum, it then becomes a spherical pendulum. The path of the bob is now some sort of a curve on a spherical surface whose radius is the length of the suspension. The pendulum has two degrees of freedom, i.e., to describe its position now requires the use of two variables. You should convince yourself of the validity of these statements by finding the position variables in each of the above cases. There are many ways of doing so, but no matter how it is done, one finds that it requires only one position variable for the simple pendulum and two position variables for the spherical pendulum.

Motion about a center of force. The motion of the pendulum bob in this experiment is a particular example of a more general type of motion, one of great importance in many branches of physics and astronomy. This is the motion of a particle about a center of force.

Suppose a particle of mass m moves under the action of a single force $\mathbf{F}$ that is always directed toward or away from a fixed point O in space. The magnitude of the force may be any function of the distance r between the fixed point O and the instantaneous position P of the particle. Let the instantaneous velocity of the particle at point P be $\mathbf{v}$. This velocity will be directed along the tangent to the path of the particle. See Fig. 14.1. What can we say about the motion of the particle moving under these prescribed conditions? Newton's laws of motion enable us to draw two conclusions about the motion. They are:

1. The path of the particle must lie in the plane defined by the force vector $\mathbf{F}$ and the velocity vector $\mathbf{v}$, i.e., a plane through point O that includes both $\mathbf{F}$ and $\mathbf{v}$. If the velocity vector $\mathbf{v}$ happens to lie along $\mathbf{F}$, then, as a special case, the path of the particle must lie along the line OP. The validity of these statements is based on the observation that the acceleration of the particle must lie along the line OP. You should supply the additional details of the proof.

2. The magnitude of the angular momentum of the particle $mr^2\omega$ about an axis through point O and perpendicular to the plane of motion must remain constant. Here, ω is the angular speed with which the line OP is turning about the axis through O. The validity of this statement is based upon the observation that the force $\mathbf{F}$, acting upon the particle, never exerts any torque on the particle about the axis through O. Here, again, you should supply the additional details of the proof.

Since the mass of the particle is constant, we conclude that the particle must move in a plane about the point O in such a manner that $r^2\omega$ remains constant. This is equivalent to saying that the particle moves so that the line OP sweeps out equal areas in equal times. The proof of this is given in the following section.

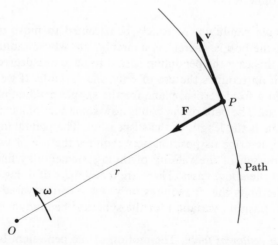

FIGURE 14.1   Central-force motion of particle.

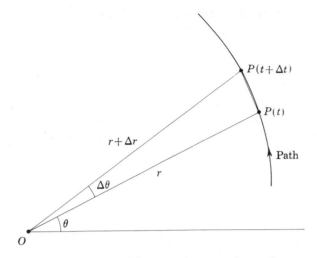

FIGURE 14.2 Central-force motion: equal area theorem.

In Fig. 14.2 let P (t) be the position of the particle at time t. At this time, line OP makes an angle θ with some base line in the plane of motion. At time $t + \Delta t$, the position of the particle is $P(t + \Delta t)$ and line OP has increased in length to the value $r + \Delta r$, and makes an angle $\theta + \Delta\theta$ with the base line. We assume that the time interval Δt is sufficiently small, thus the actual path of the particle in this small time interval very nearly coincides with its displacement. The area ΔA swept out by the line OP in time Δt is thus the area of the triangle. This area, by trigonometry, is equal to $\frac{1}{2}(r + \Delta r)(r) \sin \Delta\theta$. For small values of Δr and $\Delta\theta$, this expression reduces to $\frac{1}{2}r^2\Delta\theta$. Therefore

$$\frac{\Delta A}{\Delta t} = \frac{1}{2}r^2\frac{\Delta\theta}{\Delta t} = \frac{1}{2}r^2\omega = \text{constant} \qquad (14.1)$$

This result will be recognized as Kepler's second law for planetary motion. Actually, it is a much more general law and holds for the motion of any particle moving about a center of force.

Orbit of the particle. It is impossible to determine theoretically the actual path of the particle (orbit) until we know how the force F on the particle depends on the distance r[1]. In a great many cases (gravitational and electrical) the force is inversely proportional to the square of the distance r. In these cases it may be proved (the proof requires a knowledge of differential equations) that the orbit is a conic section (ellipse, parabola, or hyperbola) with the point O at one of the foci.

In another set of cases (elastic and mechanical) the force F is directly proportional to the distance r, and is a force of attraction. In these cases the orbit of the particle is always an ellipse (or one of its degenerate forms) with point O at the center.

[1]The symbol F in italics represents the magnitude of the corresponding force vector **F** in boldface—the usual vector convention in printing.

The motion is periodic; the period of the motion T is given by the well known relation

$$T = 2\pi \sqrt{\frac{m}{k}} \qquad (14.2)$$

where k is the force constant, i.e., F/r. Thus if the motion of the particle in its elliptical orbit is projected onto any straight line lying in the plane of the ellipse, the projected motion will always be simple harmonic motion. And finally, if rectangular coordinate axes are chosen so that they coincide with the major and minor axes of the ellipse, then the coordinates of the particle at any time t may be written

$$\left. \begin{array}{l} x = a \cos 2\pi \dfrac{t}{T} \\[2mm] y = b \sin 2\pi \dfrac{t}{T} \end{array} \right\} \qquad (14.3)$$

Equations (14.3) are the parametric equations of an ellipse with semi-major axis a and semi-minor axis b. The period of the motion is T, i.e., the time for the particle to make one complete circuit around the ellipse. The angle $2\pi(t/T)$ is the eccentric angle of the ellipse. By eliminating this angle in Eqs. (14.3), we obtain the standard equation for the ellipse

$$\frac{x^2}{a^2} + \frac{y^2}{b^2} = 1 \qquad (14.4)$$

The bob of a spherical pendulum has the motion just described, provided that its maximum angular displacement from the vertical is not too large. The argument is the same as that for a simple pendulum, i.e., the displacement must be small enough to make the restoring force on the bob very nearly proportional to its displacement. It follows that the period of the spherical pendulum is equal to the period for the simple pendulum of the same length. Furthermore, its orbit is essentially a combination of two linear harmonic motions of the same period at right angles to one another. Equations (14.3) are merely mathematical statements of these facts.

Space-time record. In this experiment we verify the theoretical conclusions concerning the motion of a spherical pendulum. To do this, we obtain a space-time record of the motion of the bob (its center) on a sheet of recording paper placed just below it. The record produced by the spark timer consists of a set of dots, equally spaced in time, whose locus is the orbit of the bob.

An analysis of this record should show (1) that the motion of the bob is essentially motion about a center of force, and (2) that the orbit of the bob is an ellipse.

Method: Place the tray (table) containing the conducting paper in its bottom immediately below the bob of the spherical pendulum. When the bob is at rest, the sparking electrode on the bottom of the bob should clear the paper by about 1 mm and should be near the center of the paper. Level the table. Make the necessary electrical connections and turn on the power for the spark timer so that it may warm up before being used.

Make a trial run without using the spark timer. Pull the bob of the pendulum back near the edge of the tray. In releasing the bob give it a moderate lateral push so that it moves over the paper in an elliptically shaped path, not a circle. The minor axis of the orbit should be about one-half to two-thirds that of the major axis. The bob should move smoothly around its orbit without any wobbling motion. At no point in the path should the sparking electrode actually touch the paper. Once you have succeeded in putting the bob into a satisfactory orbit, depress the sparking switch on the spark timer for one complete revolution of the bob. A small amount of overlap is not serious. Without moving tray or paper, bring the bob to rest, allowing it to hang in its normal equilibrium position. Again, use the spark timer to get the center of the orbit, a matter of considerable importance in the analysis. Repeat the above procedure with a new sheet of conducting paper. Each student will then have one space-time record to analyze.

There are several different ways of analyzing the data. We give one possible method. You may wish to use a different method, and may do so with the instructor's permission.

The record of motion of the bob in one complete period of motion is given on the paper as a set of dots lying along the orbit of the bob. The dot intervals correspond to equal time intervals (usually $\frac{1}{60}$ sec). In addition, the center of the orbit is given. The method of analysis that we describe requires the use of a particular set of rectangular axes; namely, one that has the x axis lying along the major (long) axis of symmetry of the orbit, and the y axis along the minor (short) axis of symmetry. These are not given directly on the record and therefore must be constructed. In order to construct the x axis draw a circle about the center of the orbit that has a diameter somewhat smaller than the approximate major axis of the orbit. This circle will intersect the orbit at four points, two of which lie near the right end of the major axis. Draw two lines from the center to these two points of intersection. Bisect the central angle formed by these lines. Use this bisector as the x axis. It should (very nearly) coincide with the major axis. Draw the y axis perpendicular to the x axis at the center of the orbit (the origin). This work must be done with great care since the verification of the elliptical nature of the orbit amounts to checking Eq. (14.4), which in turn assumes that the coordinate axes coincide with the major and minor axes of the ellipse. If they do not coincide, then Eq. (14.4) is transformed (by rotation of the axes) into a much more complex equation.

On the other hand, we need not worry about the orientation of the axes so far as areas are concerned, since these will be invariant under any rotation of the axes.

After having drawn in the axes, the next step is to choose convenient pairs of points on the orbit that represent equal time intervals, but that are still close enough together so that the orbit between any two points forming a pair is essentially a straight line. Of course, the original dots on the record satisfy this condition, but they are usually so close together and so numerous that an analysis using all of them would be exceedingly tedious. Further, it is only necessary to use pairs of points in the first quadrant since the orbit should be symmetrical with respect to the x and y axes. A convenient choice of pairs of points in this case may be 4-, 5-, or 6-dot intervals. Start near the end of the major axis (there may be no dot at the end). Choose a well defined dot on the orbit as a starting point. Label it 1. Make the chosen count

(4, 5, or 6) along the orbit and label the last dot 2. Repeat the process, starting at dot 2 and labeling the last dot 3. Continue this process until you have virtually covered that part of the orbit lying in the first quadrant (see Fig. 14.3).

Determine the x and y coordinates of each of the labeled dots by direct measurement. For this procedure, it is most convenient to use a drawing board with a T-square and right triangle. Tabulate the results.

In order to verify the equal areas in equal times theorem, determine the area of each of the alternate triangles $O12$, $O34$, $O56$, and so on as shown in Fig. 14.3. Note that these triangles are completely independent; i.e., they have no common side. The area of each triangle, e.g., $O56$, may be computed most conveniently by use of the expression $\frac{1}{2}|x_5 y_6 - x_6 y_5|$, where x_5, y_5 are the coordinates of dot 5, and x_6, y_6 the coordinates of dot 6. Tabulate the areas of these triangles. They should be the same within the limits of experimental error. Estimate the approximate indeterminate error in at least one of the area determinations.

In order to verify the elliptical nature of the orbit, calculate the value of

$$\frac{x^2}{a^2} + \frac{y^2}{b^2}$$

for each of the labeled dots. The value of this expression should be 1 for any point on the orbit, if the path is an ellipse and if the axes are properly chosen. Estimate the indeterminate error in the expression for one of the labeled dots.

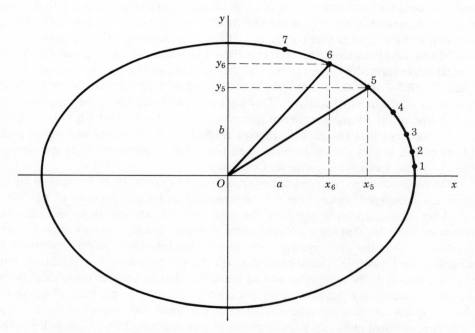

FIGURE 14.3 Spherical pendulum: space-time record of elliptical orbit.

Record: Tabulate your data and results in a systematic manner, using whatever method of analysis you wish. Do this directly on the recording paper if sufficient space is available.

QUESTIONS

1. Estimate the determinate error introduced into the value of the expression

$$\frac{x^2}{a^2} + \frac{y^2}{b^2}$$

if the coordinate axes are out of alignment with the axes of the ellipse by $1°$.

2. Show that the theoretical area of any one of the elliptical sectors that you have determined directly should be

$$\pi ab \frac{\Delta t}{T}$$

where Δt is the chosen time interval (calculus required). Compute this value and compare it with your measured values.

3. A more detailed analysis of the motion of a spherical pendulum shows that the orbit of the bob is not precisely a stationary ellipse, but rather one whose major and minor axes slowly rotate about the point O. How would you explain this in terms of the dependence of the period of a simple pendulum upon its amplitude of vibration?

OPTIONAL EXPERIMENTS

1. There are graphical methods of showing that the orbit in this experiment is an ellipse, methods that do not require construction of the major and minor axes. One of these methods is to draw six tangents to the ellipse so that a six sided polygon that circumscribes the ellipse is formed. The polygon need not be a regular one of equal sides. If the vertices of this polygon are labeled consecutively 1, 2, 3, 4, 5, 6, then lines drawn between points 1 and 4, 2 and 5, 3 and 6 will intersect at a common point. Try this construction.

2. Let major and minor auxiliary circles be drawn about point O, the former having radius a and the latter radius b. Project the recorded dots on the orbit, vertically (parallel to the y axis) onto the major circle and horizontally (parallel to the x axis) onto the minor circle. These projected points should be distributed uniformly around each circle. Explain.

PART

HEAT

Hygrometry

Object: To determine the dew point and the relative humidity of the atmosphere.

Apparatus: Dew-point apparatus, sling psychrometer, and hygrodeik.

Theory: Hygrometry is the measurement of the amount of water vapor present in a given space. According to the law of partial pressures, the total pressure exerted by a mixture of gases is equal to the sum of the individual pressures that would be exerted if each gas occupied the same volume alone. In any given volume, therefore, *the individual partial pressures are directly proportional to the amounts (i.e., the masses) of the gases present.*

The amount of water vapor that a space will contain has a maximum value that increases with increasing temperature. Introducing more vapor than this saturation value, or decreasing the temperature of a space that is already saturated, will cause the total amount of vapor to decrease, the excess being noticed in the form of condensation. The temperature at which this condensation begins is called the *dew point.*

This phenomenon of condensation is manifested almost daily in most parts of the world in the form of dew, fog, or rain. In these cases the air is nearly saturated, and a drop in temperature, or a moist wind blowing in, causes droplets to form.

A useful concept is that of *relative humidity*, which is defined as the ratio of the mass of water vapor actually present in a given volume (its *absolute* humidity) divided by the mass required to produce saturation at that temperature. Thus a relative humidity of 100% indicates saturation conditions; further, it is clear that when a mass of air reaches its dew point, its relative humidity is 100%. Tables have been prepared showing the amount of water vapor necessary for saturation at various temperatures in terms of vapor pressures. See Table D, Appendix III.

There are two general methods of determining the relative humidity. One is the dew-point method, which depends on the fact that because of local cooling, condensation will appear on a surface that is cooler than the dew point of the atmosphere. The other is the psychrometer method, which depends on the fact that the rate of evaporation of water into an atmosphere depends on the amount of water *vapor* already present. It is found that the actual vapor pressure p in millimeters of Hg is given approximately by the simple equation

$$p = p_w - 0.50(t - t_w) \tag{20.1}$$

where t and t_w are the dry- and wet-bulb temperatures respectively, and p_w is the saturated vapor pressure corresponding to t_w.

Method: *Part I. The Dew-point Method.* Fill the small, nickel-plated container about three-quarters full of ether, benzol, or other volatile liquid, and replace the cap with the thermometer. Upon subjecting the liquid in the container to a partial vacuum, some of it evaporates, cooling the rest of the liquid and the container. Record the air temperature t and, at the instant that dew begins to form on the bright surface of the container, record the temperature of the metallic surface, as indicated by the inner thermometer. Nickel surfaces are provided at either side of the container; these remain at room temperature for comparison purposes so that the slightest film on the container surface will be easily seen. Release the vacuum control, and allow the container to warm up. Record the container temperature the instant the dew disappears. If the cooling is stopped as soon as dew appears, the two values of the dew-point temperature should not differ by more than 1° C. Repeat several times and obtain the mean value of the dew-point temperature. CAUTION: Keep as far away from the container as convenient; even the faintest breath you exhale will cloud the surface. Do not confuse this effect with that of the genuine dew point. When finished, return the unused liquid to the bottle.

Since the water vapor in the atmosphere is not confined to a definite volume, its pressure will not change as the temperature is lowered. Therefore, the saturation vapor pressure at the dew point will be the same as the vapor pressure in the warmer air. Another way of saying this is that, until condensation actually starts, the amount of vapor present in a given mass of air at the dew point is the same as the amount in the same mass of warmer air. From Table D in Appendix III, the saturation vapor pressure at the dew point, i.e., the actual vapor pressure in the atmosphere, can be found. Furthermore, the saturation vapor pressure at *room* temperature can be found from the same table. The ratio of these two is equal to the relative humdity.

Part II. Psychrometer Method. The apparatus consists of two thermometers mounted side by side, the bulb of one of them being covered by a cloth that is kept wet with distilled water (see Fig. 20.1). The instrument is swung about the handle so as to pass a current of air over the bulbs at a rate of about 3 m/sec, i.e., a little over two swings per second. CAUTION: *Stand clear of furniture, walls, and other persons;* with proper care there can be no excuse for breakage!

Saturate the cloth with distilled water, and swing the instrument. Read both thermometers and again whirl the psychrometer. Continue in this manner until the read-

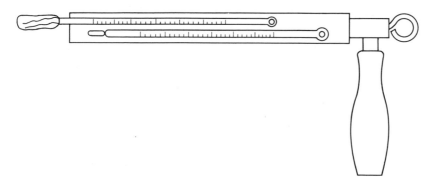

FIGURE 20.1 Sling psychrometer.

ings become steady. Record these steady values. Calculate the actual vapor pressure by use of Table D and Eq. (20.1). Determine the relative humidity.

Part III. The Hygrodeik. This is a direct-reading instrument on the psychrometer principle. Directions for its use are printed upon it. Like the sling psychrometer, it reads correctly only when air moves past it at 3 m/sec. Fan it vigorously until the readings reach a steady value. Record the wet- and dry-bulb readings, and use the chart and swinging arm to find the relative humidity, the absolute humidity, and the dew point.

Record: *Part I. Dew Point.*

	Air temp, *t*	Dew appears	Dew disappears
Average			

Vapor pressure_____
Saturated vp_____
Rel Humidity_____

Part II. Sling Psychrometer.

	Wet bulb	Dry bulb
Wet and dry readings		
Average readings		
Difference		
Actual vapor pressure		
Sat vp		
Rel humidity		

Part III. Hygrodeik.

Wet bulb_____

Dry bulb_____

Rel humid_____

Abs humid_____

Dew point_____

QUESTIONS

1. Suppose Part II of this experiment is performed in a perfectly dry room ($p = 0$). What will be the reading of the wet-bulb thermometer if the dry-bulb thermometer reads (a) 20°C, (b) 38°C? HINT: Use Eq. (20.1) and Table D.

2. What would have to be the temperature of a perfectly dry room ($p = 0$) for the value of t_w to be 0°C?

Gas

Thermometer

Object: To calibrate a constant-volume helium gas thermometer at the normal ice point and then to determine (1) the sublimation temperature of solid carbon dioxide (Dry Ice) at atmospheric pressure and (2) the steam point.

Apparatus: Constant-volume helium gas thermometer, mercury centigrade thermometer, barometer, boiler, cracked ice, and Dry Ice.

A sketch of the constant-volume gas thermometer is shown in Fig. 21.1. It consists essentially of a mercury manometer, one arm of which is open to the atmosphere while the other arm connects with a closed glass bulb containing helium gas. The glass bulb is enclosed in a brass jacket with a removable top. A stem on the top may be connected to a steam boiler for the purpose of immersing the bulb in a steam bath. Alternatively, ice or Dry Ice may be packed around the glass bulb in the brass jacket. A removable inner bucket facilitates the removal of the ice pack when desired. The height of either manometer arm is adjustable. To keep the gas in the bulb at constant volume, it is only necessary to keep the mercury level in the closed manometer arm at the fiducial or reference mark. This is a fine black ring on the stem of the bulb just below the brass jacket. The height of mercury in either arm may be read directly off a fixed scale on the apparatus (not shown in the figure).

Theory: When two systems or bodies at different temperatures are put into thermal contact with each other, heat flows from the system of higher temperature into the system of lower temperature. This flow of heat continues until the temperatures of the two systems are the same. At this point, the flow of heat ceases and the systems are said to be in thermal equilibrium with each other. Since the flow of heat, or its absence, can be detected by means other than the measurement of temperatures, this method gives us a means of determining the equality or inequality of the temperatures of two systems. It gives us no information about the amount of the inequality, however.

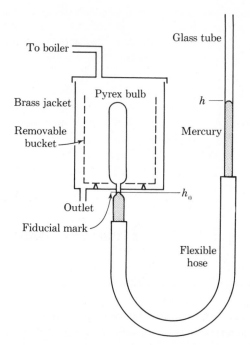

FIGURE 21.1   Constant-volume gas thermometer.

It is a matter of experience that two systems, each of which is in thermal equilibrium with a third system, are in thermal equilibrium with each other. Thus all systems in thermal equilibrium with one another have the same temperature. Temperature therefore is a property of a system that determines whether or not the system is in thermal equilibrium with other systems.

It is well known that some properties of a system are temperature independent, e.g., its mass, whereas others, such as its volume, depend upon the temperature. Temperature-dependent properties of a system are known as thermometric properties. Any one of them may be used to indicate the temperature of the system, providing that some assumed rule is given correlating the thermometric property with the temperature. The simplest rule is to assume that equal temperature changes correspond to equal changes in the value of the thermometric property, providing that the other quantities upon which the thermometric property depends are held constant. This is equivalent to saying that the relation is linear (straight-line relation). Since two points are required to fix a straight line, two fixed temperatures are also required to fix the temperature–thermometric property relation. The two fixed temperatures are usually taken to be (1) the temperature of melting ice under standard atmospheric pressure and (2) the temperature of boiling water under standard atmospheric pressure. These two fixed points are generally referred to as (1) the normal ice point and (2) the normal steam point. Unfortunately, temperature scales defined in this manner are, in general, different for each thermometric property of each substance used, except, of course, at the two fixed points. Gases under the proper conditions constitute an outstanding exception. The pressure of a gas at constant volume, or its volume at constant pressure, may be used as a temperature indicator. Under the appropriate conditions, these two thermometric properties of a gas give practically the same temperature scale for all gases (the gas temperature scale).

Using the laws of thermodynamics, it is possible to define a temperature scale, the Kelvin absolute scale, which is completely independent of the nature of the substance used as a temperature indicator. It is also possible to show that this temperature scale is identical with that given by an "ideal gas." Since all gases approach the ideal state under proper conditions, all give practically the same temperature scale under these conditions. The so-called permanent gases, such as hydrogen and helium, are very nearly ideal over wide ranges of temperature and pressure. For these reasons, the constant-volume gas thermometer (hydrogen or helium), properly calibrated, is used to define a standard temperature scale that corresponds very closely with the Kelvin scale over a wide range of temperatures. For example, the constant-volume helium thermometer gives very satisfactory temperature readings over a range from -240 to $1000°C$. From -100 to $200°C$, the deviation of the helium temperature from the Kelvin temperature is less than $0.001°C$ and even at $-240°C$ the deviation is only $0.02°C$.

In this experiment the thermometric property, used to indicate temperature, is the pressure of a fixed mass of helium gas held at *constant volume*. It is assumed that the change in pressure of this gas is directly proportional to its change in temperature. This means that the relation between the pressure P and the temperature t of this gas must be linear. Therefore

$$t = aP + b \qquad \text{(const } V) \qquad (21.1)$$

where a and b are constants. The constant b controls the zero point of the temperature scale represented by t, and the constant a controls the size of the unit degree of temperature. The constant b may therefore be absorbed into a new temperature scale $T = t - b$, and Eq. (21.1) may be written as

$$T = aP \qquad (21.2)$$

where T is called the gas temperature.

It is the gas temperature T that corresponds closely with the Kelvin[1] temperature under the proper conditions. In order to use a gas thermometer for temperature measurements, it is always necessary to calibrate the thermometer. This means that we must experimentally determine the value of constant a in Eq. (21.2) for the gas thermometer we are using. Before 1954 this calibration amounted to observing the pressures P_i and P_s given by the gas thermometer at the normal ice point temperature T_i and the *normal* steam point temperature T_s and then *assuming* that the temperature interval $T_s - T_i$ was exactly $100°$. On this basis $a = 100/(P_s - P_i)$, and Eq. (21.1) can be used to determine any other temperature T in the gas thermometer's range. Note that this calibration requires two fixed points of temperature to determine the constant a.

[1]William Thomson, Lord Kelvin, was a great British scientist of the 19th century and the originator of the thermodynamic or absolute temperature scale.

But it is clear from Eq. (21.2) that the constant a could be determined by a single fixed point of temperature with an assigned value of that temperature. In 1954 the Tenth General Conference of Weights and Measures, meeting in Paris, chose the *primary standard* of *temperature* as the temperature of the triple point of pure water and assigned to it a value of 273.16° Kelvin. The triple point of water (mixture of water, water vapor, and ice in thermal and mechanical equilibrium) is especially stable and reproducible. And the temperature value assigned to it, $T_{tr} = 273.16°K$, makes the normal ice point temperature $T_i = 273.15°K$ and the normal steam point temperature $T_s = 373.15°K$, thus preserving the old relation $T_s - T_i = 100°K$ to five significant figures.

When we use the standard triple point temperature for calibration purposes, Eq. (21.2) becomes

$$T = 273.16 \frac{P}{P_{tr}} \tag{21.3}$$

where P_{tr} is the pressure registered by the gas thermometer at the triple point and P its pressure at the temperature T.

The ideal-gas temperature which is equivalent to the Kelvin temperature may be obtained from Eq. (21.3) by taking the limiting value of the right side of Eq. (21.3) as P_{tr} goes to zero. This limiting process essentially converts the real gas in the thermometer bulb into an ideal gas. To carry out this procedure experimentally is very difficult and time consuming. But it has been done with gas thermometers using several different gases in measuring the normal steam point temperature T_s. Although the different gases gave slightly different values of T_s for the initial values of P_{tr} as the P_{tr} values were successively reduced, the values of T_s for the different gases converged to the value $T_s = 373.15°$, the normal steam point temperature on the Kelvin scale. In this experiment the helium gas thermometer was unique in that its temperature values for T_s were practically independent of the limiting process. Helium acts like an ideal gas over a wide range of temperatures. Thus with a helium gas thermometer we can use Eq. (21.3) as it stands to measure a wide range of temperatures without any appreciable corrections. But this range definitely does not include the temperature $T = 0$, i.e., absolute zero on the Kelvin scale. The meaning of absolute zero comes from the laws of thermodynamics and statistical mechanics.

It is more convenient (but less precise) to calibrate the gas thermometer using some other fixed point rather than the primary standard triple point, since the latter requires special equipment. In this experiment we use as a secondary standard the normal ice point $T_i = 273.15°K$ for calibration purposes. We can then replace Eq. (21.3) by the equation

$$T = 273.15 \frac{P}{P_i} \tag{21.4}$$

where P_i is the pressure registered by the helium gas thermometer at the normal ice point.

The normal ice point is easily obtained (a mixture of ice and water under atmospheric pressure). Strictly speaking the pressure should be 76 cm of Hg, but no pressure correction is needed, since the ice point temperature is quite insensitive to any changes that ordinarily occur in the atmospheric pressure.

The familiar centigrade temperature scale, now called the Celsius scale in honor of the Swedish astronomer Celsius who first suggested it in 1742, is obtained from the ideal-gas temperature by simply shifting its zero point from absolute zero to the normal ice point. Thus

$$t = T - 273.15 \qquad (21.5)$$

where t is the Celsuis temperature. Evidently $t_i = 0°C$, and $t_{tr} = 0.01°C$. For a more thorough discussion of temperature and its measurement see the references given below.[2,3]

In using Eq. (21.4) for the determination of temperatures we can always express the pressures in cm of Hg taken from the gas thermometer and the mercury barometer. Hence Eq. (21.4) becomes

$$T = 273.15 \frac{H_b + (h - h_0)}{H_b + (h_i - h_0)} \qquad (21.6)$$

where H_b = barometer reading in cm of Hg,

h_0 = scale reading in cm of the mercury meniscus at the reference mark,

h_i = scale reading in cm of the mercury meniscus in the open tube of the manometer when the bulb of helium gas is at the ice point temperature,

h = corresponding reading when the bulb is at temperature T.

We have assumed in Eq. (21.6) that:

1. The helium gas in the bulb has the same volume throughout the experiment and all of the gas has the temperature of the bath at thermal equilibrium.

2. The temperature of the mercury columns in the barometer and the manometer remain constant during the experiment along with the atmospheric pressure.

3. The values of h, h_0, h_i are observations *made when thermal equilibrium* has been established between the helium gas in the bulb and the bath in which it is immersed so that the two have the same temperature.

Assumptions 1 and 2 are not completely satisfied in this experiment and lead to small constant errors in the value of T—a few tenths of a degree. We can make corrections for these small errors as shown in the following section. But assumption 3 must be met in the experiment for good results. No correction is possible if you fail to wait long enough for thermal equilibrium to be established (15 to 20 min are usually required).

Errors: There are two determinate errors in this experiment for which it is possible to make corrections. The first of these arises because the volume of the gas is not held strictly constant during the course of the experiment, since the bulb expands

[2] Mark W. Zemansky, *Heat and Thermodynamics*, 5th ed. (New York: McGraw-Hill Book Company, 1968), Chap. 1.

[3] Robert Resnick and David Halliday, *Physics*, *Part I* (New York: John Wiley & Sons, Inc., 1966), Chap. 21.

or contracts with changing temperature. By a direct application of Boyle's law it is an easy matter to show that the corrected pressure P' is related to the observed pressure P by the equation

$$P' = P(1 + \gamma t) \quad \text{(correction 1)}$$

In this equation γ is the coefficient of cubical expansion of the bulb and t is the temperature of the bulb. The reference volume is V_i, the volume of the bulb at $0°C$.

The second constant error in this experiment arises because not all of the gas in the bulb is at the temperature of the bath in which the bulb is immersed. A small portion of the gas in the neck of the bulb outside the brass jacket is likely to be nearer room temperature than bath temperature. This gives an observed pressure which is too low if the bath temperature is above room temperature and too high if the contrary condition occurs. The corrected pressure P'' for this case may be worked out by use of the general gas law. Let V be the total volume of the bulb of which a large part V_b is at the bath temperature T and a small part V_r is at room temperature T_r. Then the following equations may be written:

$$PV_b = n_b RT, \qquad PV_r = n_r RT_r, \qquad P''V = (n_b - n_r)RT$$

In these equations n_b is the number of mols of gas at bath temperature and n_r the number of mols at room temperature. If n_b and n_r are eliminated from the third equation by use of the first two equations, then

$$P'' = P\left[1 + \frac{V_r}{V}\left(\frac{T}{T_r} - 1\right)\right] \quad \text{(correction 2)}$$

In addition to the two corrections just discussed it may be necessary to make a third correction when assumption (2) is not satisfied. If the room temperature changes appreciably during the course of the experiment, then a third correction is necessary. This correction factor is

$$1 - 0.000164(t_{rT} - t_{ri})$$

where t_{ri} is the room temperature in $°C$ when the ice point readings were taken and t_{rT} the corresponding room temperature when the T readings were taken.

When we combine all three of the preceding corrections into a single correction factor, we obtain

$$T_{cor.} = T\left[1 + \gamma(T - 273) + \frac{V_r}{V}\left(\frac{T}{T_{rT}} - \frac{273}{T_{ri}}\right) - 0.00016(t_{rT} - t_{ri})\right] \quad (21.7)$$

where $\gamma = 1.0 \times 10^{-5}$ cm^3 per $°C$ for a pyrex bulb. The largest correction term is likely to be $\gamma (T - 273)$, which is about 0.001 for $T = T_s$. Therefore the value of the correction factor should be rounded off at the third decimal place.

Method: Adjust the manometer arms of the gas thermometer so that, at room temperature, the mercury level in the open arm is about in the middle of the scale. Since h is a linear function of t, this initial adjustment permits a range of temperature measurements extending both below and above room temperature by about equal amounts. Read the height of the fiducial mark on the scale. Record this as h_0. All subsequent adjustments in the experiment are made leaving this arm *fixed* and adjusting the height of the open manometer arm.

The order in which the temperatures are taken and the calibration made is immaterial. There is some advantage in starting with the lowest temperature, that of Dry Ice, then going to calibration at the ice point, and finally ending up with the steam point temperature. The bulb, jacket, and bucket of the gas thermometer must be thoroughly *dry* when Dry Ice is used. Any water in the jacket will immediately freeze when Dry Ice is added. This causes trouble, so it is best to start when the jacket is most likely to be dry. It certainly will not be dry if the order is reversed.

CAUTION. After drying the jacket and bulb throughly, lower the open arm of the manometer at least 30 cm. It is essential that the mercury not rise above the fiducial mark otherwise it might enter the bulb and freeze, causing considerable delay and inconvenience. Mercury freezes at $-40°C$, a temperature much higher than Dry Ice.

Powder a quantity of Dry Ice, put the inner bucket (dry) in place, and pack it with Dry Ice. Add Dry Ice whenever necessary. Wait at least 15 min, then bring the mercury level in the closed arm up to the fiducial mark by carefully raising the open arm. Keep adjusting the level, if necessary, to prevent the mercury from rising above the fiducial mark. When the level remains steady at the mark for several minutes, thermal equilibrium has probably been reached. At this point determine the value of h and h_0, take the barometer reading H_b (see Note D, Appendix II), the room temperature t_r, and the approximate temperature of the glass tube just above the fiducial mark.

Remove the Dry Ice. Then slowly raise the open arm of the manometer as the gas pressure in the bulb builds up. Failure to do this may force mercury out of the open arm as the temperature of the bulb rises.

Next, pack the brass jacket with cracked ice until the bulb is completely covered. Wait at least 15 min, adding more ice if necessary, and keeping the mercury meniscus near the fiducial mark. After 15 min have passed, set the mercury meniscus at the fiducial and wait a few minutes to see whether or not there is any further change. When a steady state is reached, the bulb is at the ice point. Repeat the readings H_b, h, h_0, etc., that were taken at the Dry Ice point. Then remove the ice.

Finally, attach the steam boiler to the jacket and pass steam through it for at least 15 min. As the gas pressure in the bulb increases, raise the open arm of the manometer to prevent mercury overflow. When thermal equilibrium has been achieved, repeat the readings taken at the ice point. Then detach the boiler and remove any excess water in the jacket. Gradually lower the open tube as the bulb cools off.

Tabulate your data and results as shown in the Record. Calculate the value of T to four significant figures by Eq. (21.6) for the sublimation temperature of Dry Ice and for the steam temperature. By rough measurement of the dimensions of the bulb and of the stem connecting it to the manometer estimate the ratio V_r/V. Then compute the correction factor for the sublimation temperature and the steam temperature, and finally compute the corrected sublimation and steam temperatures. Compare your corrected value of the steam temperature with the standard value of 373.15°K. Account for any difference. See Appendix III, Table D.

Record: *Helium Constant-Volume Gas Thermometer.*

Data	Sublimation point (CO_2)	Ice point (H_2O) Standard	Steam point (H_2O)
H_b (cm of Hg)			
h_0 (cm of Hg), (Use average value of h_0)			
h (cm of Hg)			
$t_r °C$			
$T_r °K$			
$\dfrac{V_r}{V}$ (Estimate parts per thousand)			
Results			
$T °K$		(273.15°)	
Correction factor			
$T_{\text{cor.}} °K$			

Thermal
Radiation

Object: To make an experimental study of the transfer of heat by radiation. To verify the Stefan-Boltzmann Law and to estimate the Stefan-Boltzmann constant for black body radiation. NOTE: This is a difficult experiment and should not be attempted without some experience with electrical measurements, including the use of the thermocouple (Exp. 38).

Apparatus: Calrod electrical heater rod (steel sheath) mounted along the axis of a large pyrex glass tube, iron-constantan thermocouple, thermocouple potentiometer and accessories, ice-bath, d-c ammeter (0–5 amp), d-c voltmeter (0–150 v), mercury-in-glass thermometer (0–200°C), and vacuum pump. The calrod is about 60 cm in length and somewhat less than 1 cm in diameter. The pyrex tube is about 1 meter in length and about 15 cm in diameter.

Theory: A transfer of heat between two bodies occurs whenever the bodies are at different temperatures. The transfer is always from the hot body to the cold body and ceases when the two bodies reach the same temperature (thermal equilibrium).

There are three methods by which heat is transferred: (1) Conduction, (2) Convection, and (3) Radiation. See any good general physics textbook for a description of these methods. The methods of conduction and convection require a material medium, but the method of radiation operates regardless of the presence or absence of a medium.

In many cases of heat transfer all three methods operate simultaneously. The laws of heat transfer are quite different for the three methods. Therefore, in order to make an experimental study of any one of them, it is generally necessary to eliminate, or reduce to a minimum, the other two. In the case of thermal radiation this can best be done by placing the radiating body in an evacuated vessel.

There is overwhelming evidence that the carriers of energy in the case of radiation are electromagnetic waves of the same nature as light waves. The absolute temperature of the emitting body is one of the important factors controlling the spectrum of these waves, i.e., the energy distribution among the various wave lengths. In the special case of radiation from a body with a black surface, the absolute temperature of the surface is the sole controlling factor. The body in this case is said to be a black body, and is defined as one that completely absorbs any radiation that falls upon its surface. Theoretically, a perfectly black surface does not exist. But there are many surfaces, such as a coating of lampblack, that almost completely absorb any incident radiation upon them. A good laboratory black body is a small aperture or window in the side of a hollow enclosure. Here, any radiation falling upon the aperture is absorbed into the hollow enclosure, and only a small fraction ever gets out again. If the enclosure is maintained at a constant high temperature, the radiation that streams out through the aperture is "black body" radiation. The absolute temperature of the enclosure completely determines the character of this radiation. The relation between this temperature and the distribution of energy among the various wave lengths in the radiation is given by Planck's law.[1]

The Stefan-Boltzmann law is one of the important laws of thermal radiation, and is the law with which we are particularly concerned in this experiment. It states that the time rate of radiation per unit area from a black body is directly proportional to the fourth power of its absolute temperature. Thus

$$R = \sigma T^4 \tag{22.1}$$

where R is the thermal energy per unit area per unit time flowing out from the black body, T is the absolute temperature of the black body, and σ is the Stefan-Boltzmann constant (universal).

If the body is not black, then $R < \sigma T^4$ and $R/\sigma T^4$ is defined as the total emissivity of the body. The emissivity has the value 1 for a black body. For any other body, its value will lie somewhere between 0 and 1 and will depend not only on the nature of the surface of the body, but also on the temperature.

Equation (22.1) determines the rate at which energy is leaving the black body. But the body will also be receiving energy from its surroundings. This energy may be taken into account by modifying Eq. (22.1) to read

$$R = \sigma(T^4 - T_0^2) \tag{22.2}$$

Here, $T_0(< T)$ is the effective absolute temperature of the surroundings, and R is now the net outward flow of radiant energy per unit time per unit surface area of the black body. It is assumed that the surroundings are also black, though they may not be. However, if $T_0 \ll T$, the error introduced by making this assumption is negligible because of the fourth-power law. In fact it is often possible to neglect the T_0 correction altogether.

The Stefan-Boltzmann law was first suggested by Stefan in 1879, and then predicted by Boltzmann some years later on the basis of the laws of thermodynamics. In point of history it came much earlier than Planck's law, which of course embodies it.

In this experiment the black body radiator is a steel rod with an electrical heater

[1] James M. Cork, *Heat*, 2nd ed. (New York: John Wiley & Sons, Inc., 1942), Chap. 5.

in its core (calrod). The temperature of the rod is controlled by the electrical current in the heater. Its surface temperature is measured by use of an iron-constantan thermocouple. Transfer of heat by convection is reduced by enclosing the hot rod in a pyrex tube from which the air is pumped (see Fig. 22.1).

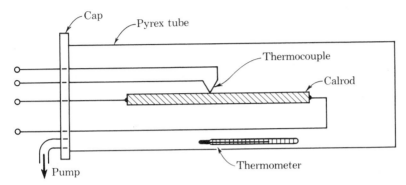

FIGURE 22.1 Thermal radiation apparatus.

Consider the condition of the hot rod when a steady state has been attained, i.e., when the temperatures of the various parts of the rod are not changing with time. Under this condition, there is a power balance between the electrical power fed into the rod and the rate of flow of heat out of the rod. The major portion of this flow of heat from the rod takes place by radiation, since convection has been practically eliminated by pumping out the system. However, there is considerable conduction of heat away from the ends of the rod through the wire leads connecting the rod to the power source. As a consequence, the surface temperature of the rod, although fairly uniform along its middle section, falls off rapidly at the end sections. In order to avoid making end corrections in this experiment, we consider the power balance in a small element of length, δL, of the rod in it middle section. The diameter of the rod is d (see Fig. 22.2). The input power is electrical and is just $I\,\delta V$, where I is the current in the element and δV is the drop of potential across the element. The output power is the rate of flow of radiant energy from the cylindrical surface of δL. This may be written as $\sigma(T^4 - T_0^4)\,\pi d\,\delta L$, i.e., R [Eq. (22.2)] times the cylindrical surface area. There is no loss of heat out through the ends of δL since the temperature gradient at these ends is zero. This is the prime reason for choosing an element δL near the middle of the rod.

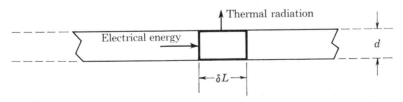

FIGURE 22.2 Power balance in Calrod element.

In the steady state for this element the power input must equal the power output. Therefore

$$I\delta V \times 10^7 = \sigma(T^4 - T_0^4)\pi d\,\delta L \qquad (22.3)$$

where I is the current in amperes in the element, δV is potential drop in volts across the element, σ is the Stefan-Boltzmann constant in cgs units, T is the absolute temperature of the surface of the element, T_0 is the absolute temperature of the surrounding glass tube, d is the diameter of the element in cm, and δL is its length in cm.

If it is assumed that the potential drop along the entire heating wire is uniform, then δV may be replaced by $(V/L)\delta L$, where V is the total potential drop across the entire heating wire and L is its total length. This assumption is very nearly valid for the calrod heater, i.e., in the range of temperatures used, the resistance of the heating wire is approximately independent of its temperature. Thus δV may be replaced by by $(V/L)\delta L$ and Eq. (22.3) may be written (after cancelling out δL) as

$$I\frac{V}{L} \times 10^7 = \sigma(T^4 - T_0^4)\pi d \qquad (22.4)$$

Equation (22.4) is based on the assumption that the calrod surface is essentially a black surface. This assumption is approximately valid, provided that either the surface of the calrod has been coated with lamp black or its surface is well oxidized by previous heating in air.

Method: Determine and record the average diameter of the calrod in the region where the hot junction of an iron-constantan thermocouple is attached. From your instructor, obtain the effective length of the heater wire in the calrod. This is less than the length of the calrod. Assemble the heater apparatus, placing the calrod in the center of the pyrex tube. Leads from the calrod and the thermocouple junction come out through a cylindrical cap that fits snugly over the open end of the pyrex tube. Place the mercury-in-glass thermometer on the bottom of the pyrex tube with its bulb just below the thermocouple junction. Connect the vacuum pump to the outlet tube in the cap.

Make the electrical connections according to the circuit diagram given in Fig. 22.3. After the system has been assembled and the proper connections made, immerse the cold junction of the thermocouple in the ice-water bath and start pumping the air out of the tube. It may be necessary to use a little vacuum wax on the cap in order to insure a vacuum tight system. With no current through the calrod, determine the emf of the thermocouple, using the potentiometer (see Experiment 38 and Appendix II-N for a description and use of the potentiometer). Remember that polarities must be correct in order to balance the potentiometer. Using Table K_2 in Appendix III, determine the room temperature of the rod. If this temperature is materially different from that given by the mercury-in-glass thermometer on the bottom of the tube, there is something wrong and the student should not proceed until this difficulty is resolved.

Heat the rod to a steady temperature state by sending a constant current of 1.50 amp through the rod. Adjust the rheostat Rh to maintain this constant current. Keep track of the temperature state of the rod by measuring the emf of the thermocouple as time goes. on. When this emf becomes reasonably steady (no continued drift up or down, only small random fluctuations), we can assume that the steady state has been reached. At this point, take three consecutive readings of the emf of the thermo-

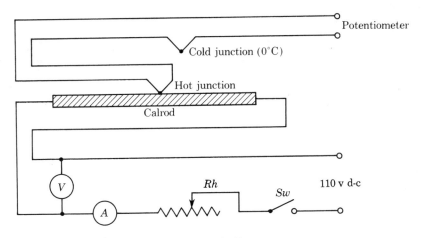

FIGURE 22.3   Circuit diagram.

couple. At the same time, read the ammeter (I), the voltmeter (V), and the temperature (T_0) given by the mercury thermometer. Record, in tabulated form, these values and their averages.

Repeat the foregoing procedure for additional values of the current in the rod: 1.75, 2.00, 2.25, and 2.50 amp. Do not use a current larger than 2.50 amp for the rod, since any larger current is likely to destroy the heater wire in the rod because of high temperature. Be sure in each case that the rod has actually reached a steady tempera-ture state before proceeding to the next higher current.

Using Table K_2 in Appendix III, determine the temperature in °C of the hot junction of the thermocouple, i.e., the surface temperature of the calrod for the se-quence of the temperature states. In each case estimate the error in this temperature. Record these temperatures and their estimated errors. Compute and record the cor-responding absolute temperatures, including that of the mercury thermometer.

From these data and results, compute corresponding values of $x(= T^4 - T_0^4)$ and $y (= (IV/\pi dL) \times 10^7)$ for each of the temperature states. Plot these values of x and y. From the theory [Eq. (22.4)], it is clear that this graph should be a straight line whose scale slope is σ, provided that the Stefan-Boltzmann law is valid. Determine and record the value of σ including its estimated error. Compare it with the accepted value of σ (= 5.67×10^{-5} erg cm^{-2} sec^{-1} deg^{-4}).

Record: Tabulate data and results.

QUESTIONS

1. Your value of σ in this experiment is quite likely to be less than the accepted value of σ. How do you explain this?

2. Develop the determinate error equation for σ, using Eq. (22.4) and neglecting the term T_0^4.

3. On the basis of your data in this experiment, explain why any error in the measurement of T_0 is likely to have an insignificant effect on the result.

Heat of
Fusion

Object: to determine the heat of fusion of ice.

Apparatus: Ice, calorimeter, watch, thermometers, balance, and paper towels. The calorimeter consists of a metal cup that may be placed inside a larger vessel consisting of an insulated water jacket that surrounds the metal cup. (See Fig. 23.1.) The purpose of the water jacket is to insulate the calorimeter cup from the effects of drafts or sudden changes in temperature of the surroundings.

Theory: The latent heat of fusion is the energy required to change a unit mass of a substance in its solid state to a liquid form, the temperature remaining constant at the melting point. When a substance makes such a change, it absorbs heat; when the reverse happens, heat is set free.

It should be noted that the heat of fusion is distinct from the specific heat of a substance involving a change in temperature.

In this experiment the latent heat of fusion of ice (the number of calories necessary to melt 1 gm of ice at its melting point) is to be found.

The method of calorimetry will be used; a mass of water will be placed in a calorimeter cup and the change in its temperature as ice is added and melted will be noted. The cup is placed in the insulating jacket. In order to counteract the effects of radiation of the cup, to and from the water jacket, the water in the cup will, at the beginning of the experiment, be a few degrees warmer than that in the water jacket, and enough ice will be added so that at the finish, the water in the cup will be an *equal* number of degrees *cooler*. Thus on the average, equal amounts of heat will be radiated from the cup to the jacket, and from the jacket to the cup, during the experiment. These amounts, then, may be ignored.

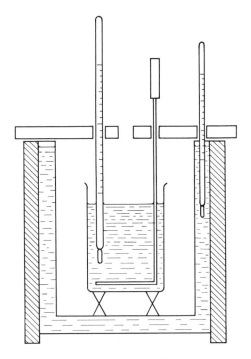

FIGURE 23.1 Calorimeter.

Let M = initial mass of the water,

M' = mass of ice added,

m = mass of the calorimeter cup plus stirrer,

s = specific heat of the cup and the stirrer (Table E, Appendix III),

e = water equivalent of the thermometer (use 0.46 times the volume of the thermometer which is immersed),

t' = original temperature of the cup and the water,

t = final temperature of the whole system,

L = latent heat of fusion of ice.

Heat lost equals heat gained. Thus

$$M(t' - t) + mt(t' - t) + e(t' - t) = LM' + M'(t - 0) \qquad (23.1)$$

From this the latent heat of fusion is

$$L = \frac{(M + ms + e)(t' - t) - M't}{M'} \qquad (23.2)$$

Method: Weigh the *dry* calorimeter cup and stirrer to the nearest tenth of a gram; add about 300 gm of water at a temperature approximately 5° above jacket temperature, and weigh again. Place the calorimeter inside the water jacket upon the insulating base provided; fasten the cover in place; and insert the thermometer so that it is immersed 2 or 3 cm deep in the water. Obtain a piece of ice weighing about 50 gm (what will be the dimensions of such a piece?) and place it on a paper towel near the calorimeter ready for use.

Take the temperature of the water to the nearest 0.1° every 30 sec for 3 min, stirring steadily. Then remove the cover, laying it on edge so as not to break the thermometer. Wipe the piece of ice with a paper towel as dry as possible (why?), placing it in the calorimeter *without touching the ice with the fingers and without splashing. Note the time at which the ice is placed in the water.* Replace the cover. This entire operation must not take more than 30 sec. Continue to stir and take readings every 30 sec. Continue in this manner for 3 min after the ice has all melted, as evidenced by the reaching of a minimum temperature. Remove the calorimeter cup and carefully weigh again to determine the mass of ice added. Be sure to make this final weighing under the same conditions as the initial weighings were made, using the same balance and weights.

Repeat the experiment.

Plot a curve of temperature vs. time for each run. The initial portion of each curve (first 3 min) should be a straight line with a slight negative slope. In order to determine the temperature at the instant the ice was added, extend this straight line to a point corresponding to the time at which the ice was added. The temperature indicated by this point is the temperature t'. Mark this point with a special symbol and pass the curve through it. The minimum point on the curve represents the temperature t.

The error in L is primarily due to errors in the measurements of $t' - t$ and M'. Therefore, the determinate-error equation is approximately

$$\frac{\Delta L}{L} = \frac{\Delta(t' - t)}{t' - t} - \frac{\Delta M'}{M'} \tag{23.3}$$

Find the two values of L and their errors. Average these values and compare with the accepted value given in Table L, Appendix III.

Before leaving the laboratory, empty the calorimeter cup, but not the water jacket. Wipe up any spilled water, and dispose of wet towels.

Record:

	TRIAL I		TRIAL II
App. No._____			
Weight of cup and stirrer	_____	$= m$	_____
Weight of cup and stirrer + water	_____		_____
Weight of cup and stirrer + water + ice	_____		_____
Material of cup and stirrer	_____		_____
Specific heat of these	_____	$= s$	_____
Initial mass of water	_____	$= M$	_____
Mass of ice added	_____	$= M'$	_____
Volume of thermometer immersed	_____		_____
Water equivalent of thermometer	_____ gm	$= e$	_____
Jacket temperature	_____		_____
Initial temperature	_____	$= t'$	_____
Final temperature	_____	$= t$	_____
Latent heat of fusion	_____	$= L$	_____
Error in L	_____	$= \Delta L$	_____

QUESTIONS

1. If there is a *constant* error of 0.5°C in the thermometer used in this experiment, what error will this introduce into the value of L? Explain.

2. If the ice is wet when placed in the calorimeter so that the mass consists of 99% ice and 1% water, what constant fractional error would be introduced into the value of L?

3. What constant fractional error in L would be introduced by neglecting to take into account the water equivalent of the thermometer? Is this error significant in this experiment?

4. Why is it necessary to be careful about the weight determinations in this experiment?

OPTIONAL EXPERIMENTS

The method of calorimetry used in this experiment for determining heat of fusion can readily be modified for determining specific heat or heat of vaporization.

1. Determine the average specific heat of each of several different metals in the temperature range from about room temperature to about 100°C. Hence show that their *molar* specific heats are very nearly the same in this temperature range.

2. Determine the heat of vaporization of water at its normal boiling point.

PART **|||**

ELECTRICITY AND MAGNETISM

Instructions for
Electricity
Laboratory

The General Instructions, Section D of the Introduction, hold equally for all experiments in the course. In addition, for electricity laboratories, the following points should be borne in mind:

The equipment in electricity laboratories is, in general, more subject to damage than that used in other parts of the laboratory course. A small mistake in procedure can result in severe damage. *Therefore, the source of electric power is not to be connected into the circuit until the instructor has authorized it.* This is especially true when the source of power is 110 v a-c or d-c from a wall outlet. A mistake here not only ruins equipment, but it can be *very dangerous* as well. Never make any circuit changes without first turning off the power. If the power source is a battery (this includes standard cells), neither terminal should be connected to the circuit before authorization. Following this rule rigorously will prevent damage to the equipment in most cases.

Most items of equipment have their ratings shown clearly. These ratings must never be exceeded. In the case of meters, the ratings are shown by the full-scale readings. Rheostats have the maximum current marked on the name plates or stamped on the ends of their mounting boards. Other instruments have ratings shown on the name plates. When using a piece of equipment for the first time, look for these ratings. Read the notes in Appendix II if the apparatus is described there. Above all, be as well informed as possible about the experiment being performed before coming to the laboratory.

Cathode-Ray
Oscilloscope

Object: To become familiar with the operation of a cathode-ray oscilloscope.

Apparatus: Cathode-ray oscilloscope, microphone, audio-frequency generator, tuning fork, and a frequency standard. *Note:* Carefully study Note L, Appendix II, before proceeding with this experiment. The construction and operation of the cathode-ray oscilloscope are discussed in that note.

Theory: *Phase Measurement and Lissajous Patterns.* Very often wave patterns have to be examined and analyzed for phase displacement with respect to some reference signal. Such measurements are easily made by using the oscilloscope. Suppose we are interested in knowing how the phase of the output signal of an audio amplifier compares with the input, and for simplicity we shall assume the input signal to be sinusoidal. If the input and output wave patterns are different, a combination of distortions may have been introduced by the amplifier. Should the patterns match except in amplitude and phase, we can then determine how much the phase of the output signal has been altered. A method for doing this is to produce Lissajous figures on the scope.

As a preliminary step, apply the same signal to the vertical and horizontal inputs of the scope, having disconnected the internal sweep circuit. Figure 30.1 shows the patterns that would appear on the screen if the outputs of both internal amplifiers (horizontal and vertical) are identical in phase and amplitude. We shall assume later that the output signals of the amplifiers differ in phase and amplitude. Combinations of two simple harmonic motions at right angles to each other that may have different phase, different amplitudes, and different frequencies produce figures known as *Lissajous figures*.

The straight line is a special case of the Lissajous figure. Assume the electric field

146

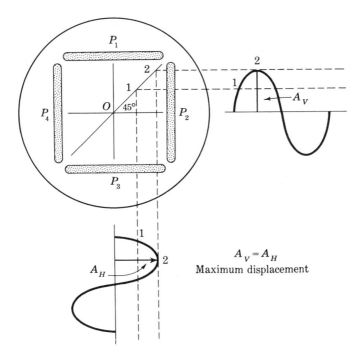

FIGURE 30.1 A combination of two simple harmonic waves in time phase but 90° out of space phase.

between plates P_1 and P_3 changes by the same amount as, and in phase with, the electric field between P_2 and P_4. This being the case, the spot will move the same distance upward as to the right. Similarly, during the negative alternation, the spot will move to the left and downward by the same amount. The net effect, therefore, is an excursion of the spot that splits the first and third quadrants. If, however, one signal is out of phase by 180° with respect to the other, the graph will be a straight line that splits the second and fourth quadrants. Suppose now that the voltage of one signal to the deflection plates is less than the other signal. The graph generated will still be a straight line, but it will have an angle other than 45°,

In the previous cases the resultant pattern has been a straight line. The following case illustrates what happens when two signals differ in phase by (1) 45°, and (2) 90°. Figure 30.2 shows two waves of the same amplitude and frequency, but the input signal e_H to the horizontal deflection plates is leading e_V by 45°. Figure 30.3 shows the pattern when the electron beam is actuated by the fields between the deflection plates under this situation. A_V and A_H are the vertical and horizontal amplitudes.

When the ellipse is centered relative to the rectangular coordinate system, the x and y intercepts and the maximum displacements are values that can be used to determine the phase relationship between the waves. Point $(A, 1)$ (Figs. 30.2 and 30.3) indicates the horizontal displacement when the vertical is zero. Point $(B, 2)$ shows the maximum displacement horizontally. Therefore, $\sin \theta = A/B$, where A

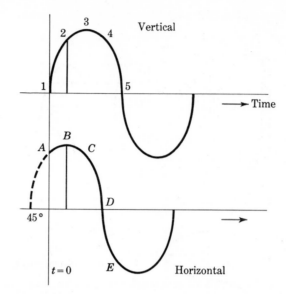

FIGURE 30.2   Two waves 45° out of time phase.

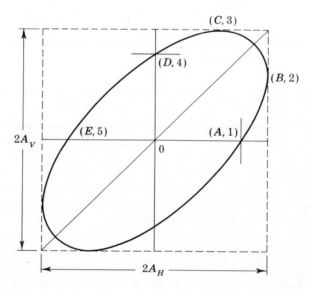

FIGURE 30.3   The figure on the scope for the two waves of Fig. 30.2.

and B represent the values of the displacements and θ is the phase angle between the two waves. If, however, at $t = 0$ the vertical displacement is maximum, the pattern becomes a circle (Fig. 30.4).

The waves producing the circle in Fig. 30.4 are (1) of equal amplitudes and (2) out of phase with each other by 90°. Had the amplitudes been different, the pattern would be an ellipse with its major and minor axes coincident with the coordinate

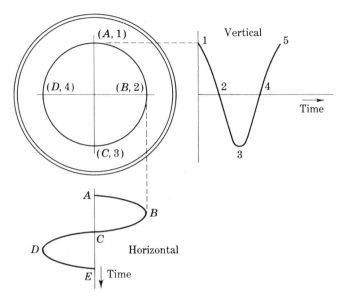

FIGURE 30.4 The figure on the scope when signals of equal amplitudes are 90° out of time and space phase.

axes. The major axis of the ellipse will be determined by the signal of greater amplitude, vertical or horizontal.

Frequency Measurement and Lissajous Patterns. In the previous section, the input signals to the vertical and horizontal plates had the same frequency. We shall now remove the frequency restriction and notice what effect, if any, there is on the wave pattern. Although the frequencies are allowed to change, we shall hold the amplitudes at approximately the same value. Figure 30.5 shows the pattern when the frequency of the horizontal input signal is one-half that of the vertical signal. The pattern is symmetrical because the waves start out in phase. An unsymmetrical Lissajous pattern results when the phase difference between the signals is different from 0°. A simple method of determining the frequency ratio is to note the number of tangency points along an imaginary horizontal line and a vertical line.[1] The ratio is commonly expressed as the vertical frequency divided by the horizontal frequency. Usually one of the frequencies is known very precisely. The Lissajous figure can be used to determine quite accurately the frequency of an unknown signal by this method. The calibration of an audio-frequency (af) generator is carried out by use of the connections shown in Fig. 30.6. The signal of known frequency is connected to the vertical (or horizontal) deflecting plates, and the signal with the unknown frequency is connected to the other set of plates (or the corresponding amplifiers). The known af frequency can be obtained from (1) 60 cycle a-c line voltage or (2) a reliable af oscillator. The a-c line frequency (60 cycles/sec) is usually held to very small tolerances.

[1]An alternate method is to count the number of crossings made by the pattern on both horizontal and vertical imaginary lines drawn through the figure.

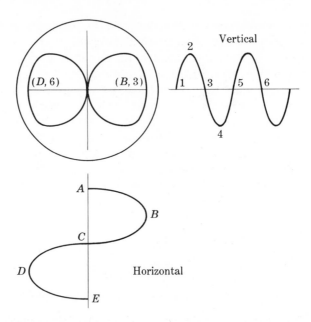

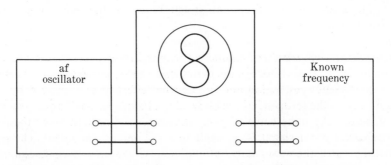

FIGURE 30.5 Lissajous patterns for two signals whose frequency ratio is 2 to 1.

FIGURE 30.6 Circuit for frequency calibration of audio oscillator.

Another af source is radio station **WWV**. This government station is maintained by the Bureau of Standards, Washington, D.C., broadcasting on eight radio frequencies (rf), 2.5, 5, 10, 15, 20, and 25 megacycles. Its rf carrier is audio modulated at 440 and 600 cycles/sec, each appearing at regular time intervals during the day. The rf and af signals are well known to the experimentalist, since they constitute the most widely-used primary standards.

In using these af signals, one needs a short-wave receiver to pick up and demodulate the rf signal. If this radio signal is used for the af source, the output (speaker coil) of the receiver may be connected to either the horizontal or vertical input of the scope.

Method: *Step 1. Wave Forms as Functions of Time.* Connect the output of the audio-frequency oscillator to the vertical input terminals of the cathode-ray oscilloscope (CRO). Note that each instrument has a terminal labeled "ground"; these should be connected to each other. Set the oscillator frequency to about 1000 cycles/sec and adjust the oscillator amplitude control and vertical gain control of the CRO until the beam vertically deflects about two-thirds of the face diameter. (Note: A distorted pattern on the screen of the scope may be due to an over-driven stage. Adjust output of af generator and gain of scope amplifier so as to produce a good pattern.) If necessary, readjust the intensity and focusing controls to obtain a good trace. Adjust the vernier frequency control carefully to "stop" the pattern. When the hand is taken off the vernier frequency control, the pattern will probably begin to "walk" along the sweep. When it does, adjust the "Synch Amplitude" control to "stop" the pattern. Study the interaction of the "Synch Amplitude" and frequency vernier controls; that is, determine how much the basic sweep frequency may be moved before the pattern "jumps out of synch" as a function of how much synchronizing signal is used. Adjust the frequency controls to obtain different numbers of cycles of the sinusoidal pattern on the scope. It should be possible to "synch in" at one, two, three, and so on, up to perhaps 15 cycles of the pattern being presented on the scope. Change the frequency of the oscillator and repeat the previous process. With a large number of cycles crowded on the scope face, increase the horizontal gain until two or three of the center cycles now occupy the scope face. Move the horizontal centering control to examine in detail other cycles of the wave. Return centering and gain controls to their original positions.

Step 2. Frequency Comparison. Connect the standard frequency oscillator (1000 cycles/sec) to the vertical (*y*) input of the scope, and connect the audio oscillator to the horizontal (*x*) input, making sure that the "ground" terminals of the two instruments are connected together. Switch the horizontal amplifier to "ext input." Adjust the oscillator frequency and all the gain controls until an ellipse is obtained (a 1:1 Lissajous figure). If possible, "stop" the ellipse. The actual oscillator frequency is now 1000 cycles/sec. Record the dial reading of the oscillator (it will not, in general, be 1000). Allow the ellipse to change phase, and note the various shapes it takes on. Adjust the oscillator to about 500 cycles/sec to obtain the 1:2 Lissajous figure. The shape is that of a "figure 8." Then, obtain Lissajous figures for frequency ratios of 1:5, 1:3, 2:3, 3:2, 3:1, and 5:1. Record the oscillator dial settings and the actual frequencies. Plot a calibration curve for the oscillator.

Step 3. Connect the input of the scope to the high-impedance winding of the output transformer of the loudspeaker. Adjust the vertical gain of the scope until a good deflection is obtained when speaking in a normal voice into the speaker. Adjust the frequency of the sweep to get a good presentation of the wave forms. The synchronization controls will not be of use here. Why? Hum a steady tone into the speaker, and examine the wave form. The synchronization control will be of assistance here. Why? Sing a note in low register into the speaker and examine this wave form. Sing a clear falsetto note and again examine the wave form. What is the difference in appearance?

Whistle a clear tone and again examine the wave form. How does this compare with the sine wave previously viewed? Sketch a typical pattern from each of the preceding procedures.

Step 4. Amplitude Measurement; Voltage Calibration. Connect the output of a sinusoidal signal generator, whose output voltage is known, to the vertical input of the scope. Adjust the pattern on the scope to read 10 v/cm, 1 v/cm, or something similar. Having calibrated the scale on the scope, measure a few unknown voltages. (Note: Once the calibration has been made do not adjust the gain controls of the vertical amplifier.) How does the measurement made with the scope compare with that obtained by using the conventional af voltmeter?

Capacitance

Object: To determine the capacitance of two capacitors (older terminology, condensers), when taken singly, and when connected in parallel and in series, by the comparison-of-deflection method with a ballistic galvanometer.

Apparatus: Ballistic galvanometer, damping key, two paper capacitors, one standard mica capacitor, two double-pole double-throw switches, and dry cell. For a description of some of these items see Appendix II, Notes F.1, 4; G.1, 3, 4.

Theory: The capacitance C of a capacitor is defined as the ratio of its charge Q to the potential difference V between its plates, that is

$$C = \frac{Q}{V} \qquad (31.1)$$

When Q is expressed in coulombs, and V is expressed in volts, then C is given in farads. A capacitance of 1 f is *extremely* large. It is customary, therefore, to express capacitance in microfarads, i.e., millionths of a farad.

A simple and direct method of comparing the capacitances of two capacitors is to compare their charges, when both have the same potential difference across their plates. Suppose, for example, that a capacitor of unknown capacitance C_x is charged to a potential V, thus accumulating a charge Q_x. Similarly, a standard capacitor of known capacitance C_0 is charged to the same potential V, thus accumulating a charge Q_0. Then, from Eq. (31.1) we have

$$\frac{C_x}{C_0} = \frac{Q_x}{Q_0} \qquad (31.2)$$

If each capacitor is discharged through a ballistic galvanometer, the two deflections (or throws) obtained will be in the same ratio as Q_x and Q_0. See Appendix

IIG.4 for the action of a ballistic galvanometer. Thus Eq. (31.2) may be replaced by the equation

$$\frac{C_x}{C_0} = \frac{D_x}{D_0} \qquad (31.3)$$

where D_x and D_0 are the two deflections. Therefore, to determine C_x, it is only necessary to observe the deflections D_x and D_0 and to know the value of C_0.

A satisfactory setup for determining C_x, as shown schematically in Fig. 31.1, consists of a galvanometer G, damping key K, battery B, condensers C_0 and C_x, and double-pole switches S_1 and S_2. By appropriate manipulation of S_1 and S_2, either capacitor may be charged to battery emf (assumed constant) and then discharged through galvanometer G.

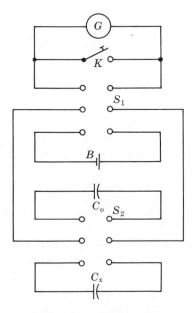

FIGURE 31.1   Wiring diagram.

Parallel and Series Connections. Two capacitors of capacitances C_1 and C_2 may be connected in parallel as shown in Fig. 31.2. The combined capacitance C_p of this parallel combination is

$$C_p = C_1 + C_2 \qquad \text{(parallel)} \qquad (31.4)$$

since the charges are added, but the potentials are the same.

Two capacitors may also be connected in series, as shown in Fig. 31.3. The combined capacitance of this series combination is given by the equation

$$\frac{1}{C_s} = \frac{1}{C_1} + \frac{1}{C_2} \qquad \text{(series)} \qquad (31.5)$$

since the potentials are added, but the charges are the same.

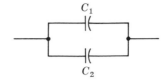

FIGURE 31.2   Parallel connection.

FIGURE 31.3   Series connection.

The determinate-error equations corresponding to Eqs. (31.3), (31.4), and (31.5), respectively, are

$$\frac{\Delta C_x}{C_x} = \frac{\Delta C_0}{C_0} + \frac{\Delta D_x}{D_x} - \frac{\Delta D_0}{D_0} \qquad (31.3a)$$

$$\Delta C_p = \Delta C_1 + \Delta C_2 \qquad (31.4a)$$

and

$$\frac{\Delta C_s}{C_s} = \frac{C_2 \Delta C_1}{(C_1 + C_2)C_1} + \frac{C_1 \Delta C_2}{(C_1 + C_2)C_2} \qquad (31.5a)$$

Time of Charging. When a capacitor is connected across the terminals of a battery, the charge Q on its plates at any time t is given by the relation

$$Q = Q_{max}(1 - e^{-t/RC}) \qquad (31.6)$$

This relation indicates that full charge is only obtained after an infinite amount of time. However, at the time $t = RC$, the capacitor is about 63% charged. If t is ten times this value, we may assume that the capacitor is fully charged. The value RC is known as the *time constant* of the circuit. Here R is the resistance in the circuit and C is the capacitance. Generally RC is quite small, e.g., 10^{-6} sec, thus instantaneous charging takes place for all practical purposes. Even if R is very large (10^6 ohms), the time constant of a 1 μf capacitor is only 1 sec. Thus one is generally safe in assuming that after 10 sec, such a capacitor connected across a battery will be fully charged.

Method: Before attempting to perform this experiment, the student should carefully read Appendix II, Notes E; G.1, 3, 4 concerning the type of galvanometer used.

Make the connections as shown in Fig. 31.1, but *do not connect* the cell until the instructor has checked your wiring. Use either one of the unknown capacitors as C_x and the standard 0.5 μf capacitor as the known C_0. With both switches S_1 and S_2 open, adjust the reading telescope and scale on the galvanometer so that the cross hairs of the telescope coincide with the zero of the scale. Excessive motion of the galvanometer coil may be stopped by closing the damping key K.

Connect C_0 into the circuit by means of switch S_2 and the cell into the circuit by means of switch S_1. The cell is now connected across C_0, thus charging it. Allow this charging process to go on for about 10 sec. Sometimes a high resistance is connected in series with the cell, in order to protect the galvanometer in case the cell is accidentally connected across the galvanometer. This may prolong the time required for completely charging the capacitor.

Check the galvanometer to see that it gives a steady zero reading. Then reverse switch S_1 so that C_0 is now connected across the galvanometer G. Observe the maximum deflection of the galvanometer in millimeters, estimating to a half millimeter. Repeat this process of charging and discharging C_0, using a longer charging time, say 20 sec. If there is no appreciable increase in the maximum deflection for the longer charging period of time, then the shorter charging period is sufficient and may be used in the remainder of the experiment. If, however, there is an appreciable increase in deflection for the longer period, it will be necessary to still further increase the charging period in order to find the proper minimum time.

After observing and recording the deflection D_0 for C_0, reverse switch S_2 so that C_x is in the circuit. Charge and discharge C_x, using the same charging period as for C_0. Observe and record the deflection D_x for C_x. Then go back to C_0. Proceed in this manner, taking *alternate* readings of D_0 and D_x until five D_0 and four D_x readings have been taken. Record these readings and compute the average values of D_0 and D_x. Estimate the error in each of these averages and record it with the average.

Determine by means of Eq. (31.3) the unknown capacitance C_x; also determine the error in C_x. Assume that the error in C_0 may amount to as much as $\pm 1\%$ of the rated value.

Replace C_x by the other unknown and repeat the above procedure for determining its capacitance.

Connect the two unknown capacitors in *parallel* and insert this combination into the circuit as C_x. Determine the capacitance of this combination and compare it with the value calculated by Eq. (31.4). In each case calculate the errors involved.

Finally, connect the two unknown capacitors in *series* and insert this combination into the circuit as C_x. Proceed as above in the determination of capacitance.

Record: Give apparatus numbers of the galvanometer, standard capacitor, and unknown capacitors. Tabulate your data. Express your results in the following form:

Item	C_x	% error
Capacitor No. 1		
Capacitor No. 2		
Parallel (exp)		
Parallel (calc)		
Series (exp)		
Series (calc)		

QUESTIONS

1. Develop the indeterminate-error equations for this experiment by use of the rules given in Sec. A.4 of the Introduction. How do they differ from Eqs. (31.3a), (31.4a), and (31.5a)?

2. If the galvanometer deflections in this experiment are too small for accurate measurement, how may they be increased without using a more sensitive galvanometer?

OPTIONAL EXPERIMENTS

1. Determine the coulomb sensitivity of a ballistic galvanometer.

2. Determine an unknown emf in terms of that of a standard cell.

3. The objective of the above experiment may be achieved by using an oscilloscope as the detector. From Eq. (31.6) we can rewrite the equation in the form $V = E(1 - e^{-t/RC})$, where V is the instantaneous voltage across the capacitor and E is the applied emf. In this alternate method we shall apply d-c voltage across an RC circuit (see Fig. 31.4) and observe the exponential rise in voltage with time.

The voltage source, however, is a square-wave generator, which is no more than a fast acting switch in series with a battery. A suitable time constant T might be 200 μsec for which 0.1 μf and 2000 ohms are chosen for capacitance and resistance respectively. It is recommended, however, that decade boxes be used for both C and R so that various combinations may be obtained.

The instantaneous voltage across the capacitor is applied to the vertical input of the scope. When the sweep frequency is synchronized to the generator frequency, the characteristic exponential growth curve of a capacitor will be displayed on the screen of the scope.

In order to measure the time constant (RC), it will be necessary to calibrate the time axis of the scope. For example, if f = 1000 cycles/sec and the sweep of the scope is synchronized to this frequency, the pattern will be one complete square-wave cycle. Because the period of the signal is 1 msec, each division on the horizontal axis will be a fractional part of 1 msec,

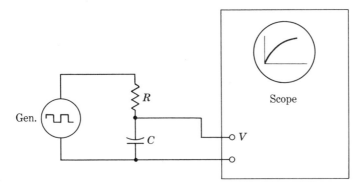

FIGURE 31.4 Voltage rise across capacitor.

depending on the horizontal amplification. For convenience it is suggested that the square wave pattern cover 10 grid divisions on the screen, in which case each division represents 100 μsec. For example, Fig. 31.5 shows one alternation spread over 5 divisions, where the time required for the voltage to rise to 0.63 of the max is the time constant (RC).

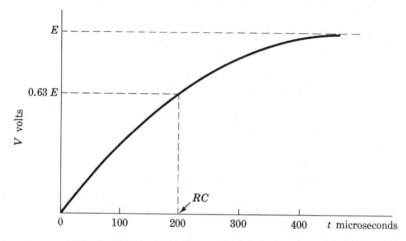

FIGURE 31.5 V vs t for a capacitor; time constant.

Joule's
Law

Object: To determine the heat equivalent of electrical *energy*, Joule's constant.

Apparatus: Calorimeter, heating coil (about 1 ohm #20 nichrome wire), clock, thermometer, ammeter (0–3 amp), voltmeter (0–3 volt), and variable power supply (0–6.3 volt) capable of furnishing a steady current of 2 amp. See Appendix II, Note H on the ammeter and voltmeter.

Theory: Energy exists in many different forms, and physical processes usually involve a conversion of one form of energy into another form. In this experiment electrical energy is converted into heat energy, and we wish to determine the number of joules of electrical energy that will produce 1 cal of heat energy. If W represents the electrical energy, expressed in joules for example, and if H represents the heat energy produced, expressed in calories, then the ratio of W to H is constant. This constant is represented by the symbol J and is frequently called Joule's constant, or the mechanical equivalent of heat. In equation form

$$J = \frac{W}{H} \tag{32.1}$$

In order to determine J, it is necessary to measure W and H. The amount of electrical energy W that is dissipated in the form of heat by a current I in a resistance R during time t is given by any one of the three equations:

$$W = VIt \tag{32.2}$$

$$= I^2Rt \tag{32.3}$$

$$= \frac{V^2}{R}t \tag{32.3a}$$

where V is the potential drop across the resistance R. If V is in volts, I in amperes, R in ohms, and t in seconds, then W is in joules.

Suppose that the heat generated in this resistance (heating coil) is measured calorimetrically by immersing the coil in a calorimeter cup filled with water. As a result, the temperature of the calorimeter cup and its contents will rise. This temperature rise may be noted on a thermometer immersed in the water, therefore, the amount of heat H given to the calorimeter may be computed.

An appropriate setup for this experiment is shown in Fig. 32.1. It consists essentially of a calorimeter with stirrer and thermometer. The calorimeter cup is partially filled with cold water. The heating coil immersed in the water is connected to the variable power supply as shown in Fig. 32.1. Note the manner in which the voltmeter and the ammeter are connected into the circuit. It is possible to eliminate either one or the other of these instruments if the resistance R of the heating coil is known. In the following discussion we shall assume that R is known and that only an ammeter is used for measuring the current I. You can easily supply the necessary modifications for the other two possibilities.

When the circuit is closed, electrical energy in the heating coil is converted into heat energy absorbed by the calorimeter. The amount of heat absorbed by the calorimeter is given by the equation

$$H = (M_w + M_c S + M_E)(T_2 - T_1) \qquad (32.4)$$

where
$\quad H =$ heat in calories,
$\quad M_w =$ mass of water in grams,
$\quad M_c =$ mass of calorimeter cup in grams,
$\quad S =$ specific heat of cup,
$\quad M_E =$ water equivalent in grams of heating coil, stirrer, and thermometer, and
$\quad T_2 - T_1 =$ rise in temperature in °C of calorimeter.

The quantities on the right side of Eq. (32.4) may be measured, thus H may be determined.

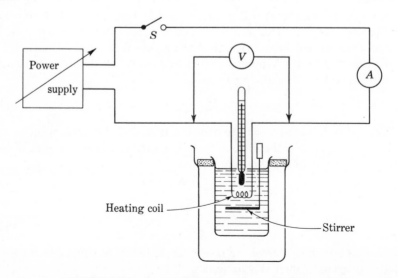

FIGURE 32.1   Joule's-law apparatus and wiring diagram.

If we substitute the values of W [Eq. (32.3)] and H [Eq. (32.4)] in Eq. (32.1), we get

$$J = \frac{I^2 Rt}{(M_w + M_c S + M_E)(T_2 - T_1)} \tag{32.5}$$

The principal errors in this experiment occur in the measurements of I and $T_2 - T_1$. The *determinate*-error equation involving these quantities is

$$\frac{\Delta J}{J} = \frac{2\Delta I}{I} - \frac{\Delta T_2}{T_2 - T_1} + \frac{\Delta T_1}{T_2 - T_1} \tag{32.5a}$$

The errors in the masses, the resistances, and the time are negligible compared to the other errors in this experiment. In addition, there is the error, ever present in calorimeter experiments, introduced by heat exchange between the calorimeter and its surroundings during the course of the experiment. Since this heat exchange is generally proportional to the time, it is advisable to reduce the time of an experimental run to a minimum. Also, it is advisable to have the initial temperature T_1 of the calorimeter as much below room temperature as the final temperature T_2 is above room temperature. In this way, the net heat exchange between the calorimeter and its surroundings during the experiment may be reduced to a minimum.

Method: Connect the apparatus as shown in Fig. 32.1.

Weigh the calorimeter cup when empty and dry. Add cold water from the tap until the cup is about three-quarters full, and weigh again. The water should be 3° or 4° below room temperature. Place the calorimeter in the metal jacket and insert the heating coil, stirrer, and thermometer. Handle the thermometer with great care. It is easily broken and hard to replace. Close the switch and quickly adjust the power supply to give a reading of precisely 2.00 amp on the ammeter. Then open the switch.

Start stirring the water and after 30 sec read the thermometer to the nearest 0.1°. Start your time and temperature measurements from this point. Read the thermometer every 30 sec thereafter, while *stirring continuously*. At the beginning of the third minute, close the switch Sw and read the ammeter as well as the thermometer. It is necessary to keep the current constant at this value during the run by continual adjustment of the rheostat. One student should perform this adjustment and record all data, while the other stirs and reads the clock and the thermometer. The temperature will now begin to rise rapidly because of the heat supplied by the coil. Stir continuously, and record readings on both ammeter and thermometer every 30 sec. When the temperature of the calorimeter has risen to a value approximately as much above room temperature as its initial temperature was below, open the switch *immediately* after reading the ammeter and thermometer. Continue to stir and to take temperature readings for an additional 2 min.

Plot a time (abscissa) vs. temperature (ordinate) curve. Indicate on this curve, along with the usual items, the room temperature and the exact times at which the current was turned on and off. The interval in seconds between these two times is the value of t in Eq. (32.5). For I, use the average ammeter reading during the run. For T_1, use the temperature of the calorimeter at the instant the current was turned on. For T_2, use the maximum temperature on the time-temperature curve. The resistance R of the coil is given. The water equivalent of the thermometer, heating coil, and stirrer may be taken as 6 gm.

Compute, by means of Eq. (32.5), the value of J in joules per calorie. Also, compute the *indeterminate* error in J by use of Eq. (32.5a). Assume that the error in the ammeter may be $\pm 1.5\%$ of full-scale reading and that the errors in T_1 and T_2 may amount to $\pm 0.05°C$. Compare your value of J with the commonly accepted value 4.18 j/cal.

Make a second run of this experiment.

Precautions:

1. Handle the thermometer with care.

2. Do not turn on the current unless the heating coil is immersed in water. It may burn out otherwise.

Record:

    App. No. Calorimeter_____

             Ammeter_____

             Heating coil_____

Resistance of heating coil:	$R = ($	$)$ ohm
Mass of calorimeter empty:	$M_c = ($	$)$ gm
Mass of calorimeter + water:	$($	$)$ gm
Mass of water:	$M_w = ($	$)$ gm
Specific heat of calorimeter:	$S = ($	$)$
Water equiv of coil, etc.:	$M_E = 6.0$ gm	
Room temperature:	$($	$)$ °C

Tabulate your time, temperature, and current data.

    From temperature vs. time curve:

Time of heating:	$t = ($	$)$ sec
Initial temp of water:	$T_1 = ($	$)$ °C
Final temp of water:	$T_2 = ($	$)$ °C
Experimental value of J:	$J = ($	$)$ j/cal
	$\Delta J = ($	$)$ j/cal
Accepted value of J		4.18 j/cal
% difference		$($ $)$

1. What constant percentage error would be introduced into your value of J by failing to take into account the water equivalent of the thermometer, heating coil, and stirrer?

2. If the thermometer you used had a constant error of $0.5°C$, what error would this introduce in your value of J?

3. What is the purpose, if any, of plotting a heating curve in this experiment? Would it not suffice to take only two temperature readings?

4. If the resistance of the heating coil was not known, how could a voltmeter be used in this experiment to complete the data? What change in Eq. (32.5) would result?

EXPERIMENT **33**

Electrolysis

Object: To determine the electrochemical equivalent of copper by use of an electrolytic cell. Therefore, to determine the value of the *faraday*, i.e., the product of Avogadro's number and the electronic charge. In the development of modern physics this is an important experiment, since it gives information about some of the fundamental properties of ions and atoms.

Apparatus: Electrolytic cell with copper sulfate solution and copper electrodes, ammeter, dropping resistance, rheostat, d-c power source, analytical balance, and a clock.

Theory: Consider two copper plates immersed in a solution of $CuSO_4$. The salt in solution is ionized, with equal numbers of Cu^{++} and SO_4^{--} ions present. If a potential difference is established between the plates, the ions will move in the solution, the positive ions toward the cathode (cations) and the negative ions toward the anode (anions). When the Cu^{++} ions reach the cathode, their charge will be neutralized by the acquisition of two electrons each. They will be deposited on the cathode as normal Cu atoms. The SO_4^{--} ions will give up their excess electrons (two for each ion) at the anode, combine with the atoms of Cu there, and then go into solution as $CuSO_4$, to be ionized again.

This description of the ionic process of electrolysis is essentially the one proposed by Arrhenius in 1887, and is still acceptable today in its general form. It gives a completely satisfactory explanation of Faraday's laws of electrolysis, stated much earlier (1833), and based upon experimental evidence. These laws are:

1. The mass of an element deposited at an electrode is directly proportional to the quantity of electrical charge that passes through the cell.

2. The mass of the element deposited by a given quantity of charge is directly

proportional to the chemical equivalent (atomic weight divided by valence) of the element.

In order to see the connection between Faraday's laws and the Arrhenius theory of electrolysis, it is only necessary to observe that, in the case of the deposition of the Cu from a $CuSO_4$ solution

$$\frac{M}{Q} = \frac{m}{q} \qquad (33.1)$$

where M = total mass of the Cu deposited, Q = the total charge passing through the cell, m = mass of each Cu atom, and q = the charge carried by each ion of Cu. But, the mass m of an atom equals its atomic weight A divided by Avogadro's number N_0. The charge q carried by each ionized atom equals its valence v (2 for cupric ion) multiplied by the electronic charge e. Thus Eq. (33.1) may be written in the form

$$M = \frac{1}{N_0 e}\left(\frac{A}{v}\right)Q = KIt \qquad (33.2)$$

where $\qquad \dfrac{A}{v}$ = chemical equivalent of the element (Cu in this case),

$\qquad\qquad N_0 e$ = the faraday, a universal constant,

$\dfrac{1}{N_0 e}\left(\dfrac{A}{v}\right) = K$ = the electrochemical equivalent of the element (Cu),

$\qquad\qquad I$ = current in the cell, and

$\qquad\qquad t$ = time during which the current exists in the cell.

In this experiment we determine the electrochemical equivalent, K, of copper by direct measurement of M, I, and t. Then, we calculate the value of the faraday by using the accepted values of A and v for Cu.

Method: The circuit for this experiment is shown in Fig. 33.1. Several precautions must be observed in this experiment. The electrolyte used is an acid solution of $CuSO_4$,

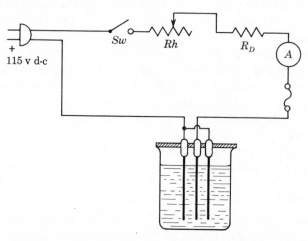

FIGURE 33.1   Wiring diagram of electrolytic cell.

commonly called "vitriol," which is very corrosive. If it is spilled on skin or clothing, large quantities of water should be used to flood the areas. Furthermore, only the copper plates used as electrodes are to be immersed in it, since it will attack and corrode the clamps holding the plates in place. When the experiment is finished, the electrodes are to be washed in running water immediately.

The electrolyte is to come within 2 cm of the clamps of the electrodes. Measure the area of the cathode (center electrode), taking into account both sides, and compute the proper current to be used—about 20 ma/cm². Make a short trial run in order to adjust the current to the proper value. The cathode should be clean and free of oxide for this run. If necessary, it may be cleaned with a piece of sandpaper. After the trial run, clean the cathode carefully under running water, avoiding any roughness that may knock off some of the flaky deposit. Dry the plate by holding it a foot or two above a heater until no sign of moisture remains. Weigh it on the balance, obtaining as accurate a weight as possible. (The *difference* in weight is the subject of interest.)

Replace the cathode in its holder and begin the run immediately. The current should be kept *constantly* adjusted to the value chosen. A 30 min run, started and stopped on the second, will suffice for the first trial. Remove and wash as before, being extremely careful not to lose any of the deposit. Dry thoroughly and again weigh carefully. Determine the mass of Cu deposited on the cathode. Using the mass and the values of current and time, determine the electrochemical equivalent. Determine the error in your value of K for Cu. Then, by taking the atomic weight of copper as 63.57 and its valence as 2, calculate the value of the faraday in coulombs. Determine the error in your value of the faraday.

If time permits, repeat the experiment.

Record: Tabulate data and results.

QUESTIONS

1. What is the greatest source of error in this experiment? Give reasons for your answer. What are some other sources? Why is it desirable to make as long a run as possible?

2. How could this coulometer be used to standardize (calibrate) an ammeter?

3. Suppose that two coulometers were placed in series, the first being a copper coulometer as used in this experiment, and the second being a silver coulometer, using $AgNO_3$ as the electrolyte. Given that the atomic weight of silver is 107.88 gm/mole, in which coulometer would the greater weight of metal be deposited in 30 min, with an average current of 2 amp? What will be the difference in the added weights?

4. Using your results in this experiment, determine the ratio of the mass of a copper ion to its charge.

EXPERIMENT **34**

Ohm's Law;
Power Transfer

Object: To determine the electromotive force (emf) and the internal resistance of an electrical power supply. To determine the electrical power transfer from this power supply to an external load resistor.

Apparatus: Power supply (voltaic cell and resistor inside a 2-terminal "black box"), d-c milliammeter (0–100 ma), d-c voltmeter (0–3 v), and a decade resistance box (0–10,000 ohms) used as a load resistor.

Theory: A conductor is said to obey Ohm's law when the current through the conductor (held at constant temperature) is strictly proportional to the applied potential difference across the conductor. Most metallic conductors obey this law, but there are conductors that do not. The former are called ohmic, and the latter non-ohmic conductors. In this experiment we are concerned with ohmic conductors.

The resistance of a conductor (ohmic) at a fixed temperature is defined by the equation

$$V = IR \tag{34.1}$$

where R is the resistance of the conductor, I is the current through the conductor, and V is the potential difference across the conductor. If V is in volts and I is in amperes, then R is in ohms. Equation (34.1) is frequently called Ohm's law for convenience, but the conditions on which it is based should be kept in mind. The resistance of a conductor depends not only upon the nature of its composition and upon its size and shape, but also upon its temperature.

In order to avoid confusion when there is more than one resistance in the circuit, subscripts should be used with Eq. (34.1), which holds only when the subscripts on V, I, and R are all the same, e.g., $V_t = I_t R_t$. In this way, the error of applying the total potential difference to only one particular resistance in the circuit, or some similar error, may be avoided.

The power P in watts, dissipated by a resistance as a result of a current through it, is given by the equation

$$P = VI \qquad (34.2)$$

where V is the potential difference in volts across the resistance, and I is the current in amperes through the resistance.

Equation (34.1) can be extended to include the entire circuit, i.e., power supply and external load resistor, by writing

$$E = IR_i + IR_e \qquad (34.3)$$

In Eq. (34.3) E and R_i are, respectively, the emf in volts and the internal resistance in ohms of the power supply, R_e is the resistance in ohms of the external load resistor, and I is the common current in amperes in both power supply and load resistor. See Fig. 34.1 for a schematic diagram of this circuit.

When we multiply Eq. (34.3) by the common current I, we obtain the power equation for the circuit. Thus EI equals the total power P in watts furnished by the voltaic cell, I^2R_i equals the power P_i in watts dissipated in the power supply, and $I^2R_e \, (= V_eI)$ equals the power P_e in watts dissipated in the external load resistor. The latter may be determined for any value of R_e in terms of any two of the values of I, R_e, and V_e, where V_e is the voltage drop across R_e.

In writing Eq. (34.3) we have assumed that putting an ammeter and a voltmeter into the circuit as shown in Fig. 34.1 does not disturb the circuit. This would only be true for an ammeter of zero resistance and a voltmeter of infinite resistance. Real meters do not satisfy these conditions, and small corrections are necessary for precise measurements. In this experiment we shall neglect these corrections.

We can always replace IR_e in Eq. (34.3) by V_e, the potential difference across R_e indicated by the voltmeter, and write this equation in the form

$$V_e = E - IR_i \qquad (34.4)$$

Since E and R_i are constants in the equation (for a given power supply), we see that V_e is a linear function of I for varying values of R_e. We can measure the corresponding values of V_e and I for a set of discrete values of R_e and then plot a V_e versus I curve

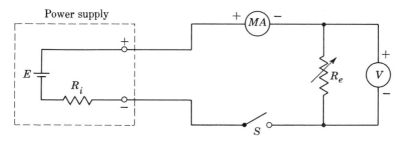

FIGURE 34.1   Schematic diagram of power supply connected to load.

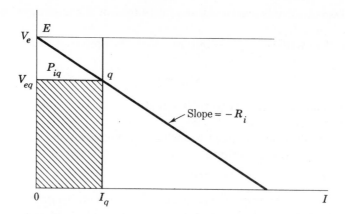

FIGURE 34.2 V_e vs I of power supply under variable load.

as shown in Fig. 34.2. This curve by Eq. (34.4) is a straight line with an intercept on the V_e axis of E (the emf of the power supply) and with a scale slope of absolute value equal to R_i (the internal resistance of the power supply). Thus from this curve we can determine both E and R_i.

But the diagram in Fig. 34.2 also shows, in terms of scale areas, the *power transfer* from the power supply to the external load resistor of variable resistance R_e. For any point q on the line with coordinates I_q, V_{eq} the scale area of the shaded rectangle $(V_{eq}I_q)$ is P_{eq}, that of the unshaded rectangle above is P_{iq}, and their sum is $P_q = EI_q$ (the total power). If we now allow the point q to move down the line (increasing I), we see that the P_{eq} rectangle increases in size until it reaches a maximum value for q at the halfway mark and then starts decreasing in size as q moves beyond that mark. Thus it can easily be shown that the maximum power that can be transferred from a given power supply to an external load resistor takes place when R_e has the same value as R_i, i.e., when the resistance of the load is matched to that of the supply.

This is a simple example of a general "power transfer theorem" of great importance in all circuit theory. We should note however that in this *maximum* power transfer the power efficiency of the transfer (P_e/P) is only 50%. For high efficiencies the external load resistance must be much greater than the internal resistance of the power supply.

Method: First read the following items in Appendix II for descriptive material about operation and precautions in the use of various instruments listed under Equipment Notes F.1, 2, 6; H. 1, 2.

Connect the circuit as shown in Fig. 34.1, remembering the precautions in the Instructions for Electricity Laboratory. Set the resistance R_e (decade box) at 100 ohms, then close switch Sw momentarily to see that the meters read in the right direction and do not go off scale. Set the resistance R_e at 35 ohms, close the switch, and take the voltmeter and milliammeter readings. Record the values of R_e in ohms, V_e in volts, and I in *amperes* (not milliamperes) in a suitable table of values.

Repeat this procedure for values of $R_e = 30, 25, 15, 10,$ and 5 ohms.

Plot a V_e versus I curve as shown in Fig. 34.2 and draw the best straight line through the plotted points, extending it to cut both the V_e and I axes. Determine and record the value E (intercept on the V_e axis) and the value of R_i (absolute scale slope of the line).

Compute and record in a fourth column of your table the value of P_e ($= V_e I$) for each value of R_e. Plot a second curve of P_e versus R_e values. Indicate the value of R_e at the peak of this curve and compare it with your value of R_i.

Record: Tabulate the values of R_e, V_e, I, P_e. List the values of E and R_i. Estimate the errors in these values except for those of R_e, which are negligible. The errors in the meters may be taken as $\pm 2\%$ of the full-scale reading.

QUESTION

1. Show that $R_e = R_i$ for maximum power transfer by the use of calculus or otherwise.

Measurement of Resistance by the Wheatstone Bridge Method

Object: To determine the resistances of two coils, when taken singly and when connected in series and in parallel, by means of (1) a slide-wire Wheatstone bridge, (2) a dial-type Wheatstone bridge.

Apparatus: Double resistance coil, dial resistance box, rheostat, switch, battery, galvanometer, slide-wire Wheatstone bridge, dial Wheatstone bridge. See Appendix II, Notes G.1, 2; F.2 on the galvanometer and on the dial box.

Theory: The resistance R of a conductor is defined as the ratio of the potential difference V across the conductor to the current I in the conductor, i.e., $R = V/I$. If V and I are expressed respectively in volts and in amperes, then R will be expressed in ohms.

There are several different methods available for measuring resistance. The most direct method is the ammeter-voltmeter method. For accurate measurements by this method, the ammeter and the voltmeter must have appropriate ranges, they must read correctly, and the resistance of one of them must be known.

The Wheatstone bridge method, used in this experiment, possesses distinct advantages over the ammeter-voltmeter method in that it is both a null and a comparison method. The unknown resistance is compared with a standard known resistance by getting a zero (null) deflection in a galvanometer connected in the bridge circuit.

The slide-wire Wheatstone bridge, as shown schematically in Fig. 35.1, consists of a uniform resistance wire, usually 1 m in length, stretched between points A and B. The unknown resistance X is connected between A and D and a dial resistance box R is connected between B and D. A battery, with switch Sw and rheostat Rh connected across AB, furnishes current for the bridge. A galvanometer G is connected between D and C. By means of a sliding contact, the position of point C is

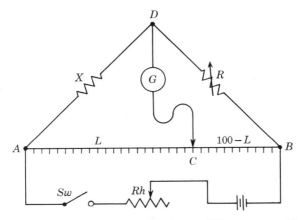

FIGURE 35.1   Schematic of slide-wire Wheatstone bridge.

variable and may be anywhere in the interval AB; i.e., one terminal of the galvanometer may be connected to any point on the bridge wire.

When the battery switch is closed, current will exist in all arms of the bridge, including the galvanometer arm. However, it is possible, under normal conditions, to find some one position C for the slider on the bridge wire such that the current in the galvanometer arm is zero; i.e., the galvanometer deflection will be zero when contact is made with the bridge wire at this point. Under this condition, the bridge is said to be balanced.

Since, for a balanced bridge, the galvanometer current is zero, it follows from Ohm's law that the potential difference between D and C must be zero. Therefore, the potential difference between A and D must equal that between A and C; and the potential difference between D and B must equal that between C and B.

Let I_1 be the current in X and in R (why is this current the same?), and let I_2 be the current in the bridge wire. Furthermore, let R_{AC} be the resistance of the bridge wire between A and C, and let R_{CB} be the resistance of the bridge wire between C and B. Then

$$I_1 X = I_2 R_{AC} \quad \text{and} \quad I_1 R = I_2 R_{CB}$$

If we divide the first equation by the second and solve for X, we get

$$X = \frac{R_{AC}}{R_{CB}} R \tag{35.1}$$

Since the bridge wire is assumed to be uniform, it follows that

$$\frac{R_{AC}}{R_{CB}} = \frac{\text{length } AC}{\text{length } CB} = \frac{L}{100 - L}$$

where L is the length in centimeters of the section AC. It is assumed that the wire is 100 cm long.

Equation (35.1) may then be written in the form

$$X = R \frac{L}{100 - L} \tag{35.2}$$

The error equation corresponding to Eq. (35.2) is

$$\frac{\Delta X}{X} = \frac{\Delta R}{R} + \frac{100 \, \Delta L}{(100 - L)(L)} \tag{35.2a}$$

It may be shown that the coefficient of ΔL in this equation has a *minimum* value when $L = 50$ cm, i.e., when the balance point is at the center of the bridge wire. Thus the fractional error in X is reduced to a minimum when the bridge is balanced in this manner. This means, of course, that R should be chosen as nearly equal to X as possible for greatest accuracy.

Equation (35.2) enables us to compute the value of X in terms of a known resistance R and the position of the balance point on the slide-wire. A very wide range of resistances may be measured by this method, i.e., resistances ranging from one to several thousand ohms, provided that one has a variable standard R. For the measurement of very low or very high resistances, the bridge method must be modified.

In a *balanced* Wheatstone bridge, it is possible to exchange the positions of the battery and the galvanometer without disturbing the balance equation. Can you prove this? However, analysis shows that the bridge is more sensitive when the galvanometer, rather than the battery, is connected between the junction of the high-resistance arms and that of the low-resistance arms. In this case X and R are the high-resistance arms. It can also be shown that the sensitivity of the bridge increases with increasing battery current. However, it is necessary to keep the current in the arms of the bridge from exceeding the current capacity of the arms.

SERIES AND PARALLEL CONNECTIONS

When two resistances are connected in series, as shown in Fig. 35.2, the combined resistance X_s is given by the equation

$$X_s = X_1 + X_2 \qquad \text{(series)} \qquad (35.3)$$

When two resistances are connected in parallel, as shown in Fig. 35.3, the combined resistance X_p is given by the equation

$$\frac{1}{X_p} = \frac{1}{X_1} + \frac{1}{X_2} \qquad \text{(parallel)} \qquad (35.4)$$

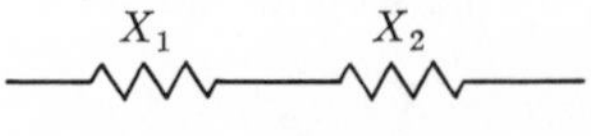

$$X_1 \qquad X_2$$

FIGURE 35.2 Series connection.

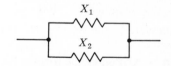

FIGURE 35.3 Parallel connection.

The corresponding error equation for this case is

$$\frac{\Delta X_p}{X_p} = \frac{X_2}{X_1 + X_2} \frac{\Delta X_1}{X_1} + \frac{X_1}{X_1 + X_2} \frac{\Delta X_2}{X_2} \qquad (35.4a)$$

In the dial Wheatstone bridge, the bridge wire is replaced by a set of ratio coils so that the ratio R_{AC}/R_{CB} is given directly on a dial as some decimal multiple of 1 such as 0.1 or 100. If this ratio is represented by Q, then Eq. (35.1) becomes

$$X = RQ \qquad (35.5)$$

By varying R and Q, a wide range of X may be measured. In this type of bridge the ratio coils and the variable resistance R are contained in a single box with dials for reading R and Q. There are three sets of terminals on the box for connecting the battery, the galvanometer, and the unknown resistance. In some types the galvanometer and the battery are incorporated in the bridge so that one needs only to connect in the unknown resistance.

Method: *Part I. The Slide-wire Bridge.* Examine the bridge and note how it may be utilized to conform with the schematic diagram in Fig. 35.1. For example, points A, B, and D in the diagram correspond in the real bridge to heavy metal strips with negligible resistance. Connections may be made at any of the terminals on these strips.

Connect the unknown resistance X_a and a dial resistance box to the bridge by means of short heavy leads. The unknown resistances consist of a set of two resistors, a and b, with independent terminals. Be certain that all the connections are tight. Loose connections are the cause of much trouble in electrical measurements. Make all the other connections called for in Fig. 35.1 *except the two connections* at the battery. Have your instructor check your setup before making the connections at the battery terminals.

After all connections have been made, set the rheostat at about 40 ohms, set the dial resistance box at 10 ohms, depress the low-sensitivity button, L, on the galvanometer, and close the battery switch. Set the slider near the one end of the slide wire, e.g., at the 5 cm mark, and momentarily depress it to make contact with the wire. Ordinarily, a large deflection of the galvanometer will occur to right or left. Raise the slider, move it up scale about 10 cm, and try again. The galvanometer deflection should be smaller, thus indicating that you are moving the slider toward the balance point. Continue this process until the galvanometer deflects in the opposite direction, indicating that you have passed the balance point. Reverse the direction of motion of the slider and, using smaller intervals, seek the balance point. In the immediate neighborhood of the balance point it will be necessary to use the high sensitivity range of the galvanometer by depressing the H button.

In carrying out the above process do not scrape the slider along the wire, and do not press it down with such force that it dents the slide-wire.

In case you are not able to find a balance point anywhere on the slide-wire, either because the galvanometer deflections are always in the same direction, or because there is no galvanometer deflection, look for an open circuit or a loose connection. If this fails, call the instructor.

After you have found this preliminary balance point, note the position of the slider with reference to the center of the bridge wire. If it is more than 10 cm away from the center, adjust R in order to bring the balance point within 10 cm of the center of the bridge wire. This adjustment is necessary if an accurate value of X_a is to be obtained, as indicated in the theory of this experiment. Determine the position of this final balance point C as accurately as possible, using the high sensitivity range of the galvanometer. Make an estimate of the error in the position of this balance point by finding how far the slider may be moved away from this balance point without giving a noticeable galvanometer deflection. Record $L(= AC)$, $100 - L$ $(= CB)$, the final reading of the dial box R, and the error in the balance- point position L. Using Eqs. (35.2) and (35.2a), determine the resistance X_a and the percentage error in X_a. The error in R may be taken as 0.25%.

Proceed in a similar manner to determine the resistance of the second resistor, the resistance of the two resistors in series, and the resistance of the two resistors in parallel. Also, determine the percentage error in each case.

Make a second determination of each of the above resistances by *reversing* the positions of X and R in the bridge. Use the same value of R for each resistance that was used in the first determination. *Note well* that the reversal of the positions of X and R in the bridge will produce a new balance point C' on the bridge wire that will fall on the opposite side of the center from that of C, and at very nearly the same distance from the center. For this reversed position of X and R, balance equation (35.2) must be modified to read

$$X = R\frac{100 - L'}{L'} \qquad (35.2b)$$

where $L' = AC'$ and $100 - L' = C'B$. It is highly advisable that the student redraw Fig. 35.1 with X and R reversed in positions and develop Eq. (35.2b).

Compute the average value of these two determinations for each resistance and use these averages as your final result for measurement of resistance with a slide-wire Wheatstone bridge.

Check the series and parallel resistances against the values calculated by use of Eqs. (35.3) and (35.4). They should check within the limits of the errors in each case.

Part II. The Dial Wheatstone Bridge. Use the dial bridge to remeasure all of the resistances used in Part I of this experiment. Set Q in these measurements to a value that gives four significant figures for X. In each case estimate the error in the setting of R for a balance and compute the corresponding percentage error in X. Assume that the error in Q is negligible.

Record: Give the apparatus numbers of the resistance a and b, slide-wire bridge, galvanometer, dial box, and dial bridge. Tabulate your data. Express your results in the following form:

Item	Slide-wire bridge		Dial bridge		% difference
	X (ave)	% error	X	% error	
Resistance a					
Resistance b					
Series (exp)					
Series (calc)					
Parallel (exp)					
Parallel (calc)					

QUESTIONS

1. Show that it is possible to interchange the positions of the battery and galvanometer in Fig. 35.1 without changing balance equation (35.2).

2. Show that the coefficient of ΔL in Eq. (35.2a) has a minimum value when $L = 50$. (Calculus required.)

3. Develop the error equation (35.4a) by applying the rules given in Sec. A.4 of the Introduction.

4. If d equals the *shift* of the *balance point* to the *left* in Fig. 35.1 when X and R are reversed in position, show that $X/R = (100 + d)/(100 - d)$.

Electric and
Magnetic
Fields

Object: (1) To map the equipotential lines and then determine the lines of force for a two-dimensional electric field. (2) To map the lines of force and then determine the traces of the equipotential surfaces for a three dimensional magnetic field.

Apparatus: Electric field mapping apparatus,[1] 4–6 v battery, leads, electrodes, conducting paper (Teledeltos),[2] probe, table galvanometer, potential divider, and voltmeter. Bar magnet, magnetic compass, board, and paper.

Theory: There are two different ways of graphically representing a steady field of force, such as an electric or magnetic field. One may use either (1) a set of lines (lines of force) that give everywhere the direction of the intensity of the field (force per unit test charge), or (2) a set of surfaces (equipotential surfaces) that give every-where the potential of the field (potential energy per unit test charge). Not all fields possess a potential. We are only concerned with those that do.

These two representations of a field are complementary in the sense that, if one of the representations is given, the other can be inferred. This fact arises because of a fundamental connection between the intensity of a field and its potential; namely, that the intensity of the field at any point is always equal to the negative space rate of change of its potential. The space rate of change of any quantity is usually called the gradient of that quantity, e.g., we speak of temperature gradient, velocity gradient, potential gradient, and so on. Thus we say that the intensity of a field equals its negative potential gradient. A direct consequence of this connection is that the lines of force for any field are everywhere perpendicular to its equipotential surfaces.

There exists an illuminating geometrical interpretation of the ideas of lines of force and equipotential surfaces (lines) in the case of a two-dimensional plane field.

[1]Apparatus designed by Frank Verbrugge.
[2]Manufactured by Western Union Telegraph Company.

This field may be a truly two-dimensional field, such as the electric field produced by a set of parallel line charges, or it may be simply a plane section of a more complicated three-dimensional field, such as those studied in this experiment. In either case we assume that the field being studied, for example, lies in the horizontal xy plane. At each point in this plane, we erect a perpendicular equal to the potential V of the field at that point. The locus of the end points of these perpendiculars forms a surface (the potential surface) in the three-dimensional space whose rectangular coordinates are x, y, and V. This surface will have peaks, valleys, ridges, and so on, similar to the surface of the earth. In order to survey the surface, we draw contour lines (lines of equal level) and gradient lines (lines of steepest descent) on it. These two sets of lines are everywhere perpendicular to each other. Why? If we now project these two sets of lines onto the xy plane, the projected contour lines of the potential surface are just the equipotential lines of the field, and the projected gradient lines are just the lines of force.

A similar geometrical interpretation exists for a three-dimensional field. Unfortunately, it is necessary in this case to introduce a *fourth dimension*, representing potential. Our surface then lies in a four-dimensional hyperspace. Such spaces can be treated mathematically, but we no longer possess any intuitive geometrical knowledge concerning them.

It is often possible to reduce the analysis of a three-dimensional field to that of a two-dimensional field, or a combination of such fields. An example is the electric field produced by an isolated system of two identical small conducting spheres, each having the same positive charge. In this case an axis of symmetry exists—a line through the centers of the two spheres. A knowledge of the field in any plane containing this axis is sufficient to give us knowledge of the entire field.

As an example, we show in Fig. 36.1 the lines of force and the traces of the equipotential surfaces in a plane of symmetry for the system just mentioned. The character of the entire field is obtained by rotating the figure about the line AB. In this case the equipotential surfaces are obviously surfaces of revolution.

The geometrical significance of the plane field pattern shown in Fig. 36.1 is obtained by drawing the potential surface to which it corresponds. This surface is

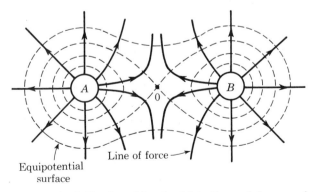

FIGURE 36.1   Electric field of positive doublet: lines of force and traces of equipotential surfaces.

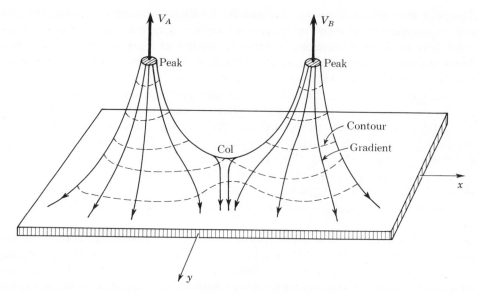

FIGURE 36.2   Relief map of potential surface.

shown in Fig. 36.2. Note the character of the contour lines (dotted) and gradient lines (full). Their projection on the xy plane gives Fig. 36.1. Note especially the col (mountain pass or saddle point) that exists between the two peaks in Fig. 36.2. The contour and gradient lines in the neighborhood of this "mountain pass" have a distinctive character that is transferred in full to the nature of the field in the neighborhood of point O in Fig. 36.1. This point O, halfway between the centers of the two spheres, is a critical or neutral point where the intensity of the field is zero. Therefore, a small test charge placed at this point would be in equilibrium, but the equilibrium would not be stable. Why not?

There are a number of important conclusions that we can draw concerning fields of force from the foregoing analysis.

1. Lines of force are everywhere perpendicular to the equipotential surfaces (or lines) for a given field.

2. Lines of force always run from regions of high potential to regions of low potential. As one moves along any line of force in the direction of the field, the potential decreases at a maximum rate, and this rate of decrease is just equal to the intensity of the field.

3. No field exists in any region where the potential is constant. Furthermore, the surface of this region is always an equipotential surface, and, therefore, the lines of force entering or leaving this surface must be perpendicular to the surface. A prime example of this is the case of a charged or uncharged electrical conductor in any electrostatic field. There is no corresponding effect of this kind in a magnetic field, since there are no "magnetic conductors."

4. In representing a field by lines of force which give the direction of the field, it is also possible to represent the magnitude of the intensity of the field by the space

density of the lines of force, i.e., the number of lines drawn per unit area, the unit area being perpendicular to the lines. For example, a field having unit intensity at some point could be represented by drawing one line of force per cm² at that point. This is the usual convention but we could just as well represent a unit field by drawing 1000 lines per cm² if we wish.

Similarly, in representing a field by equipotential surfaces, we can represent the magnitude of intensity of the field by the space density of those surfaces (i.e., the number of surfaces that cut a line of unit length drawn perpendicular to the surfaces). In this case, it is necessary that the potential difference between successive equipotential surfaces be constant. The usual convention is to make this constant equal to one, but this is not necessary.

Method: *Part I. Electric Field.*

a. *Point-point field.* The essential parts of the apparatus for this experiment are shown in Fig. 36.3. The apparatus consists of a board on which a sheet of conducting paper (left) and a corresponding sheet of blank paper (right) are placed. The conducting paper is coated with colloidal graphite. Possible substitutes for the conducting paper are conducting rubber sheets or conducting glass plates. Electrodes connected to a battery make contact with the conducting paper at points A and B, thus establishing a potential difference across the paper. As a consequence, there is a flow of charge over the paper between points A and B, e.g., along the line APB in Fig. 36.3. The equipotential lines on the paper are everywhere perpendicular to the lines of force.

The purpose of this part of the experiment is to determine these equipotential lines. To do so, a potential divider is connected across the battery, as shown. Part of the current from the battery passes through the resistance of the potential divider, thus producing a potential drop across the resistance $A_0 B_0$ equal to that between the points A and B on the conducting sheet. If the sliding contact on the potential

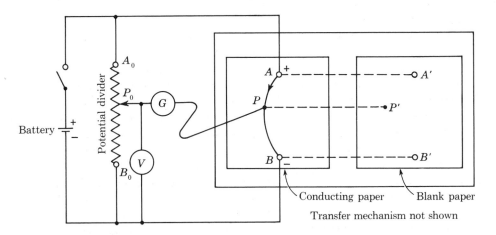

FIGURE 36.3   Schematic of electric field mapping equipment.

divider is set at any point P_0, there will be a corresponding set of points on the conducting sheet, all of which are at the same potential as the point P_0. These points on the conducting sheet must lie on an equipotential line. They may be found by moving the probe across the paper until some point P is found, at which the galvanometer G gives zero deflection when the probe makes contact with the conducting sheet. This means that points P_0 and P are at the same potential; i.e., there is no potential difference and therefore no current in the galvanometer. Once point P has been found, the probe should be moved to a neighboring point that gives zero deflection on the galvanometer, and continue from that point P_0. These points are transferred to the blank paper, and a smooth curve is drawn through them, giving a single equipotential line. The voltmeter reading V gives the numerical value of the potential on this line.

The sliding contact on the potential divider is then moved to another point and the preceding process is repeated, thus giving a second equipotential line. Step by step, the process is repeated, until a sufficient number of equipotential lines have been determined, giving a fairly accurate description of the field between A and B on the conducting sheet. In this experiment at least seven lines are required, with approximately five points on each line. There should be at least one line in the immediate neighborhood of each of the electrodes, and at least one line across the central region of the conducting sheet.

After the equipotential lines have been determined, the student should draw in the corresponding lines of force, starting at electrode A, and running to electrode B.

b. *Point-plane field.* As a second part of this experiment, the point electrode at B should be replaced by a flat metal plate in good contact with the conducting sheet. The student should determine the equipotential lines and lines of force for this field (point-plane field).

Part II. Magnetic Field. The magnetic field that is to be plotted in this experiment results from a permanent bar magnet and the earth's magnetic field. The axis of the magnet is to be placed parallel to the direction of the earth's field, but with the north pole of the magnet directed toward the south, and the south pole of the magnet directed toward the north.

Determine the north-south line at your station by means of the magnetic compass needle. Be sure that the bar magnet and all other pieces of iron are removed from the station in this process. Then place the board, with the permanent magnet in the center, in such a position at your station that the north pole of the magnet is directed toward the south. Cover the board with a large piece of paper. Outline the position of the permanent magnet, indicating its polarity and the direction of the north-south line on this paper. Place the compass near the north pole of the magnet and make dots as near each end of the needle as possible, and in line with it. Then, move the compass in the direction in which its north pole points, until the south pole is above the dot previously made at the north pole. Make another dot at the north pole of the compass in this new position. Continue in this fashion until the series of dots leads to the south pole of the bar magnet, or to the edge of the paper. Draw a smooth curve through the points, and indicate, by arrows, the direction of the field. It is clear that this line is a line of force of the magnetic field. In a similar manner, trace other lines of force until the field is clearly represented on all sides of the bar magnet. It is not necessary to start from the north pole of the magnet in each case. It is just as well to start from the south pole, or from some point away from the magnet.

Two regions will probably be found where the direction of the compass needle is indeterminate—points of neutral or critical equilibrium. These points are called neutral points, because the earth's field at these points is just balanced by the field of the bar magnet. The region in the neighborhood of these points should be mapped with considerable care. Since the field in this neighborhood is very weak, it may be necessary to gently tap the compass each time a direction is taken in order to overcome the effect of friction at the pivot of the compass.

After you have mapped the lines of force of the magnetic field (black pencil), draw in the equipotential lines (red pencil). This may be done outside of the laboratory.

Record: The record of this experiment will consist of:

1. The graphs of the two electric fields and the one magnetic field, properly titled and labeled.
2. A schematic drawing of the electric-field apparatus.
3. Answers to the questions given below.

QUESTIONS

1. Show that equipotential surfaces always cut lines of force at right angles in any field of force.

2. Can two different lines of force or two different equipotential surfaces ever cross each other? Explain.

3. Are there any neutral or critical points in the electric fields similar to those present in the magnetic field in this experiment?

4. Would it be possible to produce a field such as that shown in Fig. 36.1 by means of the electric-field apparatus that you used in this experiment? If so, how would you do it?

5. It may be shown that the strength of the field produced by a bar magnet at a point on its extended axis, a distance r from the magnet, is approximately $2M/r^3$, provided that r is fairly large compared with the length of the magnet. M is the magnetic moment of the magnet. At either one of the neutral points in Part II of this experiment, $2M/r^3$ must be equal and opposite to the strength of the earth's horizontal magnetic field. Taking the value of this field as 0.17 oersted, calculate the magnetic moment M of the magnet used in this experiment.

OPTIONAL EXPERIMENTS

1. In the electric field part of this experiment, the procedure adopted led to a direct determination of the equipotential lines of the field, after which the lines of force were drawn. With the same apparatus, devise and use a procedure that will lead to a direct determination of the lines of force instead of the equipotential line. Why is this procedure inconvenient to use?

2. In Part II of the experiment, the axis of the magnet was anti-parallel to the earth's horizontal field. Try the experiment with the axis perpendicular to the earth's horizontal field.

Potentiometer

Object: To measure, using a slide-wire potentiometer, the emf of two different cells when taken singly and then when taken in opposition. To determine the error in a laboratory voltmeter when it reads 1 v, by determining the "true" potential difference with the potentiometer.

Apparatus: Slide-wire potentiometer, storage battery, rheostat and switch, double-pole double-throw switch, standard cell, two unknown cells, table galvanometer and tap key, two dial resistance boxes, and a voltmeter. See Appendix II, Notes G.1, 2; F.6; H.1 for descriptions of the galvanometer, standard cell, and voltmeter.

Theory: Although the most direct and convenient method of measuring a potential difference is by means of a direct-reading voltmeter, this method has the distinct disadvantage that the voltmeter draws some current in the measurement. In doing so it changes, materially in many cases, the potential difference that one wishes to measure.

The potentiometer method of measuring potential difference avoids this difficulty, since the potentiometer draws no current when it is *balanced* for a reading of potential difference; i.e., the method is a null method. It is also a comparison method, because the unknown potential difference is compared with the known emf of a standard cell. As a result, very precise and accurate determinations of potential difference can be made with this method.

In one of its simplest forms, the slide-wire form, the potentiometer consists of a long uniform wire AB in which a constant current I is maintained by means of a battery connected to the ends of the wire. See Fig. 37.1. There is thus a uniform drop in potential along the wire as one goes from the positive terminal of the wire A to the negative terminal B.

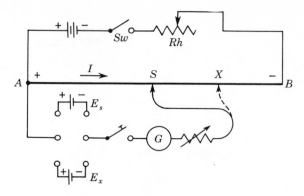

FIGURE 37.1   Slide-wire potentiometer.

Suppose that a standard cell of known emf E_s and a cell whose emf E_x is to be determined are connected, as shown, so that either positive pole may be connected through the double-pole switch to A, and either negative pole may, at the same time, be connected through the galvanometer G to some point on the slide-wire. Consider the standard cell to be in the circuit first. It will be possible to find some point S along the slide-wire such that the drop in potential along the wire from A to S is just equal to E_s, provided E_s is less than the total drop in potential from A to B. If the galvano-meter circuit is closed at this point S by means of a sliding contact, there will be *no* current in the galvanometer and standard cell, because the emf of the standard cell just "balances" the potential difference from A to S. Under these conditions, the potentiometer is said to be balanced for the standard cell and $E_s = V_{AS}$. By Ohm's law, V_{AS}, the potential drop from A to S, is equal to IR_{AS}, where R_{AS} is the resis-tance of the bridge wire between A and S. Hence

$$E_s = IR_{AS} \qquad (37.1)$$

In a similar manner we may consider E_x to be connected in the circuit. There will be a point X on the wire for which V_{AX} just equals E_x, provided $E_x < V_{AB}$. There-fore

$$E_x = IR_{AX} \qquad (37.2)$$

If we divide Eq. (37.2) by Eq. (37.1) and solve for E_x, we get

$$E_x = \frac{R_{AX}}{R_{AS}} E_s = \frac{L_x}{L_s} E_s \qquad (37.3)$$

since R is proportional to length L for a uniform wire.

Equation (37.3) enables us to determine E_x in terms of E_s and the ratio L_x/L_s. E_s is known and the two lengths, L_x and L_s, may be determined experimentally by "balancing" the potentiometer, i.e., finding points S and X for which one obtains zero galvanometer deflections.

When two cells are connected in *opposition*, as shown in Fig. 37.2, the combined emf is the *difference* between the separate emf's, i.e.,

$$E_x = E_1 - E_2 \qquad (37.4)$$

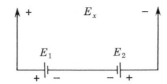

FIGURE 37.2 Cells in opposition.

The positive terminal of this combination is the positive terminal of the cell that has the larger emf. In the figure shown it is assumed that E_1 is greater than E_2.

Calibration of Voltmeter. The potentiometer may be used to calibrate a voltmeter and is often so used. It is only necessary to determine the potential difference across the terminals of a voltmeter with a potentiometer, and to compare this potential difference with the voltmeter reading. Suppose the voltmeter V is connected to a battery B through a variable resistance R, as shown in Fig. 37.3. The voltmeter reading may be varied from 0 to almost the full emf of the battery by adjustment of R. If, at the same time the potentiometer reading across V is taken, the two values may be compared.

Error Equations. The *determinate*-error equation corresponding to Eq. (37.3) is

$$\frac{\Delta E_x}{E_x} = \frac{\Delta E_s}{E_s} + \frac{\Delta L_x}{L_x} - \frac{\Delta L_s}{L_s} \tag{37.3a}$$

Generally, the error in E_s is much smaller than those in L_x and in L_s. Thus in this experiment, this error may be neglected.

Standard Cell. The standard cell used in this experiment is of the Weston unsaturated type. See Appendix II, Note F.6. Its emf is given on a tag attached to it. It should be handled with great care. To prevent changes in its emf, it should *never* be used to supply currents in excess of 10^{-4} amp.

Method: *Preliminary.* Make the connections as shown in the upper part of Fig. 37.1 (line *AB* and above). Connect the voltmeter between the terminal *A* and the slider *S*. Set the slider at point *B*. Close the battery switch and note the reading of the voltmeter. Adjust the rheostat until this reading is well over 1.5 volts. Note the

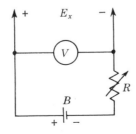

FIGURE 37.3 Voltmeter calibration.

voltmeter reading. Now set the slider at position A and note the voltmeter reading. Move the slider 20% of the length AB and again note the voltmeter reading. Repeat this process for slider positions at 40, 60, 80, and 100% of the length AB. Observe that the voltmeter readings increase uniformly with the length AS and are in fact directly proportional to this length. This means that there must be a uniform drop in potential along the wire from A to B, any portion of which may be "tapped off" by use of slider S. Disconnect the battery and remove the voltmeter from the circuit.

Part I. Emf of Cells. Complete the connections as shown in Fig. 37.1, but do not connect in any battery or cell until the instructor has checked your wiring. *This is especially important for the standard cell.* A wrong connection may ruin it. In finally making battery and cell connections be very careful to arrange the polarities to correspond with the figure. Otherwise it may not be possible to find balance points. Why not? As point A, it will be convenient to use the zero end of the meter sticks.

Set the slider at approximately 120 cm (on the second section of the slide-wire; distance measured from point A), set the battery rheostat at about half resistance, set the dial resistance box in series with the galvanometer at 9000 ohms, close the battery switch, and then close the standard-cell switch. A galvanometer deflection should occur. Try to reduce this deflection to zero by varying the resistance in the battery rheostat. When this has been done, reduce the dial-resistance-box reading to zero, then determine the exact balance point S by shifting the slider until the galvanometer deflection is zero. In this process do not rub the slider along the wire, but rather lift it up before moving it. If there is an appreciable interval of the slide-wire over which the galvanometer deflection appears to be zero, determine the end points of this interval (small but opposite deflections at the end points) and take the middle point of this interval as the balance point for the standard cell. This apparent lack of sensitivity when the standard cell is used is probably due to a high protective resistance incorporated in the standard cell.

Immediately after determining the balance point S for the standard cell, set the dial box at 9000 ohms, throw the double-pole switch for E_x, and determine the new balance point X by shifting the position of the slider. For final adjustment, reduce the dial-box setting to zero. *Do not disturb the battery rheostat in this process.*

Then switch back to the standard cell and redetermine its balance point. Any shift in the balance point of the standard cell indicates a change in the current through the potentiometer wire. Continue this process of switching back and forth between E_s and E_x until three L_s and two L_x have been determined. Use the average values of L_s and L_x for the determination of E_x.

Replace the first unknown cell by the second unknown cell in the potentiometer circuit. Determine as above the emf of this second cell.

Connect the two unknown cells in opposition and use this combination as E_x. Determine as before this value of E_x and compare it with the calculated value.

The error in the position of the balance point may be taken as one-fourth the interval over which no appreciable galvanometer deflection occurs.

Part II. Error of Voltmeter. Connect the voltmeter in series with a dial resistance box and a good dry cell as shown in Fig. 37.3. Use this combination as E_x. Adjust the dial resistance box in series with the voltmeter until the voltmeter reads exactly 1 v, i.e., as closely as one can judge. Determine E_x across the voltmeter with the potentiometer in the manner outlined in Part I. Be certain that the voltmeter is reading 1 v while this determination is being made. The error in the voltmeter is the difference between E_x (potentiometer) and the voltmeter reading 1 v. Determine this error.

Record: Give the apparatus numbers of the potentiometer, standard cell, galvanometer, dial box, and voltmeter. Tabulate your data making separate tables for each determination of an unknown emf or potential difference. Summarize your results in the following form:

Item	Use	E, volts	% error
Std. Cell	Standard emf		. . .
Cell No. 1	Measured emf		
Cell No. 2	Measured emf		
Opposition	Measured emf		
Opposition	Calculated emf		
Voltmeter	Calibrated at 1 v		

QUESTIONS

1. Why is it necessary that the battery furnishing the current for the potentiometer have a larger emf than any emf or potential difference to be measured with the potentiometer?

2. In Part II of this experiment on the calibration of a voltmeter, why is it necessary that the voltmeter be reading 1 v when the calibration is made? Suppose R in Fig. 37.3 is set so that the voltmeter reads 1 v when connected into the circuit. If the voltmeter is disconnected and the potentiometer is used to determine E_x, what will be the value of E_x?

3. Suppose an ammeter is available in this experiment along with the other apparatus already specified. How could one proceed to determine the resistance per unit length of the potentiometer wire AB in Fig. 37.1?

4. How would one proceed to make this slide-wire potentiometer a direct-reading potentiometer, i.e., one in which the position of the balance point in meters would equal the measured potential difference in volts?

OPTIONAL EXPERIMENT

1. The slide-wire potentiometer used in Exp. 37 involves the basic principles of nearly all manufactured potentiometers but is not in a very practical form for continued use because of the long slide wire. In many manufactured potentiometers the slide wire is replaced in part by a set of dial resistors and in part by a circular slide wire of several turns mounted on a rotary drum. The current through this chain of resistances is accurately controlled so that the instrument becomes a direct-reading potentiometer. The Leeds and Northrup student type potentiometer and type K potentiometer are of this character. They may be used to measure d-c voltages in various ranges to an accuracy of about 5 parts in 10,000 for the former and 5 parts in 100,000 for the latter type of instrument.

Use the former of these instruments for making the potential difference measurements in Exp. 37. See Appendix IIN.

Thermocouple

Object: To calibrate a copper-constantan thermocouple for low-temperature measurements and to determine the freezing point of mercury.

Apparatus: Copper-constantan thermocouple (24 ga. Cu; 20 ga Const), galvanometer, low-resistance shunt, two decade resistance boxes, damping key, reversing switch, 20-ohm rheostat with switch, Dewar flasks (thermos bottles), ice, solid carbon dioxide, alcohol, storage battery, voltmeter, mercury, and an iron test tube.

Theory: When a circuit is formed of two wires of different metals with the two junctions at different temperatures, an electromotive force is produced in the circuit. This is known as the thermoelectric effect and was discovered by Seebeck in 1821. Such a device, known as a thermocouple, may be employed in the measurement of temperature. When properly calibrated, the thermocouple becomes a convenient and sensitive thermometer.

A pair of metals frequently used as a thermocouple is copper and constantan. Constantan is an alloy of 60% copper and 40% nickel. This couple may be used to measure a wide range of temperatures (−200 to 350°C) when properly calibrated. Figure 38.1 shows a copper-constantan thermocouple with its two junctions, I and J, and a millivoltmeter for measuring the thermal emf. If junction I is at a higher temperature than J, then the current is in the direction indicated.

In using the thermocouple for the measurement of temperature, it is customary to keep one of the junctions at the fixed temperature of 0°C by immersing it in a mixture of ice and water. Let this be the junction I in Fig. 38.1. The other junction J is then placed in the region whose temperature is to be found. After thermal equilibrium is established, the emf produced by the thermocouple is measured. This emf is a function of the temperature of junction J. The functional relationship may usually

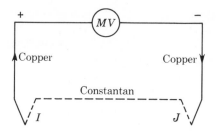

FIGURE 38.1   Thermocouple.

be represented by the empirical equation

$$E = at + bt^2 \qquad (38.1)$$

where E is the emf produced by the couple, t is the temperature of junction J, and a and b are empirical constants to be determined by experiment. The derivative of E with respect to t, i.e., dE/dt, is called the thermoelectric power of the couple and is approximately a linear function of t.

Tables have been prepared for a number of different standard thermocouples, giving the relationship between E and t. A table of values for the copper-constantan thermocouple is given in Appendix III, Table K_1. In using this table of values, based upon a standard copper-constantan thermocouple, it is generally necessary to correct the reading of the laboratory copper-constantan thermocouple in order to make it correspond with the tabulated values. It has been found that the correction is small and is very nearly proportional to the observed E of the thermocouple over a wide temperature range. This means that the corrected value of the emf, say E', is directly proportional to the observed value E. Therefore

$$E' = KE \qquad (38.2)$$

where K is very nearly equal to one. If E_0' and E_0 represent respectively the corrected value and the observed value of the emf at some known temperature t_0, then K may be determined and Eq. (38.2) may be written in the form

$$E' = \left(\frac{E_0'}{E_0}\right) E \qquad (38.3)$$

In this experiment the calibration temperature t_0 is the sublimation temperature of solid CO_2 (dry ice). It has the value $-78.5°C$, at which temperature the tabulated value of the emf of a standard copper-constantan couple is $E_0' = 2.72$ mv. E_0 in Eq. (38.3) is determined by immersing the junction J of the laboratory couple in a mixture of solid CO_2 and alcohol, and observing the emf developed. The couple could then be used to measure any temperature between -78.5 and $0°C$ by observing E at that temperature, calculating E' by means of Eq. (38.3) and using Table K_1 to find the temperature which corresponds to E'. It will be used in this experiment to find the freezing point of mercury.

In Fig. 38.1 a millivoltmeter is shown for the purpose of measuring the emf of the thermocouple. This is not a very satisfactory method of measuring the emf of the couple, since the millivoltmeter draws some current and measures the potential difference (P.D.) across the terminals of the couple rather than its total emf.

A much better procedure is to replace the millivoltmeter with a potentiometer, such as a Leeds and Northrup student potentiometer. Even if these are not available, it is possible to construct a simple potentiometer that enables one to measure the emf of the couple without drawing current from it.

In Fig. 38.2, a satisfactory circuit for achieving this result is shown. The battery B supplies current through a rheostat Rh to a low resistance S in series with a decade resistance box R. A voltmeter V reads the P.D. across S and R. The emf of the thermocouple TC is balanced against the P.D. across the low resistance S, as indicated by a null reading of the galvanometer G included in the circuit. A resistance box A is included in the galvanometer circuit to prevent excessive galvanometer deflections when the system is not balanced. Balance is achieved by adjusting R until the galvanometer deflection is zero. As the balance point is approached, A (initially several thousand ohms) is gradually reduced to zero for maximum sensitivity. When a balance is reached, it is evident that E for the couple is given by the equation

$$E = V\frac{S}{R + S} \tag{38.4}$$

where V is the reading of the voltmeter. Since R is in general quite large compared to S, it is possible to write Eq. (38.4) in the form

$$E = V\frac{S}{R} \tag{38.5}$$

The final working equation for this experiment may be obtained by combining Eqs. (38.3) and (38.5). We get

$$E' = E_0'\frac{V}{V_0}\frac{R_0}{R} \tag{38.6}$$

where the symbols with the subscript correspond to values at the calibration tem-

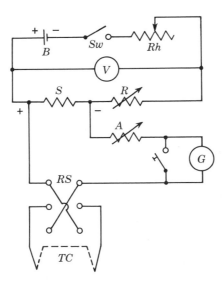

FIGURE 38.2   Circuit diagram for thermocouple.

perature t_0, and those without the subscript correspond to values at the unknown temperature.

The errors in this experiment are rather difficult to estimate without a more complete experimental analysis because of the nature of the assumption concerning the correction to be made on the observed emf of the laboratory thermocouple. For precise work, the couple should be calibrated at a number of different known temperatures in the range in which the couple is to be used.

As far as Eq. (38.6) is concerned, the indeterminate error in E' may be computed in the ordinary manner. It should be noted that since V and V_0 are practically equal to each other in this experiment, any constant error in the voltmeter reading will contribute little or nothing to the error in E'. The value of E_0' taken from the table is probably good to within ± 0.005 mv. Finally, the readings R and R_0 from the decade resistance box are accurate to within $\pm 0.25\%$.

Once the error in E' has been determined, the error in the temperature to which it corresponds may be obtained by use of Table K_1.

Method: Connect the apparatus according to Fig. 38.2, except for the battery. After the instructor has checked the circuit, connect the battery, but leave all switches opened. Set the rheostat at about 20 ohms. Set the decade resistance boxes at 9000 ohms. Fill a thermos bottle with a mixture of ice and water. Place both junctions of the thermocouple in this mixture, and wait 3 or 4 min for thermal equilibrium to be established. Then close the reversing switch, but not the battery switch. Change the box A setting step by step from 9000 to 0 through the intermediate settings of 6000 and 3000. Stop the adjustment if there is any appreciable galvanometer deflection. Such deflection under these circumstances indicates trouble. It should be corrected before proceeding with the experiment.

Part I. Calibration of Thermocouple. Half fill a second thermos bottle with alcohol or acetone, and *slowly* add solid CO_2 until the flask is almost full. The mixture will bubble over if the CO_2 is added too rapidly. Take one of the junctions of the thermocouple out of the mixture of ice and water (leave the other in the ice), wipe it dry, and then place it in the CO_2 mixture. After thermal equilibrium has been established, the two junctions will be at 0° and −78.5°C, respectively.

With the box A set at 9000, close the battery switch and the reversing switch. See that the voltmeter is reading. Then balance the emf of the thermocouple against the P.D. across S by reducing R from its original setting of 9000 ohms. Balance is indicated by zero deflection of the galvanometer. If a decrease in R does not reduce the galvanometer deflection, reverse switch RS and try again. *For a final balance, use the full sensitivity of the galvanometer.* Before taking this reading, be sure to stir both the ice mixture and the CO_2 mixture. Record the decade-box reading R_0 to the nearest ohm and the voltmeter reading V_0 at the balance point. Take four more readings of R_0 and V_0, stirring the mixtures before each reading. Any constant change of R_0 in these five readings indicates a lack of thermal equilibrium. In this case the process should be continued until a constant value of R_0 is obtained. Stirring the mixture before each reading is important. Set the A box to 9000, and open the battery and reversing switches after these readings.

Compute the average value of R_0 and V_0. For the error in average R_0, use either the mean deviation or 0.25%, whichever is larger. The value of E_0' at $-78.5°C$ may be obtained from Table K_1.

Part II. Freezing Point of Mercury. Remove the junction from the CO_2 mixture and wipe it off with a paper towel. Place it in the tube containing the mercury. Then, carefully place this tube in the mixture of CO_2 and alcohol. Do not allow the mixture to boil over in this process. Since mercury freezes at about $-40°C$, it should freeze in this mixture in 5 or 10 min and eventually reach the temperature of the mixture.

After the mercury seems to be frozen, close the battery switch and reversing switch; then balance the circuit as in Part I by varying R. Stir both mixtures before taking readings. Repeat this process as time goes on until a steady value of R is obtained. This will indicate that the mercury is in thermal equilibrium with the CO_2 mixture. The value of R should be nearly equal to R_0 found in Part I of the experiment, provided that the voltmeter reading V has not changed appreciably. At this point, the tube containing the mercury and thermocouple junction should be removed from the CO_2 mixture.

As soon as the tube of mercury is taken out of the CO_2 mixture, its temperature will rise very rapidly until the mercury reaches its melting point. At this point, the melting mercury absorbs its latent heat of fusion, thus keeping the temperature constant during the process of fusion. After all of the mercury has melted, its temperature will again rise until it reaches room temperature. Thus the melting point of mercury may be determined by use of the heating curve (temperature versus time) for mercury.

In order to obtain this heating curve it is necessary to determine the temperature of the mercury as a function of the time. The total time interval, of course, must be large enough to include both the solid and liquid states of the mercury. This may be done in the following manner.

Immediately after the tube of solid mercury is taken from its CO_2 bath, its temperature should be taken every 30 sec by means of the thermocouple. This can be done by taking a set of R and V values at half-minute intervals until the temperature of the mercury is well above its melting temperature. Since the mercury heats up very rapidly at first, it is advisable to keep the circuit in *continual* adjustment for balance by constant adjustment of R. In this process the ice mixture should be stirred continuously to ensure a constant temperature of $0°C$ for the other junction of the couple. It will be noted that while the mercury is melting the value of R will remain practically constant, indicating a constant temperature for the mercury. After melting has taken place, the temperature of the mercury will again rise, requiring an increasing value of R for balance. Continue taking R and V values every half minute until the value of R attains the maximum resistance of the decade box. At this point, the switch RS should be reversed and readings discontinued for 2 or 3 min (keep track of the time). After this time has elapsed, take five or more readings at half-minute intervals. Then, open the battery switch and the reversing switch.

By means of Eq. (38.6) and Table K_1, compute the set of temperatures for the mercury. Plot these temperatures (ordinate) against the times (abscissa). Using this heating curve, determine the melting temperature of mercury to the nearest half degree. Determine the error in this temperature. Compare this observed temperature with the accepted melting point of mercury ($-38.9°C$).

Record: Record the apparatus numbers of the important components of your equipment. Tabulate your data. Summarize your results.

QUESTIONS

1. What will be the value of R in Part II of this experiment when the mercury reaches the temperature of $0°C$?

2. Why was it necessary to reverse RS at a certain point in the heating curve and discontinue readings for a short time?

3. How could one determine the latent heat of fusion of mercury by use of the heating curve? What additional data would be necessary for this determination?

4. Suppose one junction of this thermocouple were held at room temperature, say $20°C$, instead of at the ice point. Would it still be possible to measure temperature with this couple by using Table K_1? Explain.

OPTIONAL EXPERIMENT

1. Perform Exp. 38 using a direct reading potentiometer such as the Leeds and Northrup student potentiometer. For a description of this potentiometer see Appendix IIN.

Electromagnetic
Induction

Object: To study some of the phenomena of electromagnetic induction; particularly to note the emf induced in a secondary coil when the magnetic flux linking it is changed. This magnetic flux is produced either by a current-carrying primary coil or by a permanent magnet.

Apparatus: Table galvanometer, primary coil (200 turns, No. 20 wire), secondary coil (350 turns, No. 24 wire), battery, switch, permanent bar magnet, brass rod, iron rod, and high resistance (20,000 ohms).

Theory: Whenever the magnetic flux (magnetic lines of force) linking a coil of wire *changes*, there is an emf induced in the coil that is proportional to the *time rate* of *change* of this magnetic-flux linkage. The magnetic flux linking the coil may arise from any source, e.g., permanent magnet, earth's field, current in another coil, or current in the coil itself. But the induced emf is independent of the nature of the source; it depends only upon the time rate of change of magnetic-flux linkage, whatever the source. This is known as Faraday's law of electromagnetic induction (1831) and is the foundation principle on which much of our modern-day electrical machinery is based.

   The direction of the induced emf is given by Lenz's law; that is, the induced emf is always in a direction that will set up electrical conditions—induced currents—that tend to *oppose* the *change* that is bringing them about. This is in complete accord with the law of the conservation of energy.

   A simple diagram will serve to illustrate these principles. In Fig. 39.1, a loop of wire with some magnetic flux linking it is shown. As long as this magnetic flux is constant, there is no induced emf in the wire. Suppose, however, that this magnetic flux is *increasing* with time. There will then be an emf induced around the wire pro-

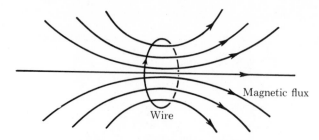

FIGURE 39.1   Magnetic flux linkage.

ducing an induced current. The magnitude of this emf will be directly proportional to the time rate of change of the magnetic flux. The direction of this induced emf and current must, by Lenz's law, be such as to oppose the increasing magnetic flux, i.e., the induced emf and current must be around the loop in a counterclockwise direction as one views the coil from the rear, looking in the direction of the magnetic flux. In this case the magnetic field set up by the induced current will be in a direction *opposite* to that of the *increase* of magnetic flux. On the other hand, if the magnetic flux linking the coil is *decreasing* with time, the induced current will set up a field that opposes the *decrease* in this magnetic flux. This means that the induced current is in the direction opposite to that in the first case.

In this experiment the induced emf in the secondary coil will be detected by connecting it to a galvanometer and observing the deflection. The magnetic flux linking this secondary coil will be produced either by use of a current-carrying primary coil, or by use of a permanent bar magnet.

Method: First determine the relation between the direction of the current in the galvanometer and the direction of its deflection. To do this, connect the battery in series with the galvanometer and a very high resistance (20,000 ohms) as shown in Fig. 39.2. Since the direction of the current furnished by the battery is known, the direction of the current through the galvanometer and the corresponding deflection can be determined.

Then, connect the secondary coil (wound with fine wire) to the galvanometer. Connect the primary coil (wound with coarse wire) to the battery through a switch. Keep the switch closed only while readings are being made. The direction of current in the primary is determined by the battery. The direction of any current in the secondary may be deduced from the observed direction of the galvanometer deflection.

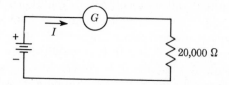

FIGURE 39.2   Correlation of current and galvanometer deflection.

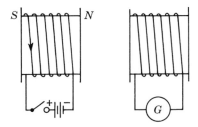

FIGURE 39.3   Orientation of coils.

Step 1. Line the coils up as shown in Fig. 39.3 so that the windings in both coils are in the same direction (clockwise or counterclockwise). Close the battery switch. Move the secondary quickly away from the primary. Bring it quickly back. Record the directions of the currents in the primary and secondary coils under these two conditions. To do this, draw diagrams similar to Fig. 39.3 and put in arrows indicating directions of motion and currents. Repeat this process, moving the primary instead of the secondary. Again, record directions of motion and currents in this case.

Step 2. Make and break the current in the primary and record the corresponding directions of the primary and secondary currents. Determine the galvanometer throws, while the primary current is made and broken, when the primary and secondary coils are separated by 0, 1, 2, 3, 4, 5, and 10 cm. Plot a curve of galvanometer throws against distance between coils.

Step 3. Open the primary circuit. Place the secondary as shown in Fig. 39.4. Thrust the *N* pole of the bar magnet into the secondary coil and note the direction of the galvanometer throw. Pull the *N* pole out and observe the throw. Perform the same operations using the *S* pole of the magnet. Record the direction of motion of the magnet and that of the induced current by means of a diagram similar to Fig. 39.4.

Step 4. Line the two coils up as in Fig. 39.5, with a distance of 2 cm between the ends. Place a brass rod through the cores of both coils and repeat the first part of Step 2. Record your results. Replace the brass rod with an iron rod and repeat. Record your results. Be sure to include the sizes of the throws of the galvanometer in this step.

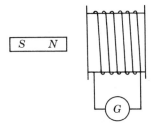

FIGURE 39.4   Orientation of coil and bar magnet.

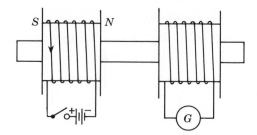

FIGURE 39.5   Magnetic coupling; variable core.

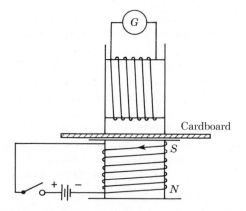

FIGURE 39.6   Magnetic coupling; variable orientation.

Step 5. Place the secondary on the primary, as shown in Fig. 39.6. Determine whether or not there is any position for which the galvanometer throw is zero when the primary current is made and broken.

<h3 style="text-align:center">QUESTIONS</h3>

1. In Step 1 what is the source of energy that appears in the secondary circuit?

2. In Step 2 what is the source of energy that appears in the secondary circuit when the primary circuit is broken? HINT: A magnetic field possesses energy. What is the source of energy in the secondary circuit when the primary circuit is made? Explain the observed difference in galvanometer deflections when the distance of separation is 2 and 10 cm.

3. In Step 4 explain the difference in galvanometer deflections for the brass and iron rods. Compare these deflections with the corresponding one obtained in Step 2.

4. In Step 5 explain in terms of magnetic-flux linkage why, for a certain orientation of the secondary coil, there is no appreciable induced current in the secondary. Use a sketch.

Earth

Inductor

Object: To determine the horizontal and vertical components of the earth's magnetic field using an earth inductor.

Apparatus: Earth inductor, ballistic galvanometer, damping key, dial resistance box, ammeter, rheostat, battery, switch, and magnetic compass. See Appendix II, Notes G.1, 3, 4.

Theory: The horizontal and vertical components of the earth's magnetic field may be measured by turning a coil (earth inductor) in the earth's field in such a manner that it cuts the horizontal or vertical component of the magnetic flux. If this coil is connected to a ballistic galvanometer, the charge flowing through the coil and galvanometer will be proportional to the magnetic flux cut.

The earth inductor (Cenco) consists of a frame supporting a coil. There are two separate windings on this coil—the current winding and the inductor winding. The instrument is arranged so that the coil may be rotated, by spring action, through 180° about an axis either in a horizontal or vertical plane. The binding posts for the current coil are mounted on the ring, and those for the inductor coil are mounted on the framework. A data plate giving the electrical and geometrical characteristics of the coil is located on the frame.

The connections to the earth inductor are shown in Fig. 40.1. The inductor winding is connected to a ballistic galvanometer G through the decade resistance box DR. A damping key DK is placed across the galvanometer to control its deflections.

The current winding is connected to a battery through an ammeter A, switch Sw, and rheostat Rh.

Suppose the earth inductor is placed in a horizontal plane with its axis of rotation parallel to the H field of the earth, i.e., with its axis of rotation lying parallel to the

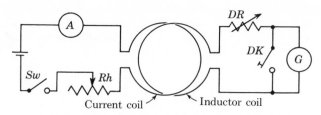

FIGURE 40.1 Wiring diagram for earth inductor.

direction of a compass needle. In this position it encompasses a maximum amount of the vertical flux of the earth's field, the V field. If the coil is suddenly rotated through 180°, all of the vertical magnetic flux will be cut *twice* without any disturbance caused by the H field. (In this process the current winding must be disconnected for mechanical reasons.) Let A represent the average area of the coil, V represent the vertical intensity of the earth's field, and N_i represent the number of turns in the inductor winding. Then ϕ_V, the total magnetic-flux change in this process, will be

$$\phi_V = 2AN_iV \tag{40.1}$$

By Faraday's law of electromagnetic induction, the induced emf, E_i, in volts will be

$$E_i = -10^{-8}\frac{d\phi_V}{dt} \tag{40.2}$$

If we substitute I_iR for E_i in Eq. (40.2) and integrate with respect to t over the time of the process, we get

$$RQ_V = -10^{-8}\phi_V = -2 \times 10^{-8}AN_iV \tag{40.3}$$

where R is the resistance of the galvanometer circuit. The throw of the galvanometer d_V is proportional to the charge Q_V passing through it, i.e., $Q_V = Kd_V$. Therefore, Eq. (40.3), after solving for V, becomes

$$V = \frac{-10^8RKd_V}{2AN_i} \qquad \text{(gauss)} \tag{40.4}$$

In order to determine H, the horizontal component of the earth's field, we may place the earth inductor on its side so that its axis of rotation is vertical, and its plane is perpendicular to the magnetic meridian. In this case the inductor encompasses the maximum amount of horizontal magnetic flux. When the coil is turned through 180°, this flux is cut twice without interference from the vertical flux. A throw d_H of the galvanometer results, which is related to H by means of the equation

$$H = \frac{-10^8RKd_H}{2AN_i} \qquad \text{(gauss)} \tag{40.5}$$

Equations (40.4) and (40.5) are sufficient to determine V and H, provided that K is known. This is generally not the case, and it is necessary to carry out an auxiliary experiment in order to determine it. This may be done as follows: The current winding is now connected as shown in Fig. 40.2 and serves as a primary circuit. Any *change* in the primary current will induce an emf in the secondary circuit (inductor winding), thus producing a throw of the galvanometer. The emf, E_s, induced in the secondary because of a changing current I_p in the primary, is given by the relation

$$E_s = -M\frac{dI_p}{dt} \qquad (40.6)$$

where M is the mutual inductance between the two circuits. E_s may be replaced by RI_s in Eq. (40.6) and the equation integrated over the time of change in the customary manner. We get

$$RQ_s = RKd_s = -M\delta I_p \qquad (40.7)$$

where δI_p is the *change* in the primary current and RK is the same combination of secondary circuit resistance and galvanometer constant as appears in Eqs. (40.4) and (40.5).

If we solve Eq. (40.7) for RK and substitute in Eqs. (40.4) and (40.5), we finally get

$$V = \frac{10^8}{2}\frac{M\,\delta I_p}{AN_i}\frac{d_V}{d_s} \qquad (40.8)$$

and

$$H = \frac{10^8}{2}\frac{M\,\delta I_p}{AN_i}\frac{d_H}{d_s} \qquad (40.9)$$

These Eqs. (40.8) and (40.9) enable us to determine V and H in terms of the measured quantities d_V, d_H, d_s, δI_p, and the quantities M, A, and N_i, obtainable from the data plate of the apparatus.

The approximate determinate-error equations for the experiment are

$$\frac{\Delta V}{V} = \frac{\Delta\delta I_p}{\delta I_p} + \frac{\Delta d_V}{d_V} - \frac{\Delta d_s}{d_s} \qquad (40.8a)$$

and

$$\frac{\Delta H}{H} = \frac{\Delta\delta I_p}{\delta I_p} + \frac{\Delta d_H}{d_H} - \frac{\Delta d_s}{d_s} \qquad (40.9a)$$

The small errors in M, A, and N_i may be neglected.

Method: Determine the direction of the magnetic meridian at the place where the inductor is to be used. A large compass may be used for this purpose, and a chalk line should be drawn on the table to show this direction.

Connect the inductor winding (terminals on frame) to the galvanometer circuit as shown in Fig. 40.2. Do not make the connections to the current windings (terminals on coil) yet. Place the inductor so that its face is horizontal and its axis of rotation is in line with the magnetic meridian (chalk line). Set the dial box at about 1000 ohms resistance. Then release the spring catch, allowing the coil to flip through 180°. Observe the throw d_V of the galvanometer. It should be about 125 mm. If it is much more or much less than 125 mm, adjust the decade resistance until the proper throw is obtained. The exact value of the decade resistance is immaterial, since it does not enter the final equations. Make five determinations of the galvanometer throw d_V.

Then, turn the inductor on its side with the face of the coil perpendicular to the magnetic meridian and with the axis of rotation in a vertical position. Place the inductor so that the d_H deflection is in the same direction as the d_V deflection. The inductor is now in position to "cut" the horizontal magnetic flux of the earth. Make five determinations of the throw of the galvanometer for this case, i.e., d_H. *Do not change the decade resistance in this process* since the galvanometer constant K is a function of this resistance.

Finally, connect the battery circuit to the terminals of the current winding on the ring. Set the rheostat Rh so that the ammeter reads about 0.50 amp when the battery switch is closed. The galvanometer will deflect in opposite directions when the switch is closed and when it is opened. Choose the operation (close or open) for your record that leads to a galvanometer throw in the *same* direction as for d_H and d_V. Make five determinations of the galvanometer throw d_s for this operation. Read the ammeter for each throw. This reading will equal the change δI_p in primary current.

Record the values of $d_V, d_H, d_s,$ and I_p. Use their averages to compute V and H by means of Eqs. (40.8) and (40.9). The values of M and N_i are given on the data plate of the inductor. Also, this plate gives the internal and external diameters of the coil. Use the average of these values in computing A. Note that the value of M is given in millihenrys (thousandths of a henry). It must be converted to henrys before being used in the equation.

In calculating the errors in V and H, use the mean deviations in d_V, d_H, d_s; or $\frac{1}{2}$ mm, whichever is the larger. The error in the ammeter may be taken as $\pm 1\%$ of full-scale reading.

Determine the angle of dip θ by means of the equation

$$\tan \theta = \frac{V}{H}$$

This is the angle that the total intensity of the earth's field makes with the horizontal.

Record:

App. No. Earth inductor_____
Galvanometer_____
Ammeter_____
Dial box_____
Dial-box setting (   ) ohms

Rdg	d_V, mm	d_H, mm	d_s, mm	I_p, amp
1	110.5	35.0	140.5	0.500
2				
3				
4				
5				
Average				

Inductor-coil Data:

$N_i = 1000$ turns D (average) = (   ) cm
D (inside) = (   ) cm A (average) = (   ) cm^2
D (outside) = (   ) cm $M = ($) mh

Results:

$V = ($	) gauss	$\Delta H = ($	)
$\Delta V = ($	)	$\theta = ($	)
$H = ($	)	Total intensity $= ($	)

QUESTIONS

1. Under what condition is the throw of the galvanometer in this experiment independent of the time of flow of charge through it? Would you expect d_V or d_H to change materially if the inductor coil were rotated very slowly in this experiment? Explain.

2. Explain carefully how the earth inductor could be used to determine the magnetic north-south direction and therefore act as a magnetic compass. (Principle of the earth inductor compass.)

OPTIONAL EXPERIMENT

1. The principles involved in the use of the earth inductor for determining the intensity of the earth's magnetic field can be applied to the measurement of the magnetic flux density of any magnetic field. The circuit is usually modified (for strong fields) by replacing the inductor coil with a small search coil of known area and number of turns. For calibration purposes, a magnetic-flux standard is connected in the galvanometer circuit. The current coil in Fig. 40.2 is eliminated. By proper manipulation of the search coil and the standard, the magnetic flux density of the unknown field may be determined. Use this method to determine the strength of the magnetic field produced by an appropriate electromagnet or permanent magnet.

Mutual Inductance;
Measurement of Magnetic Flux

Object: To calibrate a ballistic galvanometer for the measurement of magnetic flux using a standard primary-secondary air solenoid. To measure the magnetic flux of an old Hibbert standard (or any suitable substitute).

Apparatus: Ballistic galvanometer, standard solenoid, Hibbert standard, damping key, single-pole switch, reversing switch, battery, ammeter, and rheostat. See Appendix II, Notes G.1, 3, 4.

The standard solenoid consists of a long primary coil (50–100 cm) of several hundred turns of wire and a short secondary coil (5–10 cm) of several hundred turns around the middle of the primary. The mutual inductance of such an inductor may be determined accurately by formula involving the dimensions of the solenoid and the number of turns in the primary and secondary windings. But the formula is only valid provided there are no ferromagnetic substances in or near the solenoid.

The Hibbert standard (when calibrated) is a convenient device to use in conjunction with a search coil and ballistic galvanometer for measuring magnetic flux but is no longer on the market. This standard consists essentially of a coil that may be dropped through a radial magnetic field of known magnetic flux, which is produced by a permanent magnet formed by the shell and core of the standard. The coil slides on a shaft that passes through the center of the field. At the top of the shaft is a catch, which holds the coil out of the field and, when released, allows the coil to fall under the action of gravity into the field. In its initial position, there is no magnetic flux from the Hibbert standard linking the coil. In its final position, the entire flux of the standard links the coil. See the sketch, Fig. 41.1, which shows the cross section of the Hibbert standard.

Theory: When two coils are placed so that a current in one (the primary) produces a magnetic field so that some part of this field links the other (the secondary), the coils are said to be coupled magnetically and to possess mutual inductance.

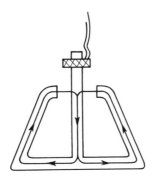

FIGURE 41.1   Cross section of Hibbert standard.

The mutual inductance M between two coils is given by the equation

$$E_s = -M\frac{dI_p}{dt} \tag{41.1}$$

that is essentially a form of Faraday's law of electromagnetic induction. In this equation E_s is the emf induced in the secondary coil and dI_p/dt is the time rate at which the current I_p in the primary coil is changing. If E_s is given in volts, I_p in amperes, and t in seconds, then M will be expressed in henrys. A pair of coils then will have a mutual inductance of 1 h if the current in the primary, changing at the rate of 1 amp/sec, induces an emf of 1 v in the secondary. This unit of mutual inductance, the henry, is a large one, and it is customary to express mutual inductance in a thousandth of a henry, the millihenry, or even in a millionth, the microhenry.

In laboratory work Eq. (41.1) is not convenient to use, involving, as it does, a rate of change of current. It may be put into a more convenient form by use of Ohm's law, when the secondary coil is part of a closed circuit. In this case the induced emf, E_s, in the secondary produces a current I_s in the secondary circuit, which is given by the equation

$$E_s = I_s R_s \tag{41.2}$$

where R_s is the total resistance of the secondary circuit. If we eliminate E_s between Eqs. (41.1) and (41.2), multiply the resulting equation by dt, and integrate over the time during which the primary current is changing, we get

$$R_s Q_s = -M\,\delta I_p \tag{41.3}$$

where $Q_s\ (= \int I_s\,dt)$ is the total charge circulating in the secondary circuit, and δI_p is the total change in the primary current. Q_s may conveniently be measured by connecting a ballistic galvanometer in the secondary circuit; δI_p may be measured by using an ammeter in the primary circuit.

A circuit that enables us to carry out these operations, and hence determine Q_s, is shown in Fig. 41.2. The chief element in this circuit is a primary-secondary air solenoid. It consists of a long primary coil with a short secondary coil wound around its center. The primary coil is connected to a battery through a reversing switch RS, an ammeter A, and a variable rheostat Rh. The secondary coil is connected to a ballistic galvano-

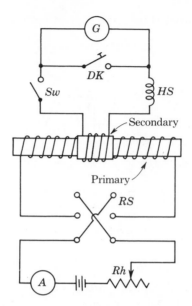

FIGURE 41.2   Circuit diagram for determination of magnetic flux.

meter G through switch Sw and a Hibbert standard HS. A damping key DK is connected across the galvanometer.

In order to determine Q_s it is only necessary to cause an abrupt change in the primary current I_p and observe the corresponding deflection d of the ballistic galvanometer. This change in I_p may be made by closing, opening, or reversing the switch RS. δI_p will be the difference between the current in the primary just before the change has occurred and that just after the change has been made. The galvanometer deflection d will be proportional to Q_s, the quantity of charge that circulates in the secondary circuit as a result of this change. Equation (41.3) may then be written

$$R_s K d = -M\,\delta I_p \tag{41.4}$$

where K is the ballistic constant of the galvanometer, i.e., the ratio of Q_s to d. This process amounts to a calibration of the ballistic galvanometer.

We may now determine the magnetic flux of the Hibbert standard by allowing its coil to drop into the cavity of the standard and noting the corresponding galvanometer deflection d_H. By use of Faraday's law we can show that

$$R_s K d_H = -10^{-8} N_H \phi_H \tag{41.5}$$

where $R_s K$ has the same value as in Eq. (41.4), N_H equals the number of turns in the Hibbert coil (given), and ϕ_H equals the magnetic flux in maxwells of the Hibbert standard. By dividing Eq. (41.5) by Eq. (41.4) and solving for ϕ_H, we get

$$\phi_H = 10^8 \frac{M\,\delta I_p}{N_H}\frac{d_H}{d} \tag{41.6}$$

where all quantities on the right side are either known or measured.

The formula for the mutual inductance M of the standard solenoid is

$$M = \frac{\pi^2}{10^9} \frac{D_p^2 N_s N_p}{L_p} \left[1 - \frac{1}{2}\left(\frac{D_p}{L_p}\right)^2 \right] \qquad \text{(henrys)} \qquad (41.7)$$

where D_p = *average* diameter of the primary coil in centimeters (core diameter + wire diameter),

N_s = number of turns in the secondary,

N_p = the number of turns in the primary, and

L_p = length of primary in centimeters.

The bracket factor in Eq. (41.7) is the so-called end correction, and approaches unity when $D_p \ll L_p$.

Error Equations. The *determinate*-error equation corresponding to Eq. (41.6) may be written

$$\frac{\Delta M}{M} = \frac{\Delta d}{d} - \frac{\Delta d_H}{d_H} - \frac{\Delta \delta I_p}{\delta I_p} + \frac{\Delta (N_H \phi_H)}{N_H \phi_H} \qquad (41.8)$$

The determinate-error equation corresponding to Eq. (41.7) is very nearly

$$\frac{\Delta M}{M} = \frac{2\Delta D_p}{D_p} - \frac{\Delta L_p}{L_p} \qquad (41.9)$$

since N_s and N_p are known exactly and the error in the correction term may safely be ignored. Why?

Method: Connect the apparatus as shown in Fig. 41.2 keeping all switches open during the process. *Keep* the *standard solenoid* as *far away from* the *other apparatus* as possible in order *to avoid* any *magnetic materials.*

Adjust the galvanometer scale so that the initial reading on the scale is zero when the galvanometer coil is at rest. Since the determination of ϕ_H depends primarily upon the ratio of two galvanometer throws—d when the *primary current* is *reversed* and d_H when the *flux standard* is *used*—it is highly advisable to have these two throws in the *same direction* and of *about the same magnitude*. Therefore, a preliminary trial should be made with the view of satisfying the above conditions either by adjustment of the flux standard, or by adjustment of the primary current. If the Hibbert flux standard (fixed) is used, adjustment of the primary current should be made by use of the battery rheostat to meet these conditions.

After this preliminary trial has been run, and adjustments made so that d and d_H are approximately equal in magnitude and occur in the same direction for a noted change of the reversing switch RS, the principal part of the experiment may be performed.

With the primary circuit open, but with the galvanometer switch Sw closed, determine the throw (maximum deflection) d_H of the galvanometer when the Hibbert coil is dropped. Make five determinations of d_H. Use their average in the determination of M and use the mean deviation or $\frac{1}{2}$ mm as the error in d_H, whichever is the larger.

Next close the reversing switch RS, choosing its closed position so that on *reversing* it, the resulting galvanometer deflection is in the *same direction* as the deflection d_H. Take five readings of the galvanometer throw d and the corresponding reading of the ammeter. After each reading, the reversing switch must be returned to its initial position and the galvanometer brought to rest at its zero position. The change of current δI_p in the primary circuit in this process will be $2I_p$, where I_p is the ammeter reading. From these data (d_H, d, I_p) and the values of M and N_H, compute by use of Eq. (41.6) the value of ϕ_H. Estimate its error by use of the error equations. The error in the ammeter readings may be taken as $\pm 1\%$ of full-scale reading. The error in the galvanometer readings may be taken as their mean deviation or $\frac{1}{2}$ mm, whichever is larger.

Record: Give the apparatus numbers of the galvanometer, flux standard, solenoid, and ammeter. Tabulate your data. Summarize your results.

QUESTIONS

1. Develop the error equations.

2. What percentage error would be introduced into your value of M(theo) if the end correction were neglected? Is this error significant?

3. If it were desired to determine the numerical value of the ballistic galvanometer constant K in this experiment, what additional information would be required?

4. If a piece of soft iron happened to be near the end of the primary-secondary solenoid, what effect would this have on the results of this experiment?

Alternating-current
Series Circuit

Object: To study current, potential difference or voltage, impedance, power consumption, and phase relationships in an R-L-C series a-c circuit.

Apparatus: a-c circuit board, a-c ammeter (0–1), a-c voltmeter (0–150), a-c wattmeter (0–150), and a variable transformer. The circuit board consists of the following elements in series: switch, fuse, condenser bank (8 μf paper), choke (inductance, 0.7 h), and a 60 w lamp (resistance). Jacks are placed at the points M, N, O, P, as indicated in Fig. 42.1, so that measurements may be made across any or all of the elements without disconnecting the circuit.

Theory: 1. *Phase Relationships.* When a pure resistance carries an alternating current, the resulting potential difference (V_R) is also alternating, and in phase with the current. The same current in a pure inductance results in an alternating voltage

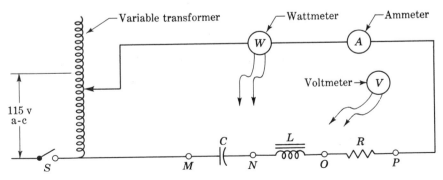

FIGURE 42.1 Wiring diagram of series a-c board.

across the inductance that leads the current by 90°. In a capacitance, the voltage lags the current by 90°. The angle by which the voltage leads or lags the current is called the phase difference angle. This is illustrated in Fig. 42.2 in which the voltage and current curves are shown separately in (a) and (b), and the more general case of a combination is shown in (c).

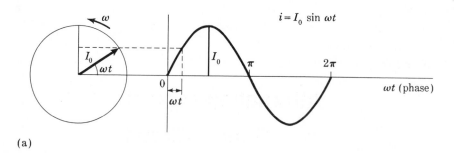

(a)

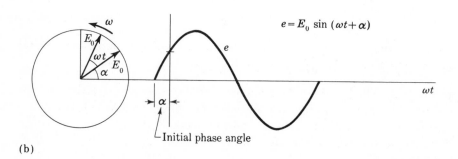

(b)

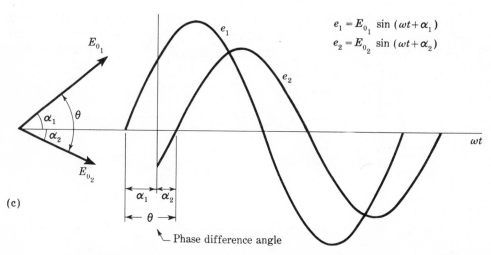

(c)

FIGURE 42.2 Phase relationship of voltage and current.

2. *Power.* The calculations of power, using effective values of current and voltage (the values indicated by most a-c electric meters), are identical for a resistive element with those in d-c theory, that is,

$$P_R = V_R I$$

However, in elements that include inductance or capacitance, the phase angle between the current and the voltage must be taken into account, i.e.,

$$P = VI \cos \theta \qquad (42.1)$$

where $\cos \theta$ is called the power factor.

In the purely resistive element the voltage and the current are "in phase," that is, there is no phase difference between them, and the term $\cos \theta$ becomes unity. On the other hand, in a perfect capacitor or a perfect inductor, the current respectively leads or lags the impressed voltage by 90°, and the power consumption as given by Eq. (42.1) is seen to be zero in each case. See Appendix II, Note H.4 for a discussion of the wattmeter.

3. *Impedance.* A generalized form of Ohm's law may be applied to a-c circuits, i.e.,

$$Z = \frac{V}{I} \qquad (42.2)$$

where Z is given in ohms if V is in volts and I is in amperes. There are two basic types of impedance; the pure resistance which does not cause a phase difference between V and I, and the pure capacitance or the pure inductance which shifts the phase 90°. This latter type is called a reactance and is designated by the symbol X. It is measured in ohms and has an algebraic sign associated with it. A positive reactance (a pure inductance) is one in which the voltage leads the current by 90°; a negative reactance (a pure capacitance) is one in which the voltage lags the current by 90°.

In general, an impedance consists of a combination of resistance and reactance. The impedance of the ordinary inductance coil combines a resistance with a positive reactance, and therefore has a phase angle greater than 0° but less than 90°. Since resistance and reactance voltages are out of phase by 90°, they are taken as the component of the impedance voltage. From the sketch of Fig. 42.3 and Eqs. (42.1) and

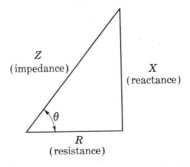

FIGURE 42.3 Impedance triangle.

(42.2), it may be seen that power is consumed only by the resistive component of an impedance.

4. *Calculation of Reactances.* The reactance of an inductor depends on the size of the inductance and on the frequency of the current; i.e.,

$$X_L = 2\pi f L \tag{42.3}$$

where X_L is in ohms, L in henrys, and f in cycles per second.

The reactance of a capacitor is similarly given by

$$X_C = -\frac{1}{2\pi f C} \tag{42.4}$$

where X_C is in ohms, C in farads, and f in cycles per second.

The reactive component of any impedance (see Fig. 42.3) is given by

$$\overset{\ast}{X} = Z \sin \theta \tag{42.5}$$

whereas the resistive component is given by

$$R = Z \cos \theta \tag{42.6}$$

5. *Series Circuit.* In a series circuit such as the one to be studied, the current (Fig. 42.4) in each element of the circuit must be identical with that in all the other elements. Clearly then, if the phase relationships between current and voltage in the individual elements are to hold, the individual potential differences must differ in phase, each from the others.

It will be convenient, therefore, to use the common current as the reference of phase. It is obvious from the foregoing that the voltage across the resistor may be expected to be in phase with the current, the voltage across the inductor to lead the current, and the voltage across the capacitor to lag the current. Furthermore, in general, the voltage impressed across the entire circuit will not be in phase with the current in the circuit. It will, in fact, be the vector sum of the individual potential differences in the circuit. (Compare this with the d-c case; can it be said that the d-c circuit is a

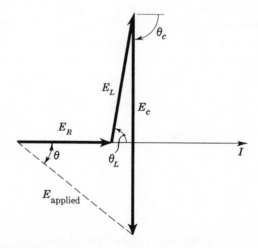

FIGURE 42.4   Voltage vector diagram of series a-c circuit.

special case of the a-c circuit? In what way?) Since the voltages add vectorially, it is possible that an individual voltage may be larger in magnitude than the vector sum of all the potential differences, i.e., the applied emf (see Fig. 42.4).

It is clear, from Secs. 3 and 4, that the algebraic sum of the reactances in a series a-c circuit is the total reactance of the circuit. Similarly, the simple sum of all the resistive components in the circuit must add up to the resistive component of the entire circuit. The total impedance Z of the circuit will then be given by

$$Z = \sqrt{R^2 + X^2} \tag{42.7}$$

where R and X are respectively the total resistance and the total reactance in the circuit. Likewise, the phase difference angle θ for the circuit is given by

$$\tan \theta = \frac{X}{R} \tag{42.8}$$

Finally, the power factor for the circuit is given by

$$\text{P.F.} = \cos \theta = \frac{R}{Z} \tag{42.9}$$

The error equations in this experiment are complicated by the presence of trigonometric functions; therefore, no attempt will be made to derive all the error equations. However, there is one place where the errors in P, V, and I have a pronounced effect on the result. This occurs when one attempts to determine the angle θ from the power factor $\cos \theta$, when this power factor is very nearly equal to one, i.e., when θ is very nearly equal to zero. In this case it may be shown that $\Delta\theta$ is given approximately by the equation

$$\Delta\theta = \pm \sqrt{2\left(\frac{\Delta I}{I} + \frac{\Delta V}{V} + \frac{\Delta P}{P}\right)} \qquad \text{for } \theta \cong 0 \tag{42.1a}$$

Method: 1. Check the zero readings of all the instruments. They must lie flat on their backs. Avoid parallax (see Appendix II, Note E) by using the mirror behind the scale, and lining up the needle with its image in the mirror before taking a reading.

2. Use the smaller capacitance available in the condenser bank if an adjustment is provided.

3. Adjust the current to 0.400 amp with the voltmeter and wattmeter potential leads *disconnected*. Later, when these are connected, the ammeter reading will change, but this may be ignored since the current in the *circuit element* will remain very close to 0.400 amp. (Why?)

4. Use the 150 v scale on both the voltmeter and the wattmeter.

5. Connect the voltmeter leads across R, L, C, and MP (whole circuit) in succession, recording each reading. The current should be checked before and after each reading. If the current has changed after removing the voltmeter leads, the reading should be discarded and another one taken after readjusting the current.

6. Repeat instruction 5 for the wattmeter potential leads.

7. Change the current slightly with the rheostat and then readjust it to its original value of 0.400 amp. Then repeat instructions 5 and 6.

8. Repeat instruction 7.

9. *Computations.* Average the three groups of readings on each instrument and enter the averages in columns 2, 3, and 4 of the record.

(a) A determinate error is introduced into the data because the wattmeter measures not only the power dissipated in the element being measured, but also that consumed by the voltage winding of the wattmeter itself. This latter amount should be subtracted from the wattmeter reading to obtain the power actually dissipated in the measured element alone. The correction may be taken as E^2/R_{wm}, where the wattmeter resistance, R_{wm}, is printed on the dial of the instrument. Compute the actual power consumed in each element of the circuit and in the total circuit.

(b) Compute the power factor for each element of the circuit and for the total circuit, using Eq. (42.1). Assume that the power factor of the condenser is zero. It is too small to be measured by these methods.

(c) Compute the phase angle θ for each element and for the total circuit. Its sign is known for each element, but not for the total circuit.

(d) Compute the impedance Z in ohms of each element and of the total circuit, using Eq. (42.2).

(e) Compute the reactance X and the resistance R of each element and of the total (see Fig. 42.2). The sign of X for each element is known, but it is not known for the total.

(f) Compute the size of the inductance L in henrys of the choke and the size of the capacitance C in farads of the condenser bank, using Eqs. (42.3) and (42.4). The frequency is 60 cps.

(g) Compare the corrected total power dissipated in the entire circuit with that of the corrected sum of those in the three elements.

(h) Compare the total resistance with the sum of the separate resistances.

(i) Compare the total reactance with the algebraic sum of the separate reactances. Choose the sign of the total reactance to correspond with the sign of the algebraic sum of the separate reactances.

(j) Compare the total impressed voltage with the vector sum of the separate potential differences. To carry out this procedure, construct a vector diagram (see Fig. 42.3). Use a full sheet of graph paper for accurate work.

(k) Compare the calculated values for the capacitance of the condenser and the resistance of the lamp bulb with the expected values. Account for differences in each case.

(l) Using Eq. (42.7), compute the expected value of Z for the entire circuit and compare this with the measured value.

(m) Using Eq. (42.8), compute the expected value of the phase angle θ of the entire circuit and compare this with the measured value and with the value obtained from the vector diagram of item j.

(n) Using Eq. (42.9), compute the expected P.F. for the entire circuit and compare this with the measured value.

Record: (*Partial sample*)

App. No. a-c circuit board_____

Wattmeter_____

Voltmeter_____

Ammeter_____

Resistance of wattmeter 10,050 ohms

Size of condenser_____μf

Size of lamp bulb_____watts,_____volts

Quantity / Element	I, ave read amps	E, ave read volts	P, ave read watts	P, actual watts	Power factor	θ, deg	Z, ohms	R comp of Z ohms	X comp of Z ohms	Size	Expected size
Measured — Lamp	0.400	79.7	32.6	32.0	1.00	0.0	199	199	0	199 ohm	
Measured — Choke	0.400	106.2	6.6	5.5	0.129	82.6	266	34	264	0.700 henry	⊗
Measured — Condenser	0.400	110.8	1.0	0.0	0.000	−90.0	277	0	−277	9.58×10^{-6} farad	
Measured — Total	0.400	95.3	38.8	37.9	0.994	±6.3	239	238	±26	⊗	⊗
Calc Tot	(0.400)	94.2*	⊗	37.5	0.998	−3.2 / −3.4*	236	233	−13	⊗	⊗
%Diff in Total	⊗	1.2	⊗	1.1	0.4	⊗	1.2	2.1	⊗	⊗	⊗

* Indicates value obtained from the vector diagram.
⊗ Indicates a space which is not to be filled.

215

QUESTION

1. In this experiment a large percentage difference frequently occurs between X_{total} as measured and X_{total} as calculated. Explain. [HINT: See Eq. (42.1a).]

Resonance

Object: To study the phenomenon of resonance in an *R-L-C* series a-c circuit by varying the frequency.

Apparatus: Capacitor $\cong .05\ \mu f$, resistors $\cong 50$ and 200 ohms, rf choke $\cong 5$ mh, variable af generator, and oscilloscope.

Theory: In Fig. 43.1 we show the circuit for this experiment. It consists primarily of elements *R*, *L*, and *C* connected in series to the output terminals of a variable audio-frequency generator, the power source for the circuit. Since we wish to measure the current in the circuit for constant output voltage as a function of frequency, it is necessary to have some means of measuring the current at different frequencies without permitting the output voltage of the generator to change. This may conveniently

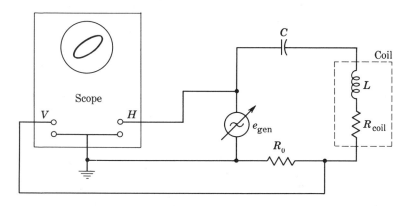

FIGURE 43.1 Schematic for series *R-L-C* circuit.

be done by using an oscilloscope to indicate both the voltage output of the generator (horizontal deflection, d_H) and the voltage across the known resistor R_0 (vertical deflection, d_V). The pattern on the scope for horizontal deflection only is a horizontal line whose half-length is proportional to the peak output voltage of the generator, for vertical deflection only a vertical line whose half-length is the peak voltage across the resistor R_0, and for both deflections simultaneously a Lissajous ellipse.

Because $X_L = 2\pi f L$ and $X_C = -1/(2\pi f C)$, a plot of the respective reactance versus frequency yields the graphs as shown in Fig. 43.2. At resonant frequency f_r

$$2\pi f_r L = \frac{1}{2\pi f_r C}$$

$$\therefore \quad f_r = \frac{1}{2\pi\sqrt{LC}} \tag{43.1}$$

In general the current varies with the impedance, i.e.,

$$I_{\max} = \frac{E_{\max}}{Z}$$

where $I_{\max}$ and $E_{\max}$ are the instantaneous maximum values and Z is the impedance given by the expression

$$Z = \sqrt{(R_0 + R_{\mathrm{coil}})^2 + (X_L + X_C)^2} \tag{43.2}$$

In this experiment we shall be measuring the deflection on the screen of the scope,

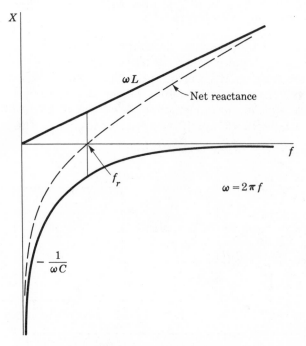

FIGURE 43.2 Reactance curves.

consequently,

$$d_V = \frac{Kd_H}{\left[R^2 + \left(\omega L - \frac{1}{\omega C}\right)^2\right]^{1/2}} \qquad (\omega = 2\pi f) \qquad (43.3)$$

At resonance where $X_L + X_C = 0$, Z is minimum and is equal to the resistance in the external circuit. Fig. 43.3 shows a typical curve of current (actually d_V) plotted against frequency, holding the source voltage constant.

Because the frequency of the applied voltage is allowed to vary, the phase difference between the current and the applied voltage changes, becoming zero at resonance. In general the phase difference angle can be determined from the relation

$$\tan \theta = \frac{X}{R} \qquad (43.4)$$

where X is the net reactance and R includes both R_0 and R_{coil}.

There is one other property of the resonance peak in addition to its position and height. That property is the "sharpness" of the peak. In specifying the sharpness of resonance it is customary to determine a frequency band width above which the current exceeds $0.707I_{\text{max}}$. At resonance the current is a maximum and is determined from

$$I_{\text{max}_r} = \frac{E_{\text{max}}}{R} \qquad (43.5)$$

On both sides of f_r, f_{lo} and f_{hi} represent frequencies at which the net reactance is equal in magnitude to the resistance. At these points, therefore,

$$Z = \sqrt{R^2 + R^2} = \sqrt{2}\,R \qquad (43.6)$$

and

$$I_{\text{max}} = \frac{E_{\text{max}}}{Z} = \frac{E_{\text{max}}}{\sqrt{2}\,R} = 0.707 I_{\text{max}_r}$$

$$\left(\frac{I_{\text{max}}}{I_{\text{max}_r}}\right)^2 = \left(\frac{1}{\sqrt{2}}\right)^2 = \frac{P}{P_r} \qquad (43.7)$$

It should be apparent from Eq. (43.7) that the power at f_{lo} and f_{hi} is one-half the power at resonance, and for this reason these points are often referred to as the half-power points.

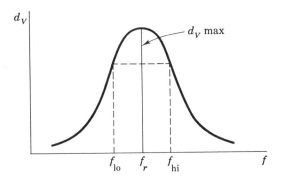

FIGURE 43.3 Resonance curve.

At the half-power points

$$\omega_{hi}L - \frac{1}{\omega_{hi}C} = R \tag{43.8}$$

and

$$\omega_{lo}L - \frac{1}{\omega_{lo}C} = -R \tag{43.9}$$

In separate operations by subtracting and adding Eqs. (43.8) and (43.9) it is easy to show that

$$f_r^2 = f_{hi}f_{lo} \quad \text{and} \quad f_{hi} - f_{lo} + \frac{1}{(2\pi)^2 LC}\frac{f_{hi} - f_{lo}}{f_{hi}f_{lo}} = \frac{2R}{2\pi L} \tag{43.10}$$

from which we obtain

$$\frac{(f_{hi} - f_{lo})}{f_r} = \frac{R}{2\pi f_r L} = \frac{1}{Q} \tag{43.11}$$

where the symbol Q is used to express or determine the degree of sharpness or selectivity.

Method: After having connected the elements as shown in Fig. 43.1, adjust the output of the signal generator and the horizontal gain of the oscilloscope such that the pattern covers approximately 6 cm of the screen. Vary the frequency through resonance and make the necessary change in the vertical gain to give a reasonable vertical displacement. At resonance the Lissajous pattern on the scope should be a straight line, since the applied voltage is now in phase with the current in the series circuit. On both sides of resonance the pattern on the scope is the general case of the ellipse.

Step 1. For each frequency measure the maximum vertical displacement and the y intercepts. Tabulate your data. At each frequency make certain that the voltage output of the generator is adjusted to a fixed predetermined value.

Step 2. Plot two resonance curves ($d_{V_{max}}$ versus f) on a single graph paper; one for $R_0 = 50$ ohms and the other for $R_0 = 200$ ohms. Indicate the resonant frequency and determine the value of Q.

Step 3. Determine the phase difference angle at (a) resonance and at (b) each of the half-power points.

QUESTIONS

1. If direct voltage measurements are to be made with the scope, explain the procedure. (See Appendix II, Note L for scope calibration.)

2. Show that $Q = (1/R)(\sqrt{L/C})$.

3. Derive Eq. (43.11).

4. Determine the value R_{coil}.

EXPERIMENT **44**

Vacuum
Tube

A. VACUUM TUBE CHARACTERISTICS

Object: To study the characteristics of a triode.

Apparatus: Triode board, ammeter (0–50 ma), voltmeter (0–400 v), vacuum tube 12AU7 (or 6AU7, 6C4, 6B10, 6AG11, 6SN7GTB, 6SL7GT), and power supply (6 or 12 v filament, 0–12 v d-c grid, and 0–400 v d-c plate). See Appendix II, Note F.

Theory: The triode consists of (1) a cathode (electron source), (2) a plate (the positive electrode to which the electrons are attracted), and (3) the grid (the gate which regulates the flow of electrons from the cathode to the plate).

The operation of the vacuum tube depends on the fact that certain metals when heated emit electrons quite freely from their surfaces. The cathode, either the directly heated or indirectly heated, will "boil" off electrons to form a cloud surrounding the cathode. The process will not, however, go on indefinitely, since the cloud of electrons (the "space charge") quickly becomes so dense that it is able to repel subsequent electrons back to the cathode. Equilibrium is reached when as many electrons return to the cathode per second as are emitted.

In the triode a grid or mesh of wire is inserted in the region of maximum space charge (see Fig. 44.1). Because of its proximity to the space charge, the grid has a greater effect on the motion of the electrons in this region than does the plate. If the grid is negative with respect to the filament, it repels the electrons back toward the filament, forming a smaller but more compact charge nearer the filament; thus, fewer electrons get through to the plate. By making the grid sufficiently negative, the plate current i_b can be entirely cut off.

If, however, the grid voltage E_c is held constant while varying the plate voltage

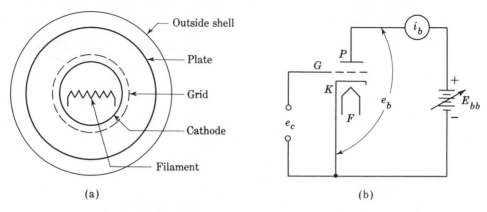

FIGURE 44.1   Triode: (a) cross section of tube, and (b) wiring connections of triode board.

e_b, we obtain an $i_b - e_b$ curve. By plotting such curves for several values of grid voltage a family of curves is obtained, as shown in Fig. 44.2.

Similarly if the plate voltage is held constant, a series of observations of plate current versus grid voltage will give a grid characteristic curve. A family of such

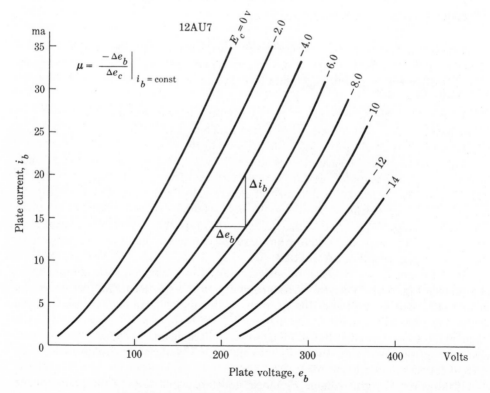

FIGURE 44.2   Family of plate characteristic curves.

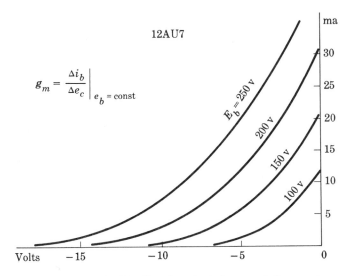

12AU7

$$g_m = \frac{\Delta i_b}{\Delta e_c}\bigg|_{e_b = \text{const}}$$

$E_b = 250$ v

200 v

150 v

100 v

ma

30

25

20

15

10

5

Volts -15 -10 -5 0

FIGURE 44.3   Family of mutual characteristic curves.

characteristics for different values of E_b is called transfer characteristics (see Fig. 44.3). They are so called because action on the grid is transferred to results in the plate circuit. The slope of a typical curve at any definite point on the curve (definite i_b and e_b) shows the rate of change of the plate current with respect to the plate voltage for the value of the grid voltage E_c corresponding to that curve. Suppose, for example, that this is 6.5 ma/45 v. The slope of the corresponding curve in Fig. 44.2 (same E_b) at the corresponding point (same i_f) shows the rate of change of the plate current with respect to the grid voltage. Suppose that this is 6.5 ma/2 v. In this example it is necessary to change the plate voltage 22.5 times as much as the grid voltage, in order to produce the same change in the plate current. Thus, in this example, the grid is 22.5 times as effective as the plate in controlling the plate current. The quantity 22.5 is called the *amplification factor* of the tube and is represented by the symbol μ. This factor may be obtained approximately from a single set of characteristics by noting that in order to hold i_b constant during a small increase of the grid voltage Δe_c, the plate voltage will have to be decreased by a larger amount Δe_b. The ratio of the latter to the former will equal the amplification factor as defined by the equation

$$\mu = \frac{-\Delta e_b}{\Delta e_c}\bigg|_{i_b = \text{const}}$$

Method: The filament of the tube is surrounded by a metal cylinder called the cathode, which when heated by the filament emits electrons. The filament in this case is not a good emitter.

Do not exceed the maximum value of plate current as listed by the manufacturer. Note that this limitation means that all the spaces in the record will not be filled. With zero or positive grid voltage, less plate voltage is required to attain the maximum allowable plate current. On the other hand, with larger negative grid voltages, even the maximum available plate voltage will not be enough to cause the maximum current.

Step 1. Using the test voltmeter, measure the filament a-c voltage. It should be close to the listed value on the tube. (NOTE: The filament of the 12AU7 may be connected to either a 6.3 or 12 v source. This is made possible by the center connection brought out at pin No. 9. In this experiment the 6.3 v source was used.)

Step 2. Connect the grid power supply so that $E_c = 0$ v. (Note that merely disconnecting the grid is not sufficient, for then it "floats" in potential. The grid catches electrons that it cannot dispose of and rapidly becomes very negative.) Take a series of readings of the plate current for values of plate voltage (see Record), being careful not to exceed the tolerable value. Repeat for values of E_c in 2 v steps to minus 12 v.

Step 3. Plot the family of plate characteristics curves.

Step 4. From your family of curves in Step 3 determine the amplification factor at which I_b is held constant at 15 ma.

Record:

E_b (VOLTS) \ E_c (VOLTS)	0	−2	−4	−6	−8	−10	−12
0							
25							
50							
75							
100							
125							
150							
200							
250							
300							

B. VACUUM-TUBE AMPLIFIER

Object: To study the characteristics of a simple vacuum-tube amplifier.

Apparatus: Amplifier test board, d-c voltmeter (0–500 v), 12 v power supply, d-c voltmeter (0–15 v), milliammeter (0–50 ma), oscilloscope, VTVM, and resistors (22,000 and 100,000 ohms).

Theory: A vacuum-tube amplifier consists of one or more stages of amplification. In any given application, enough stages of amplification are used to amplify the incoming signal to the point where the final output voltage is capable of actuating a suitable device, such as a voltmeter, loudspeaker, impulse counter, and so on. The over all amplification obtained in such an amplifier is equal, of course, to the product of the individual stage gains and may range from factors of many billions down to values of less than unity, depending on the application. The present experiment concerns a single-stage amplifier and some of the concepts and characteristics associated with it.

A principal matter of interest is the gain of the stage. The gain may, of course, be measured. Calculation or prediction of its value is often an involved procedure, if exact values are desired, but good approximate values can be predicted rather easily under normal circumstances. To do this, we use an "equivalent circuit," (for small a-c signals) which depicts the tube as a source of emf with an internal resistance and places this source in the remainder of the circuit as it actually is, except that the fixed supply voltages are eliminated. The equivalent circuit resulting from this treatment is solved by use of Kirchhoff's laws, and in this way the behavior of the actual circuit is usually quite well predicted.

In Fig. 44.4 a simple vacuum-tube amplifier stage and its equivalent circuit is shown. The actual circuit (a) consists of the tube, a load resistor R_L, a plate supply voltage E_{bb}, and a grid bias voltage E_{cc}. A signal voltage e_g, to be amplified, is inserted as shown, and the output voltage $e_{out}(e_o)$ is taken between plate and cathode. The action of the circuit is as follows: with no signal voltage, $e_g = 0$; the d-c value of the plate voltage is determined by the value of the grid bias, the plate supply voltage, and the load resistor. If there is any plate current i_b, then the plate will be at a lower potential than the supply voltage because of the "IR drop" in R_L. Thus E_b will have a no-signal, or quiescent, value E_{bo}. We are interested, however, only in the change in voltage across the terminals B and D in response to an incoming signal e_g. It is this change in e_b that we define as e_o. If a positive signal e_g is applied, then the grid of the tube is driven more positive (less negative) with respect to the cathode. This causes an increased plate current. The increased current in the load resistor R_L causes an increased IR drop across it, and thus the plate of the tube goes to a lower potential. The difference between the quiescent value E_{bo} and the new value of e_b is e_o. Since the plate voltage decreases for a positive input signal, we see that the circuit reverses the sign of a signal as well as amplifying it. It will be instructive for the student to verify this by applying the above reasoning to a negative input signal.

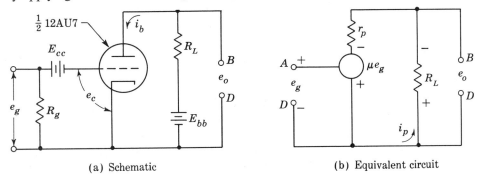

(a) Schematic   (b) Equivalent circuit

FIGURE 44.4 Triode amplifier circuit: (a) schematic, and (b) equivalent circuit.

In the equivalent circuit (b) the tube has been replaced by a voltage source μe_g and a resistance r_p. μ is the amplification factor of the tube; r_p is the plate resistance. Since the equivalent circuit concerns only changes resulting from an input signal, the plate and grid batteries have been left out of the circuit. A positive input signal e_g causes an output current i_p in the direction shown. As it is drawn, the equivalent circuit may now be analyzed by ordinary Kirchhoff methods. This is a single-loop circuit, and its equation for voltages around the loop may be written as follows:

$$\mu e_g - i_p R_L - i_p r_p = 0 \tag{44.1}$$

The output voltage e_o is simply the IR drop across R_L, and since the voltage is measured with respect to point D, convention states that i_p, as shown, is in a negative direction. Thus

$$e_o = -i_p R_L \tag{44.2}$$

Solving Eq. (44.1) for i_p we get

$$i_p = \frac{\mu e_g}{r_p + R_L}$$

We substitute this in Eq. (44.2) and find that

$$e_o = -\frac{\mu R_L}{r_p + R_L} e_g \tag{44.3}$$

The quantity $-\mu R_L/(r_p + R_L)$ is called the gain of the stage, since it is the ratio of the output voltage to the input voltage. It is directly proportional to the amplification factor of the tube and depends also on the plate resistance of the tube and on the load resistance. If the load resistance is high compared to the plate resistance, then the gain is seen to reach the value $-\mu$ in the limiting case. In all other cases, the gain of the stage is less than the amplification factor of the tube.

The analysis thus far assumes that the output voltage is applied to an infinite resistance; i.e., the external resistance across points B to D is infinite. If the external circuit has a finite resistance, this resistance must be inserted in the equivalent circuit between B and D. It is then simply in parallel with resistance R_L, and the formula of Eq. (44.3) must be modified to

$$e_o = -\frac{\mu R_b}{r_p + R_b} e_g \tag{44.4}$$

where R_b is the net load resistance and is equal to the parallel resistance of R_L and the following circuit.

Method: *Part. I. Determination of Stage Gain with Load.* Familiarize yourself with the connections of the vacuum-tube test board, and then set up the circuit of Fig. 44.5. With the plate-supply voltage E_{bb} set at 300 v take a set of plate voltage readings e_b for grid bias values varying from 0 to -8 v in steps of 1 v. Select a grid bias such as -5 v as the reference or operating point. The corresponding values of plate voltage and plate current are known as the quiescent values respectively. Subtract the quiescent value from each reading of the total plate voltage to obtain the value of e_o for each signal voltage. Plot the output voltage against the input signal volt-

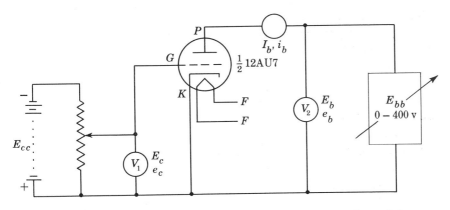

FIGURE 44.5 Triode amplifier; determination of operating point.

age. Determine the slope of the resulting line and calculate the actual stage gain. By use of Eq. (44.4) calculate the theoretical gain of the amplifier using the values $\mu = 17$ and $r_p = 7000$ ohms for the 12AU7 tube.

Part II. Determination of Frequency Characteristics. Gain of amplifier. Connect the circuit as shown in Fig. 44.6. As in Part I use a grid bias supply E_{cc} of -5 v and a plate supply E_{bb} of 300 v. Then, as illustrated in Fig. 44.6, e_g is the a-c signal applied to the grid of the tube and e_o is the output signal. The ratio of e_o/e_g gives the gain by direct measurement, and this may be accomplished by the use of either or both (a) a VTVM and (b) an oscilloscope.

If the vacuum-tube voltmeter (VTVM) were used, adjust the input signal e_g to a reasonable value such as 2 v (rms measured on the instrument). With this signal applied at the input, measure the output voltage across R_L, i.e., between point C and ground. The ratio e_o/e_g is the measured value of the gain of the single-stage amplifier.

In method (b) an oscilloscope may be used to measure the gain. Here again it would be advisable to apply a suitable voltage to the grid, its value being determined by

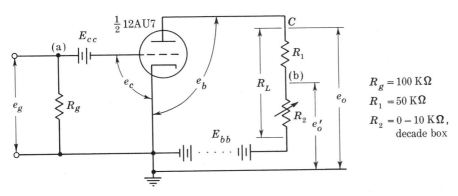

FIGURE 44.6 Schematic of simple triode amplifier.

calibrating the scope. The voltage e_o' which is a fraction of the total output voltage e_o is made equal to the applied voltage e_g by adjusting the value of R_2 (decade box). The ratio of R_L/R_2 is the gain of the stage.

QUESTIONS

1. The oscilloscope or vacuum-tube voltmeter used in Part I does not have an infinite input impedance. If its input impedance is 1 megohm (10^6 ohms), what percentage error in the stage gain is introduced by ignoring its loading effect on the plate circuit?

2. Over what range of frequencies is the experimental vacuum-tube amplifier "flat"? For the purposes of this question, assume that a drop of 3% or less is satisfactorily flat. Answer this for each gain-frequency curve plotted.

3. In most oscilloscopes the amplifier for the horizontal deflection plates is a different type from that for the vertical deflection plates, and thus the two amplifiers will have different frequency characteristics. Suppose that the horizontal amplifier were poorer in quality than the vertical amplifier. What effect would this have on the results in Part II? Describe how, using the same input signal to both, the oscilloscope amplifiers could be "calibrated" at some frequency and then used to get an accurate measure of the phase shift in the experimental amplifier.

Transistor

Object: To study the characteristics of a transistor. To construct a simple transistor amplifier.

Apparatus: Transistor Board, voltmeter (0–12v), VTVM, ammeter (0–30 ma, 0–1ma), power supply (0–12v), dry cell (1.5 v), audio generator, oscilloscope, and Transistor 2N104 (or 2N109, 2N1303, 2N1302, EN2222, 2N2156, 2N4143, 2N3053, 2N5130, EN2369A, 2N3055, etc.).

Theory: The simplest of the semiconductor devices is a single junction, which acts as a rectifier, though not perfect. The diode (two-element tube) permits current to flow during one alternation of an a-c signal and offers an open circuit during the negative alternation. The semiconductor (Fig. 45.1) offers low resistance during the positive half cycle and high resistance during the negative alternation. Nonetheless the resistance can be made so low in some semiconductor junctions that the inverse current is negligible, and as a consequence, high efficiency of rectification can be achieved.

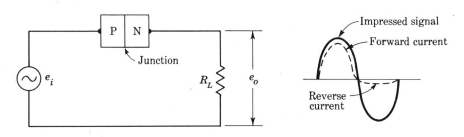

FIGURE 45.1   Diode action.

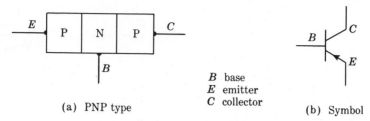

(a) PNP type

B base
E emitter
C collector

(b) Symbol

FIGURE 45.2   Triode transistor.

Triode Action. The common transistor consists of 2 junctions (Fig. 45.2). It is essentially a current generator with a characteristically low impedance. The vacuum tube, as was noted in Exp. 44, is a voltage generator with a high impedance. The operation of the two devices is similar so an understanding of one is helpful in studying the other.

In Fig. 45.3 forward current takes place at the emitter-base junction. Similarly, if the collector of the PNP transistor is made positive with respect to the base, forward current takes place at the collector-base junction. However, if the voltage is reversed across the junction, the current is reversed.

The action of the NPN transistor is different only in that the applied d-c voltages are opposite to that of the PNP.

Many factors influence the layout for a circuit such as input impedance, current gain, current and voltage requirements, etc. Without elaborating on the various circuits that are available, we shall consider (a) the characteristics of a PNP transistor, and (b) the common-emitter connection in a very simple low-frequency amplifier. In the transistor amplifier we shall be interested in the current gain, namely,

$$\beta = \frac{\Delta I_c}{\Delta I_b}\bigg|_{V_c = \text{const}}$$

Method: *Part I. Static characteristics.* The performance of the transistor may be studied by plotting a family of curves similar to the characteristic curves of the vacuum tube.

Step 1. To determine experimentally the current amplification α.

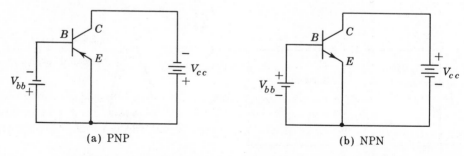

(a) PNP

(b) NPN

FIGURE 45.3   Conventional dc polarities for transistor types.

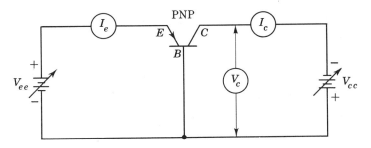

FIGURE 45.4 Common-base connection.

In Fig. 45.4:
(a) Set I_e at 5 ma, and vary V_c from 0 to 10 v in steps of 2 v.
(b) Repeat the above step for I_e = 10, 15, 20, 25, 30 ma.
 Plot a graph[1] showing I_c versus V_c. From the graph determine $\Delta I_c/\Delta I_e$ at a fixed value of V_c. This ratio is the current amplification for the common-base circuit and is called alpha (α).

Step 2. To determine experimentally the current amplification β.
 In Fig. 45.5:
(a) Set I_B at 250μa and vary V_c from 0 to 10 v in steps of 2 v. Record the corresponding values of I_c.
(b) Repeat the above step for I_b = 500, 750, 1000, 1200 μa.

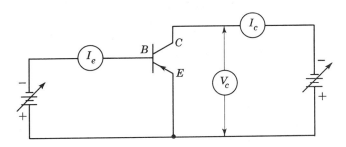

FIGURE 45.5 Common-emitter connection.

 Plot a graph showing I_c versus V_c. From the graph determine the ratio $\Delta I_c/\Delta I_b$, at a fixed value of V_c. The current amplification factor for the common-emitter circuit is labeled β.
 In the above steps α and β were determined from data obtained by connecting the transistor in two different ways. They are, however, related and that relationship may be expressed as $\beta = \alpha/(1 - \alpha)$. Your experimental results should have shown $\alpha < 1$ and $\beta \gg 1$.

[1]See Appendix II, Fig. I.9 for typical family of curves.

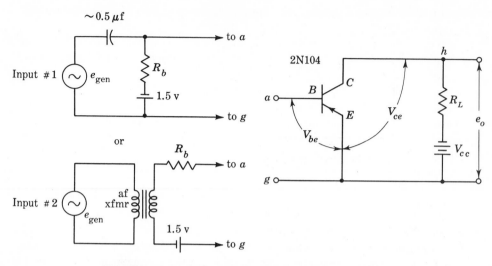

FIGURE 45.6   *R-C* coupled common-emitter amplifier.

Part II. Simple transistor amplifier. In Part I it was observed that β was much greater than unity. Thus in the common-emitter circuit (Fig. 45.6) we would expect $(e_o/e_i) \gg 1$ because of the current gain in the transistor. From the collector characteristics family of curves one can easily approximate the operating point of a transistor. Briefly, the supply voltage and the resistive load are chosen to permit a uniform excursion of the signal on both sides of the operating point. Referring to Appendix II Fig. I.9, and 2N104 collector characteristics, a supply voltage (V_{cc}) of 10 v and quiescent values of $I_b = 0.65$ ma and $I_c = 24$ ma might represent a reasonable choice. The load line then is drawn through V_{cc} at $I_c = 0$ and the operating point. From the voltage drop $I_c R_L$ one can determine the load resistor. In the figure $I_c R_L = 4.4$ v,

Typical values	
2N104	2N1303
$V_{bb} = 1.5$ v	$V_{bb} = 1.5$ v
$V_{cc} = 10$ v	$V_{cc} = 10$ v
$I_b = .65$ ma	$I_b = .20$ ma
$I_c = 24$ ma	$I_c = 10$ ma
$R_b = 2000$ ohms	$R_b = 2400$ ohms
$R_c = 400$ ohms	$R_c = 400$ ohms

hence $R_L = 4.4$ v/$(24 \times 10^{-3}$ amp$) \cong 200$ ohms. Similarly, because I_b was selected from the family of curves and a dry cell was chosen for the supply voltage V_{bb}, $R = 1.5$ v/$(650 \times 10^{-6}$ amp$)$ or $R_b \cong 2000$ ohms.

Step 1. Phase reversal. Connect the vertical input of a scope to point *a* (Fig. 45.6), the horizontal input to point *h*, and the ground to point *g*. From the Lissajous pattern on the screen of the scope determine the phase difference between the output and the input signals.

Before the output and the input signals are applied to the scope determine the change in phase that is inherent with the scope. This is done by applying the same signal to both inputs of the scope. The pattern, if the angle is 0° or 180°, should be a straight line which cuts the 1st and 3rd quadrants or the 2nd and 4th quadrants.

Because the phase difference angle is 0° for a common signal to both vertical and horizontal inputs of the scope, the pattern should be a straight line in either set of quadrants. If a phase reversal does take place in the transistor the straight-line pattern will shift to the opposite set of quadrants.

The amount of angular shift can be determined by following the directions in Exp. 30.

Step 2. Frequency response. Vary the frequency of the input signal e_i and, with a *VTVM*, measure the input and output voltages for the following frequencies: 100, 300, 500, 1000, 3000, 5000, 10K, 15K, 20K cycles/sec. (Suggestion: Hold e_i at a pre-determined value for each frequency.) On semilog paper plot gain versus frequency.

Diode; Power Supply

Object: To study the rectification property of a diode; and to study the filtering action of a power supply.

Apparatus: Vacuum-tube power-supply board, transistor power-supply board, multimeter, oscilloscope, and VTVM.

Theory: In the vacuum-tube diode electrons are copiously emitted from the heated filament or cathode. The source of the electron is the cathode, and if a second electrode within the tube is made positive, the electron flow is from the cathode to the positive electrode (plate). If the polarity is reversed the electrical current becomes zero since the plate is not heated. For this reason the diode acts as a valve and the output current is unidirectional.

In Fig. 46.1 the double diode is housed in a single envelope (such as the 5AS4, 5U4, 6X4, 6202, and many others). If one assumes an instantaneous polarity as shown in the figure, the *conventional current* is shown by the arrows. That is, there is a current in diode #1 circuit but not in the other. The reason for this is that electrons flow from the cathode to the plate only when the plate is positive relative to the cathode. However, during alternation II plate P_2 is positive relative to K, and then there is a current in the lower section of the transformer which is connected to diode #2. You will note that the direction of the current in R_L is the same for rectification by both diodes. Because both alternations of the a-c input voltage are utilized, the resultant is full-wave rectification.

Figure 46.2 shows a corresponding circuit for the transistor diode. This diode (see Appendix II, Note I) allows a high current in one direction but a negligible reverse current. For this reason the transistor diode performs the function of rectification.

The rectified output in each of the above cases is unidirectional but very "rough."

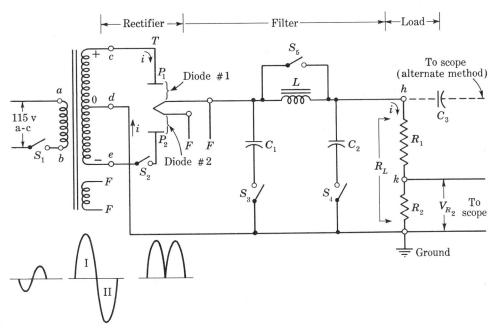

FIGURE 46.1 Schematic of vacuum-tube rectifier and filter.

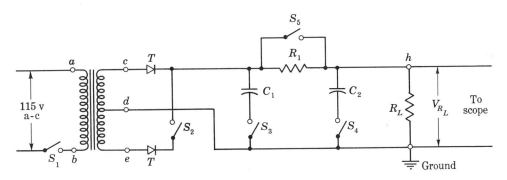

FIGURE 46.2 Schematic of transistorized d-c power supply.

For many applications one needs a smooth d-c output which may be achieved by adding a filter. By proper choice of the number and size of the filter elements the fluctuation (ripple) in the output voltage may be reduced to relatively small values. Usually the amount of ripple in the output of a power supply is useful information and is defined as

$$\text{Ripple factor} = \frac{E_r}{\overline{E_{dc}}}$$

where E_r is the rms value of the ripple voltage and $\overline{E_{dc}}$ is the average d-c output voltage.

Method: *Part I. Vacuum tube rectification and filtering.*

1. Make sure that switches (a) S_2, S_5 are closed, and (b) S_1, S_3, S_4 are open.

2. Energize the circuit by closing switch S_1.

3. With an a-c voltmeter measure V_{ab}, V_{cd}, V_{ce}, and V_{FF}. The secondary voltage may be several hundred volts, so be careful to avoid shock. Also, remember that the meter reading is the rms value.

4. *Rectification.* Connect the output (V_{R_2}) to an oscilloscope. Compare the patterns when S_2 is (a) open and (b) closed.

5. *Filtering.* With the output of the power supply applied to the scope, close switch S_3 and note the pattern on the scope. Do the same by adding elements L and C_2 in that order. Maximum filtering will be achieved when S_3 and S_4 are closed and S_5 is opened. Adjust the gain of the scope to obtain a suitable pattern.

6. Measure the ripple voltages with a calibrated scope or VTVM.

7. Determine the ripple factor.

Part II. Transistor power supply. Repeat the steps in Part I; i.e.,

1. Observe half- and full-wave rectification.

2. Note the filtering action as elements are added to the filtering section.

3. Determine the ripple factor.

Thermionic Emission:
Richardson's Equation

Object: To study the relation between the saturation current and the filament temperature of a diode vacuum tube—Richardson's equation. To determine the work function of tungsten.

Apparatus: Mounted Ferranti[1] guard ring diode (GRD7), filament d-c ammeter (0–2.5 amp), plate d-c voltmeter (0–150 v), plate d-c milliammeter (0–15 ma), adjustable filament and plate power supplies (0–6.3 v) and (0–150 v), traveling microscope, rheostats, and switches.

The GRD7 tube consists essentially of a straight tungsten filament (cathode) lying along the axis of three cylindrical coaxial *nonmagnetic* electrodes (anode and guard rings). The guard rings prevent "fringing" of the electric field between anode and cathode. The nonmagnetic materials used in the construction of the tube permit it to be used in magnetron experiments. A peephole in the anode permits direct observation of the center of the filament for temperature determination. See Appendix II, Note M_2 for a complete description of this tube.

Theory: A metal is characterized by the existence in its structure of a large number of free electrons (about one per atom) that are not permanently attached to the atoms, but are virtually free to move about in the metal. A potential barrier at the metal surface tends to prevent these free electrons from escaping at normal temperatures. However, when the metal is heated to a sufficiently high temperature, some of these free electrons acquire sufficient additional energy to carry them over the potential barrier. Thus at high temperatures the metal emits an appreciable number of

---

[1]The Ferranti GRD7 tube and accessories may be obtained from Macalaster Scientific Corporation, Route 111 and Everette Turnpike, Nashua, New Hampshire, 03060. This tube supplants the G.E. FP400 tube, no longer available.

these electrons, which may be drawn away from the heated metal by a suitable electric field.

In the case of the ordinary diode this field is produced by impressing a suitable potential difference between the filament (heated metal) and the plate. If the plate is held at a large enough positive potential with respect to the filament so that all of the electrons emitted by the filament are drawn to the plate, then the plate current is controlled primarily by the filament temperature and is practically independent of the plate potential. Under these conditions, the plate current is said to be saturated and is purely a function of the temperature of the filament and its surface area.

The relation between the saturated plate current i_b and the absolute temperature T of the filament is given by the equation

$$i_b = AT^2 e^{-w_0/kT} \qquad \text{(Richardson's equation)} \qquad (47.1)$$

where A is a constant directly proportional to the surface area of the filament, k is Boltzmann's constant, and w_0 is the work function of the metal filament. The work function w_0 is the minimum energy required to transfer an electron from an interior to an exterior point of the filament. It is possible to verify Richardson's equation and, at the same time, determine w_0, by observing the saturation currents i_b for different filament temperatures T.

The theoretical derivation of Richardson's equation is a difficult matter. It may be deduced either by use of the laws of thermodynamics or by use of Fermi-Dirac statistics applied to electrons within a metal. A rigorous development on either of these bases lies outside the province of this discussion. However, a suggestive thermodynamical argument for the equation follows.

The emission of electrons by a hot metal corresponds in many respects to the sublimation of a solid. In both cases particles (electrons or molecules) are transferred from the solid to the vapor state until, under the proper conditions, equilibrium between the solid and vapor states is realized. When this occurs, the relation between the vapor pressure P and the absolute temperature T of the system is given by Clapeyron's equation

$$\frac{dP}{dT} = \frac{L}{T\Delta V} \qquad (47.2)$$

Here, L is the molar latent heat of sublimation of the solid, and ΔV is the difference between the molar volume of the vapor and that of the solid. If it is assumed that ΔV is essentially the volume of the vapor because of the low density of the vapor as compared to that of the solid, and also that the vapor approximates an ideal gas, then

$$\Delta V = \frac{RT}{P} \qquad (47.3)$$

The molar latent heat of sublimation L is the amount of energy required to transfer 1 mol of particles from the solid to the vapor state. L may be regarded as the sum of three energies: the energy L_0 required by 1 mol of the particles to surmount the potential barrier at the surface of the solid; the kinetic energy $\frac{3}{2}RT$ of 1 mol of the particles in the vapor state; and the work RT done by 1 mol of the particles against the prevailing vapor pressure. Thus L may be written

$$L = L_0 + \tfrac{5}{2}RT \qquad (47.4)$$

In the case of ordinary sublimation L_0 is a function of T, since the specific heat of a solid is a function of T. However, in the case of electron sublimation it may be shown that L_0 is practically independent of T up to extremely high temperatures, as evidenced by the fact that the free electrons in a metal do not contribute appreciably to its specific heat. The behavior of free electrons in a metal is described by Fermi-Dirac statistics, which give these electrons an enormous energy even at absolute zero, but, which prevent most of them from absorbing any additional thermal energy. Therefore, the free electrons *inside* a metal do not act at all like an ideal gas.

If we substitute the values ΔV[Eq. (47.3)] and L[Eq. (47.4)] in Eq. (47.2) and then integrate this equation, we get

$$P = P_0 T^{5/2} e^{-L_0/RT} \tag{47.5}$$

where P_0 is an undetermined constant of integration. Equation (47.5) gives the vapor pressure P of the electron gas in equilibrium with the metal at any temperature T.

Since this electron gas is in equilibrium with the metal, the number of electrons leaving the metal during any time must equal the number of electrons returning to the metal from the vapor in this same time. This latter quantity may be calculated from elementary kinetic theory and turns out to be proportional to $P/T^{1/2}$. If the electron vapor is not allowed to accumulate in the neighborhood of the metal, but is pulled away by an electric field, this in no way affects the emission of electrons from the metal; i.e., the same number per unit time are emitted. The saturated plate current i_b is proportional to this number. Thus

$$i_b \propto \frac{P}{T^{1/2}} \tag{47.6}$$

The value of P is given by Eq. (47.5). That value, when substituted in (47.6), gives

$$i_b \propto T^2 e^{-L_0/RT} \tag{47.7}$$

Relation (47.7) is essentially equivalent to Richardson's Eq. (47.1). It is only necessary to introduce the proportionality constant A and to set $L_0 = N_0 w_0$ and $R = N_0 k$, where N_0 is Avogadro's number.

Thermodynamics alone can determine neither the integration constant P_0 in Eq. (47.5), nor the proportionality constant A in Eq. (47.1). Their evaluations require a use of statistical mechanics and kinetic theory.

Temperature of Filament: It is necessary to determine the absolute temperature of the emitter (tungsten filament) in order to verify Richardson's equation. This is a difficult task, even though several different methods are available, because the temperature of the filament is not uniform throughout its entire length.

An optical pyrometer or a radiation pyrometer may be used to determine the mid-point filament temperature. Both of these instruments operate on the basis of radiation principles. The first instrument matches a band of radiation from the hot body against a similar band from a calibrated lamp filament. The second instrument measures the total radiation from the hot body and then "deduces" the temperature. Both of these instruments are difficult to use in this experiment because of the small size of the "filament viewing hole" in the anode of the GRD7 tube.

Another method that is frequently used in determining the temperature of a pure tungsten filament is the resistance method. The ratio of the "hot" to the "cold" resistance of the filament may be correlated with the "hot" temperature of the filament. This method gives an average temperature of the hot filament, which may be much lower than its maximum temperature. Because of the exponential form of Richardson's equation, however, it is better to use the maximum temperature of the filament (at its mid-point), rather than its average temperature.

A third method—the one used in this experiment—determines the mid-point temperature of the filament by utilizing a relation between this temperature, the filament current, and the diameter of the heated filament.

Consider a small mid-point section δl of the filament, as shown in Fig. 47.1. Let its absolute temperature be T, its diameter be d, and its resistance be δR (ohms). In the steady state we may assume that the total electrical energy per unit time fed into this section is converted into heat and then radiated out through the exterior surface of the section. The electrical energy input per unit time is $i_f^2 \, \delta R \times 10^7$ ergs/sec, whereas the total energy radiated per unit time is, by the Stefan-Boltzmann law, $\epsilon(T)\sigma T^4 \pi d \, \delta l$ ergs/sec, where $\epsilon(T)$ is the total emissivity of the tungsten at temperature T and σ is the Stefan-Boltzmann constant. Thus

$$i_f^2 \, \delta R \times 10^7 = \epsilon(T)\sigma T^4 \pi d \, \delta l \qquad (47.8)$$

In this equation δR may be replaced by $\rho(T)4\delta l/\pi d^2$, where $\rho(T)$ is the resistivity of tungsten at temperature T. Making this substitution for δR in Eq. (47.8) and rearranging terms, we get

$$\frac{i_f^2}{d^3} = \frac{\pi^2}{4 \times 10^7} \sigma \frac{\epsilon(T)}{\rho(T)} T^4 \qquad (47.9)$$

The right side of Eq. (47.9) is a function of T alone and, in the case of tungsten, a known function. Thus it is possible to correlate T with the value of $i_f/d^{3/2}$ over a wide range of temperatures by means of a table of values.

Errors: The important errors in this experiment mostly arise from the difficulty of determining the effective temperature of the filament. As has been pointed out before, the filament temperature is not uniform. The filament is much cooler at the ends that in the middle. It is better to use the mid-point temperature rather than the average temperature in this experiment. The reason for this has already been given.

One may estimate the effect of temperature errors on the value of the work function w_0 in the following manner. Equation (47.1) involves two unknowns, A and w_0. If these are to be determined, then at least two sets of values of i_b and T must be measured. Let these be i_{b1}, T_1 and i_{b2}, T_2, and let Eq. (47.1) be written for each of these sets. By eliminating A between these two equations, we get

$$\frac{i_{b1}}{i_{b2}} = \left(\frac{T_1}{T_2}\right)^2 e^{(-w_0/k)(1/T_1 - 1/T_2)} \qquad (47.10)$$

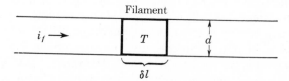

FIGURE 47.1   Section of filament.

Solving Eq. (47.10) for w_0, we get

$$w_0 = k \frac{T_1 T_2}{T_2 - T_1} \ln\left[\frac{i_{b2}}{i_{b1}}\left(\frac{T_1}{T_2}\right)^2\right] \tag{47.11}$$

The approximate determinate-error equation corresponding to Eq. (47.11) is

$$\frac{\Delta w_0}{w_0} = \frac{\Delta T_1}{T_1} + \frac{\Delta T_2}{T_2} - \frac{\Delta(T_2 - T_1)}{T_2 - T_1} \tag{47.12}$$

In arriving at this equation we have omitted terms involving the natural logarithm, since these are generally negligible. Why?

Equation (47.12) clearly indicates the effect of temperature errors on the value of w_0. If the temperature errors are determinate, then the first two terms of the right member of Eq. (47.12) become important; but if the errors are indeterminate, then the third term becomes predominant. The fact that errors in i_b do not appear in Eq. (47.12) simply indicates that these errors do not contribute much to the error in w_0. Hence it is much more important to measure T accurately, than to measure i_b accurately.

Method: The circuit is shown in Fig. 47.2. The *filament current* for the tube is furnished by an *adjustable* power supply (may be storage battery plus rheostats). The ammeter in the filament circuit gives the filament current i_f. This *current should never exceed 2.25 amp* in order to prevent damage to the tungsten filament. The ammeter must be of good quality with a maximum error of 1% or less for reliable temperature determinations of the filament.

The plate voltage V_b is given by the voltmeter and the plate current i_b by the milliammeter. Note that the current in the guard rings bypasses the milliammeter.

Adjust the filament power supply for *low* filament current and the plate power supply for *low* plate voltage before closing switches S_f and S_b. Note the readings of the meters. Carefully increase the filament current to 1.90 amp. Then increase the plate potential until the plate current is saturated; i.e., it no longer increases appreciably with increasing plate potential. Record the values of the *saturated* plate current i_b and the plate voltage V_b for $i_f = 1.90$ amp. Repeat this procedure for filament

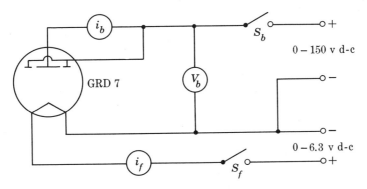

FIGURE 47.2 Circuit diagram; GRD 7 tube.

currents of 2.00, 2.10, and 2.20 amp. Do not allow the filament current to exceed 2.25 amp in the process.

In order to determine the temperature of the center of the filament at the various filament currents, it is necessary to measure the diameter of the hot filament. It is sufficient to do this at only one of the filament currents, e.g., $i_f = 1.90$ amp, since the diameter does not change appreciably for the other filament currents. Set the filament current back to 1.90 amp and take five measurements of the filament diameter (3 significant figures) with the traveling microscope. The hot filament is observed through the peephole in the anode. It may be necessary to use a red glass filter for viewing the bright filament through the microscope in order to see clearly the edges of the filament.

Computations: Compute the absolute temperature T of the mid-point of the filament for each of the four filament currents by use of Table R, Appendix III. Use the average filament diameter for each value of the filament current in making the computations. Interpolate in Table R to get the temperature to the nearest 10°K. Calculate the error in each of these temperature values owing to the estimated errors in i_f and d. Also calculate the errors in the values of $1/T$.

If Richardson's equation is valid, the graph of $\log_{10}(T^2/i_b)$ plotted against $1/T$ should give a straight line whose slope is $w_0/2.30k$. Verify this statement by use of Eq. (47.1). Make such a graph, using your values of i_b and T. Indicate the error intervals in $1/T$ on this graph by drawing a small horizontal line segment of length $2\Delta(1/T)$ through each plotted point. Draw the best possible straight line for these plotted points. This line should cut all of the small line segments representing the error intervals in $1/T$. Determine the slope of this line and then, compute the work function w_0 of tungsten in ergs and in electron volts. The value of k is given in Table L, Appendix III. Compare your value of w_0 with that given in Table N, Appendix III, for tungsten.

Determine the error in w_0, using Eq. (47.12). Use your lowest and highest filament temperatures for this calculation.

Record: List apparatus and apparatus numbers. Tabulate i_f, d, i_b, $i_f/d^{3/2}$, T, ΔT, $1/T$, $\Delta(1/T)$, and $\log_{10}(T^2/i_b)$. Give your value of w_0, its error Δw_0, and the accepted value of w_0.

OPTIONAL EXPERIMENT

1. Instead of using the method outlined in Exp. 47 for determining the temperature of the filament, use an optical pyrometer.

Measurement of e/m:
Magnetron Method

Object: To determine the value of *e/m* for electrons by the magnetron method.

Apparatus: Mounted Ferranti GRD7 tube, d-c ammeter (0 to 3 amp), two-range d-c plate voltmeter (0 to 30 v and 0 to 150 v), milliameters (0 to 5 ma and 0 to 15 ma), adjustable filament and plate power supplies, rheostats and switches.

In order to use the tube as a magnetron, it must be placed in a suitable magnetic field. This field is produced by a current carrying solenoid with a hollow core large enough to accommodate the GRD7 tube placed lengthwise inside the solenoid. The anode of the tube should be at the center of the solenoid, with the filament of the tube in alignment with the axis of the solenoid. A very convenient solenoid for this purpose (6 in., 440 turns #90) may be obtained from Macalaster Scientific Corporation (see footnote 1, Exp. 47).

Theory: If a diode consisting of a hot filament lying along the axis of a cylindrical plate is placed in a magnetic field so that the filament is parallel to the field, it is found that the plate current of the diode is a function of the strength of the magnetic field as well as of the plate potential and filament temperature. As the field is increased from zero, the plate current remains practically constant until a critical point is reached beyond which the plate current rapidly diminishes to zero.

The reason for this action follows. The electrons emitted by the filament are acted upon by crossed electric and magnetic fields that cause the electrons to move in curved paths closely approximating circular orbits. These orbits, starting at the filament, lie in planes perpendicular to the direction of the magnetic field and intersect the plate for small magnetic field strengths. Thus all electrons leaving the filament reach the plate under these conditions. As the strength of the magnetic field is increased, however, the radius of curvature of the electron orbits decreases until the orbits just barely reach the plate, i.e., are tangent to it. Any further increase in the magnetic field produces orbits that fail to reach the plate. In this latter case, electrons leaving the filament fail to reach the plate and return to the filament. This failure, of course, reduces the plate current to zero. Figure 48.1(a) shows the crossed electric and magnetic fields in the diode. Figure 48.1(b) illustrates the change in a typical electron orbit as the magnetic field is increased. Figure 48.1(c) shows the corresponding change in the plate current of the tube.

In actual practice it is found that the plate current does not drop off to zero as sharply as Fig. 48.1(c) indicates. This fact is probably due in part to a lack of alignment between filament and field and between filament and plate, and in part to the drop in potential along the filament and to the nonuniformity of the magnetic field.

The critical value of the magnetic field B_c for which the plate current drops to zero may be derived in the following manner. An electron leaving the filament with zero kinetic energy acquires kinetic energy because of the electric field between filament and plate. Because of the configuration of this field, almost the entire potential

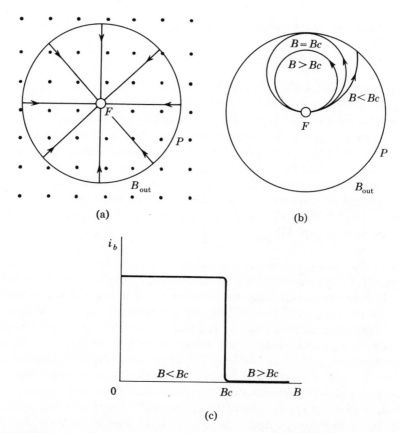

FIGURE 48.1   Magnetron tube; (a) crossed electric and magnetic fields, (b) electron orbits, and (c) plate current variation with applied magnetic field.

difference between filament and plate occurs very close to the filament. Therefore, the electron acquires practically its entire kinetic energy before it has moved very far from the filament. Its kinetic energy $\frac{1}{2}mv^2$ is given by the equation

$$\tfrac{1}{2}mv^2 = eE_b \tag{48.1}$$

where E_b is the plate potential (filament potential assumed to be zero). The magnetic field *does no work* on the electron and therefore cannot contribute to its kinetic energy.

The magnetic field, however, does exert a centripetal force on the electron, causing it to move in a circular orbit (we neglect the effect of the electric field except near the filament). This orbit has a radius of curvature R given by the well-known equation

$$Bev = m\frac{v^2}{R} \tag{48.2}$$

provided the field B is perpendicular to v, as it is in this case. If this electron just fails to reach the plate, then it is clear that $R = b/2$, where b is the radius of the plate. The electron orbit is a circle similar to that shown in Fig. 48.1(b) for $B = B_c$. Thus

$$B_c ev = m\frac{v^2}{b/2} \tag{48.3}$$

If v is eliminated between Eqs. (48.1) and (48.2), and the resulting equation is solved for e/m, we get

$$\frac{e}{m} = \frac{8E_b}{B_c^2 b^2} \tag{48.4}$$

Equations (48.1) through (48.4) are valid for any consistent set of electrical units, e.g., mks or emu. When E_b is in volts, B_c in gauss, b in cm and e/m in coulombs/gm, Eq. (48.4) becomes

$$\frac{e}{m} = \frac{8E_b}{B_c^2 b^2} \times 10^9 \tag{48.5}$$

Eq. (48.4) has been derived by assuming that the electric field and the magnetic field act consecutively instead of simultaneously upon an electron coming from the filament. A more detailed analysis in which this assumption is not made leads to the same result, provided the radius of the filament is negligible compared to that of the plate.

In using Eq. (48.4) to obtain the value of e/m, two alternative procedures are available. In the one just described, we set the plate potential E_b at some arbitrary positive value and then increase the strength of the magnetic field to the cut-off point of the plate current i_b, where $B = B_c$. But we could just as well have fixed the value of B_c and then increased the plate potential E_b' from zero to a threshold where the plate current i_b suddenly jumps from zero to its maximum value. In this latter case i_b is a function of E_b' for fixed B_c instead of the reverse and is represented by a curve that is just the reverse of Fig. 48.1(c); i.e., we get a step-up rather than a step-down curve. For reasons already given, in neither case is the break in the curve as sharp as we would like. The best that we can do, therefore, is to use the mid-point value of i_b to determine the critical values of B_c or E_b'.

In this experiment we shall use the second procedure rather than the first for reasons of convenience and accuracy. It is easier to vary the plate voltage E_b' than to vary the magnetic field strength B, since the former involves small plate currents while the latter involves large solenoid currents of several amperes. Also it is clear from Eq. (48.5) that the threshhold value of E'_b need not be determined as accurately as the solenoid current that determines B_c, since the equation involves the first power of E_b but the second power of B_c.

Some means must be used to determine the magnetic field produced at the center of the solenoid by a given magnetizing current I in the solenoid. This field may be calculated by means of the equation

$$H = \frac{4\pi NI}{10}\left(1 - \frac{1}{2}\frac{D^2}{L^2}\right) = KI, \qquad D \ll L \tag{48.6}$$

where H = strength of field (center of solenoid) in oersteds,
$\quad N$ = total number of turns,
$\quad L$ = length of solenoid in centimeters,
$\quad I$ = solenoid current in amperes, and
$\quad D$ = average diameter of solenoid in centimeters.

Equation (48.6) is only valid provided there are no magnetic materials in or near the solenoid. With this proviso, we can equate the value of H in oersteds from Eq. (48.6) to B_c in gauss in Eq. (48.5).

We now modify Eq. (48.5) to bring it into line with the measured electrical quantities. We can replace B_c by KI_c, where I_c is the current in the solenoid in amperes. Also E_b, the true plate voltage, should be written as $V_b + \delta$, where V_b is the threshold value of the plate voltage V'_p as given by voltmeter readings and δ is a small unknown residual voltage due to contact potential and other small effects. Thus Eq. (48.5) can be written in the form

$$\frac{e}{m} = A\frac{(V_b + \delta)}{I_c^2} \tag{48.6}$$

where the constant $A[= (8 \times 10^9)/(K^2b^2)]$ is computed and where V_b is in volts, I_c is in amperes, and e/m is in coulombs/gm.

If we now consider I_c^2 as the independent variable and the corresponding threshold value V_b as the dependent variable, we can plot V_b (ordinate) versus I_c^2 (abscissa) for several different values of I_c. By Eq. (48.6) these points should lie on a straight line with a scale slope of $e/(mA)$. From this scale slope we can determine the value of e/m, since A is a known constant.

Method: Wiring connections are shown in Fig. 47.2 and Fig. 48.2.

Place the solenoid over the mounted GRD7 tube so that the center of the anode is at the center of the solenoid and the filament is in alignment with the axis of the solenoid. This is automatically done if the Macalaster solenoid is used. Operate the tube at a constant filament current of about 2.2 amp; do not exceed 2.25 amp. Start with zero plate potential. See to it that there are *no magnetic materials in* or *near* the *solenoid*. Set the *current* in the *solenoid* at 2.20 amp as nearly as you can. This current should be kept constant and its value determined as accurately as is possible. Be sure to eliminate any zero error in the ammeter. Now slowly increase the plate potential V'_b observing, at the same time, the plate current i_b registered by the milliam-

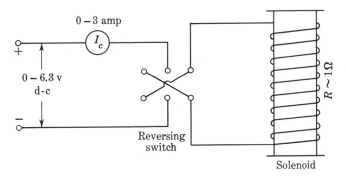

FIGURE 48.2   Circuit for solenoid.

meter. The value of i_b will be negligible until you are near the threshold value V_b, then rise rapidly to a maximum as you pass through the threshold, and finally remain practically constant for any further increase in the plate potential. Observe and record a set of V_b' versus i_b values in the vicinity of the threshold in steps of 1 v starting when i_b just begins to rise and stopping when the value of i_b reaches a plateau.

Repeat the above process after reversing the solenoid current by means of the reversing switch RSw. There should be no appreciable change in the ammeter reading in this process. This reversal tends to offset the effect of the earth's magnetic field (vertical component).

Repeat the foregoing procedures for solenoid currents of 2.50 amp and 2.80 amp.

Computations: Average i_b values (direct and reversed solenoid current) and plot the averaged i_b values versus V_b' values for each of the three solenoid currents. In each case determine the threshold value V_b (half-way up the step).

You now have three pairs of corresponding values of V_b and I_c. Plot V_b (ordinate) versus I_c^2 (abscissa) for these three pairs of values. They should lie on a straight line. Draw the best straight line through the plotted points and determine its scale slope. Finally determine the value of e/m by using this scale slope in conjunction with the computed value of K, Eq. (48.6), and the value of b (cm) for the GRD7 tube (Appendix II, Note M_2).

Record: List the pertinent values for the GRD7 tube and the solenoid. Tabulate i_b (average) versus V_b' for each of the three values of I_c; also tabulate V_b versus I_c^2. Record filament current, solenoid constant K, tube constant b, and constant A. Give your value of e/m with its estimated error.

Measurement of e/m:
Deflection Method

Object: To determine the velocity of electrons in an electron beam by the magnetic deflection method; and to determine the ratio of the charge to the mass of an electron.

Apparatus: Welch apparatus consisting of a Bainbridge type of electron tube and a pair of Helmholtz coils (or similar Cenco or Leybold apparatus). Necessary auxiliary electrical equipment.

Theory: In this experiment electrons emitted by a hot filament are accelerated by an electric field and then deflected by a magnetic field into a circular orbit. By measuring the accelerating potential of the electric field, the intensity of the magnetic field, and the radius of the orbit, it is possible to compute both the velocity of the electrons and the ratio of charge to mass for the electron.

The entire action of the electrons takes place within a large evacuated glass bulb (the electron tube) containing a small amount of mercury vapor. The essential components of this tube are shown in cross section in Fig. 49.1. The source of the electrons is a heated tungsten filament F that lies along the axis of a cylindrical carbon anode with a narrow slit in it at S. When a potential difference (about 50 v) is applied between F and C, electrons from F are accelerated toward C. Some of these emerge from the slit S and enter the main body of the tube in the form of a narrow, ribbon-like electron beam. Ionization of the mercury vapor in the tube by the electrons makes the beam visible.

If there is no transverse magnetic field across the beam, the electrons in the beam move along the straight line SA, striking the tube and being partially reflected. If a transverse magnetic field produced by a current in the Helmholtz coils is applied to the beam, the beam is deflected by this field into a circular arc. As the intensity of the

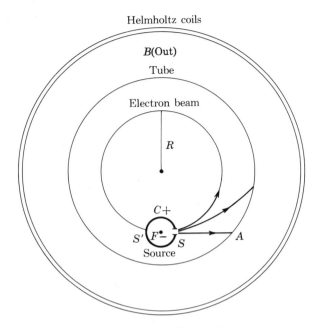

FIGURE 49.1 Cross section of Bainbridge electron tube.

magnetic field is increased, the radius of the circular arc decreases until the beam is bent into a complete circle of some radius R whose value may be determined.

The magnetic field in the tube is produced and controlled by the current in the Helmholtz coils. These coils, properly oriented with respect to the tube position, provide a fairly uniform magnetic field throughout the tube. Usually, the earth's magnetic field must also be taken into account in setting up the apparatus.

The fundamental working equations in this experiment are as follows:

The velocity of the electrons in the beam is given by the equation

$$eV \times 10^7 = \tfrac{1}{2}mv^2 \qquad (49.1)$$

where e = electronic charge in coulombs,
V = potential difference between F and C in volts,
m = mass of electron in grams,
v = velocity of the electrons in cm/sec.

This equation developed in most general physics textbooks assumes (1) that the thermal kinetic energy of the electrons coming from the hot filament is negligible compared to the kinetic energy given the electrons by the electric field between F and C, and (2) that the final velocity v of the electrons is small compared to the velocity of light, i.e., no relativity corrections are necessary. Both of these assumptions are valid in this experiment.

The deflection of the electron beam in the magnetic field is given by the equation

$$\frac{Bev}{10} = m\frac{v^2}{R} \qquad (49.2)$$

where B is the flux density of the magnetic field in gauss, R is the radius of the circular beam in cm. The other symbols have the values assigned in Eq. (49.1). This equation is also developed in most general physics textbooks.

Finally, the strength of the magnetic field produced by the Helmholtz coils at a point midway between their centers is given by the equation

$$H = \frac{32\pi N}{a\sqrt{125}} \frac{I}{10} \tag{49.3}$$

where H = strength of the magnetic field in oersteds,
  N = number of turns of wire on each coil,
  a = mean radius of each coil in cm,
  I = current in each turn of the coils in amperes.

This equation is not usually developed in general physics textbooks, but see problem 27-10 in *Physics*, by Furry, Purcell, and Street[1] for a suggested proof. Equation (49.3) gives the intensity of the magnetic field at the midpoint of the centers of the two coils. If the coils are sufficiently large, however, the region about the midpoint has about the same intensity.

The student should note that Eq. (49.2) involves B, whereas Eq. (49.3) involves H. In the electromagnetic system of units (the system we are virtually using in these equations) $B = H$, provided that the medium is nonmagnetic ($\mu = 1$). The factors 10^7 in Eq. (49.1) and $1/10$ in Eqs. (49.2) and (49.3) are simply conversion factors. They would not appear if all electrical quantities were expressed in electromagnetic units, e.g., abcoulombs, abvolts, and abamperes. Equations (49.1), (49.2), and (49.3) may also be written in the mks system of units. This alternative is left as an optional procedure.

Eliminating e and m from Eqs. (49.1) and (49.2), the value of v may be determined for any given electron beam. Eliminating v from Eqs. (49.1) and (49.2), the value of e/m for the electron may be determined. This value of e/m should be the same for all electron beams used in the experiment.

Method and Results: For details concerning the procedures in this experiment, the student should consult the operations manual published by the manufacturer of the equipment. Several different types of equipment are available for this experiment. All involve the same principles, but differ as to detail. We have described the Welch apparatus. Somewhat similar apparatus is made by the Central Scientific Company and by the Leybold Company of Germany.

Determine the velocity of the electrons in several different electron beams of varying R. In each of these cases determine the value of e/m for the electron. Make reasonable estimates of the errors involved. Compare your value of e/m with the accepted value.

[1]W. H. Furry, E. M. Purcell, and J. C. Street, *Physics* (New York: McGraw-Hill Book Company, 1952), p. 635.

PART IV

WAVE MOTION
AND SOUND

Vibrating
Wire

Object: To generate stationary transverse waves in a wire. To determine the frequency of vibration of this wire in terms of the tension and mass per unit length of the wire and the wave length.

Apparatus: Music wire (0.016 gauge), clamps, pulley, weight holder and weights, analytical balance, 100 cm specimen of the wire, driving electromagnet, stroboscope, and meter stick. See Appendix II, Note K on the analytical balance.

Theory: The velocity v in centimeters per second with which a transverse wave pulse travels along a flexible wire is given by the equation

$$v = \sqrt{\frac{F}{m}} \qquad (60.1)$$

where F is the tension in the wire in dynes, and m is its mass per unit length in grams per centimeter.

A train of running waves may be generated in the wire by driving some small portion of it back and forth at right angles to the direction of the wire. This action may be accomplished in the case of an iron or steel wire by driving it with a small electromagnet placed close to the wire and energized with a 60 cycles/sec alternating current. The wire in the neighborhood of the iron core of the electromagnet is magnetized by induction and pulled toward the iron core 120 times/sec. In this manner a train of waves of frequency 120/sec will be sent along the wire with a velocity given by Eq. (60.1). The wave length λ in centimeters of this wave train will be given by the well-known equation

$$v = n\lambda \qquad (60.2)$$

where n is the frequency of the wave train.

We may eliminate v between Eqs. (60.1) and (60.2) and solve for n, thus getting

$$n = \frac{1}{\lambda} \sqrt{\frac{F}{m}} \qquad (60.3)$$

This equation enables us to determine n in terms of F, m, and λ.

Although it is an easy matter to measure F and m, it is generally not feasible to measure λ in a running wave train. Under appropriate conditions, however, it is possible to set up so-called stationary or standing waves in the wire, which enable us to make a precise determination of λ.

If the wire is fixed at both ends (as it is in this experiment), then the waves generated at any point along the wire will run along the wire to the fixed ends, where they will be reflected back along the wire. The actual configuration of the wire at any instant is a combination of incident and reflected wave trains, and may be very complex. However, under the proper conditions, the incident and reflected waves may combine in such a way as to produce stationary or standing waves. *This occurs when the distance between the fixed ends of the wire is just an integral multiple of a half wave length of the running waves.* The waves then "fit" into the length of the wire, so to speak, strongly reenforcing each other at certain points along the wire called loops or antinodes, and completely annulling each other at intermediate points called nodes. Under these conditions, the wire vibrates in segments, as shown in Fig. 60.1. In this figure the wire is shown vibrating in three segments. The fixed ends of the wire must of necessity be nodes (positions of zero displacement). Nodes and loops alternate in position. It may be shown that the distance between two successive nodes (or two successive loops) is just equal to a half wave length. Therefore, when these stationary waves are produced, it is an easy matter to pick out the nodal positions and thus measure the wave length of the generated wave train.

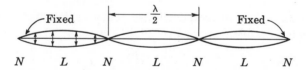

FIGURE 60.1   Standing-wave pattern in a wire.

Method: Set up the apparatus as shown in Fig. 60.2. Connect the electromagnet to the 110 v source of 60 cycle current. Place it near one end of the wire as shown, with its soft iron core a few millimeters away from the wire. Without adding any weights to the weight holder, pull down on the weight holder with your hand, gradually increasing the tension in the wire. If this is done carefully, and if the apparatus is working properly, the wire will suddenly vibrate strongly (usually in five segments) for a certain tension. If the tension is increased further, the vibration will cease until the tension reaches a second critical value at which vibrations will again occur, with the wire now vibrating in four instead of five segments. Keep increasing the tension until the wire vibrates in three segments, and then finally in two segments. It is not advisable to attempt to make the wire vibrate in a single segment because of the very large tension required.

After you have succeeded in making the wire vibrate in five, four, three, and two segments by applying force on the weight holder with your hand, repeat the same

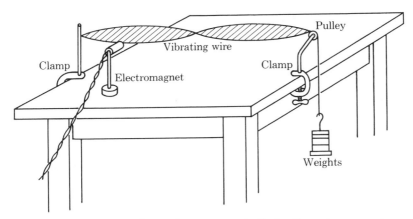

FIGURE 60.2   Experimental arrangement of vibrating wire apparatus.

performance by adding weights to the weight holder. Record the weights (weight holder included) that give the best set of stationary waves for the case of five, four, three, and two vibrating segments. In each case estimate the amount of weight that may be changed without appreciably reducing the stationary wave pattern. This change will represent the indeterminate error of the tension in the wire. In addition, examine each pattern with a stroboscope and determine the frequency.

Carefully measure the length of the wire between the clamp and the pulley to the nearest millimeter. Weigh the 100 cm sample of the wire on the analytical balance to the nearest milligram and determine the mass per unit length of this sample.

From these data, compute for the case of each stationary wave pattern (1) the wave length, (2) the velocity of the wave train, and (3) the frequency of the wave train. Also, compute the indeterminate errors in these quantities, using the estimated errors in the measured quantities. The frequency of the wave train in each case should be 120 vib/sec. The error in the frequency determination should be less than 2% in this experiment.

Record: (Sample.)
    App. No. _____
    Mass of 100 cm sample = 1.015 ± 0.001 gm
    Length of wire = 156.9 ± 0.1 cm
    Standard frequency = 120.0 vib/sec

Obs	No. of loops	Load, gm	Wave length, cm	Velocity, cm/sec	Frequency, vib/sec	Stroboscope frequency, vib/sec
1.	5	590 ± 10	62.76 ± 0.04	7550 ± 70	120 ± 1	
2.	4	...	...	...	...	
3.	3	...	...	...	...	
4.	2	...	...	...	...	
					Ave	

$$\lambda = \frac{(2)(156.9)}{5} = 62.76 \text{ cm}; \qquad \frac{100\,\Delta\lambda}{\lambda} = \frac{10}{157} = 0.064\%$$

$$\Delta\lambda = \pm 0.04 \text{ cm}$$

$$v = \sqrt{\frac{(590)(980)}{0.01015}} = 7550 \text{ cm/sec}; \qquad \frac{100\,\Delta v}{v} = \frac{1}{2}\left(\frac{1000}{590} + \frac{0.1}{1.0}\right) = 0.9\%$$

$$\Delta v = \pm 70 \text{ cm/sec}$$

$$n = \frac{7550}{62.76} = 120.3 \text{ vib/sec}; \qquad \frac{100\,\Delta n}{n} = \frac{7000}{7550} + \frac{4}{63} = 0.99\%$$

$$\Delta n = \pm 1 \text{ vib/sec}$$

Percent difference between standard frequency and average measured frequency = _____

QUESTIONS

1. Show that Eq. (60.1) is dimensionally correct.

2. Develop the error equation corresponding to Eq. (60.3).

3. If the diameter of the steel wire used in this experiment were 1% larger than the diameter of the 100 cm sample whose mass was determined, what constant percentage error would be introduced into the value of n?

4. Show by diagram, or otherwise, that the distance between two successive nodes in a stationary wave is just a half wave length of the running wave.

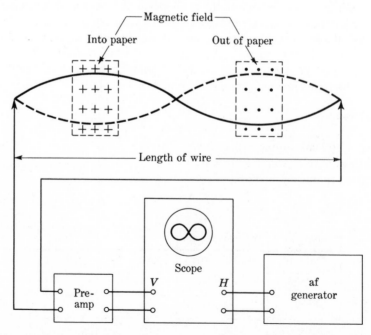

FIGURE 60.3 Schematic of apparatus using scope to determine frequencies of standing waves.

OPTIONAL EXPERIMENT

1. Instead of supplying energy to the vibrating wire for particular tensions (at a fixed frequency), one can remove energy from the vibrating string (at many different frequencies) and apply a voltage to an oscilloscope. This is accomplished by connecting the extremities of the vibrating wire to a scope. Then as the wire vibrates in a magnetic field (horseshoe magnet resting on sonometer), the induced voltage can be measured by the scope. Because the signal is small, it is necessary to use a pre-amplifier.

Fig. 60.3 shows the placement of two magnets such that the second harmonic will be detected. In general the magnet (s) is placed at the antinode of a wave pattern. In so doing the largest emf is induced due to the fact that the wire makes the greatest excursions at these points, and at the same time, the time rate of cutting the magnetic lines is the greatest.

You will note, too, that an af signal is applied to the horizontal of the scope. With a variable af generator it is then possible to produce Lissajous patterns on the scope, from which one can accurately determine the frequency of the harmonic being investigated.

Velocity of
Sound:
Resonance Tube

Object: To generate stationary sound waves in air. To determine the velocity of sound in air at room temperature from measurements of wave length for a given frequency. To compute the velocity of sound at 0°C.

Apparatus: Resonance-tube apparatus, tuning fork, rubber hammer, thermometer, and a meter stick.

Theory: The velocity of a compressional wave pulse in any medium is given by the equation

$$v = \sqrt{\frac{E}{\rho}} \tag{61.1}$$

where v = velocity of the compressional wave,
$\quad E$ = volume modulus of elasticity of the medium,
$\quad \rho$ = density of the medium.

Eq. (61.1) is derived in many general physics textbooks. See, for example, Sears and Zemansky, *College Physics*.[1] Note that the velocity v of a compressional wave pulse depends only upon the constants E and ρ of the medium. It does not depend upon the way in which the wave got started, nor does it depend upon the shape of the wave (within reasonable limits). Equation (61.1) is valid for any medium, whether it be a gas, a liquid, or a solid. However, in developing Eq. (61.1) it is assumed that the medium is confined in a tube with rigid walls; i.e., there is no lateral expansion of the medium. This condition is generally satisfied for a gas or a liquid, but not for a solid, e.g., a solid rod. In the latter case it is necessary to replace E by Y, Young's modulus for the rod.

[1]F. W. Sears and M. W. Zemansky, *College Physics*, 3rd ed. (Reading, Mass.: Addison-Wesley Publishing Co., 1960), p. 402.

In this experiment sound waves (compressional waves) are generated by a vibrating tuning fork. The medium is air at room temperature enclosed in a glass tube, sealed at its lower end by a column of water (see Fig. 61.1).

When the tuning fork is set into vibration, a train of waves consisting of alternate compressions and rarefactions in air is sent down the tube. This wave train is reflected at the water surface with a phase change of 180° and passes back up the tube. At the open end of the tube, it is again reflected, but with no phase change in this case. The resultant waves in the tube are a combination of incident and reflected wave trains and may be very complex, just as in the case of transverse waves in a wire. But just as in the case of the wire, it is possible to produce stationary waves under the proper conditions in the case of an air column. The air column will then vibrate strongly in segments, with a frequency equal to the frequency of the tuning fork. This phenomenon occurs when the length of the air column is of such value that an *odd* multiple of quarter-waves "fits" the air column, since there must be a node at the lower end of the air column (at the surface of the water) and a loop or antinode near the open end of the tube. Under these conditions, the air column resonates with the tuning fork and the intensity of sound from the system is considerably increased. This phenomenon of resonance enables us to determine when stationary waves are being produced in the air column.

The length of the air column may be varied by changing the level of the water in the tube. Thus the positions of the water surface at which resonance occurs may be determined.

It has been pointed out that an antinode must exist near the top of the tube. Analysis shows that its position is slightly above the top of the tube (about 0.6 the

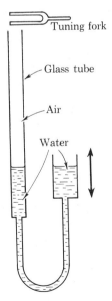

FIGURE 61.1 Air column with adjustable water level.

radius of the tube). If the water level in the tube is lowered from the top of the tube, resonance will occur when the position of the water level corresponds with the position of the first node. This position is sharply defined and may be accurately determined. As the water level is further lowered, a second resonance point may be found that corresponds to the second node. In some cases additional nodes may be found, depending upon the relation between the wave length and the tube length. The inter-nodal distance is just a half wave length. See Fig. 61.2. It is not advisable to try to determine the wave length using only the first antinode and the first node, since this distance ($\lambda/4$) cannot be accurately determined because of a lack of knowledge concerning the exact position of the first antinode. However, the distance between the successive resonance points (successive nodes) may be accurately determined and represents the value of $\lambda/2$.

Once the wave length λ has been determined by a measurement of the inter-nodal distance, we may determine the velocity of sound in air at room temperature by means of the equation

$$v = n\lambda \tag{61.2}$$

where n is the frequency of the tuning fork.

In order to compute the velocity of sound in air at 0°C it is necessary to make use of Eq. (61.1) in a modified form. If it is assumed that air obeys the general gas law and, furthermore, that the compressions and rarefactions occurring in air when a sound wave passes through it are adiabatic in character rather than isothermal, then $E = \gamma P$, where γ (1.402 for air) is the ratio of the specific heat at constant pressure to that at constant volume, and P is the pressure. The relation $E = \gamma P$ may be derived from the formula $PV^\gamma =$ constant for an adiabatic change in the volume of a gas,

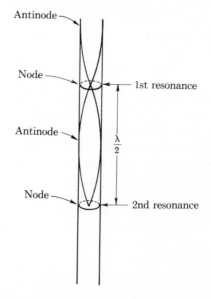

FIGURE 61.2   Standing wave pattern of an air column closed at one end.

by taking the logarithmic derivative of this latter expression. We get

$$\frac{dP}{P} + \gamma \frac{dV}{V} = 0 \qquad \text{or} \qquad -V \frac{dP}{dV} = \gamma P$$

But by definition $E = -V \, dP/dV$, therefore

$$E = \gamma P$$

Equation (61.1) for air may then be written in the form

$$v = \sqrt{\frac{\gamma P}{\rho}} \tag{61.3}$$

By the general gas law, $P/\rho = (R/\mu)T$. Thus, Eq. (61.3) becomes, after substituting for P/ρ

$$v = \sqrt{\frac{\gamma R}{\mu} T} \tag{61.4}$$

where R = universal gas constant,
μ = molecular weight of air,
T = absolute temperature.

It is clear from Eq. (61.4) that the velocity of sound in any gas is directly proportional to the square root of the *absolute* temperature. Therefore, the velocity of sound in air at any temperature may easily be computed if it is known at some one temperature, e.g., room temperature, by means of the formula

$$\frac{v_1}{v_2} = \sqrt{\frac{T_1}{T_2}} \tag{61.5}$$

Method: Raise the water level in the glass tube until it is near the top of the tube. Start the tuning fork vibrating by striking it gently with a soft rubber hammer. (PRECAUTION: Never strike a tuning fork with a hard hammer. It may ruin the fork.) Slowly lower the water level while listening for resonance to occur. The amplification of the sound at resonance is quite pronounced, even though the fork itself is emitting a scarcely audible sound. Once you have determined the approximate position of the first resonance point, make several trials by running the water surface up and down. When the point of maximum intensity is located, mark its position with one of the spring brass rings on the tube. Then lower the water surface and, in a similar manner, locate and mark the next resonance point. Continue this process to the bottom of the tube. In each case make an estimate of the error involved in locating the resonance point.

With a meter stick held against the tube, read and record the positions of the resonance points to the nearest millimeter. By subtraction, determine the distance between successive resonance points, i.e., $\lambda/2$. Estimate the error in this value of $\lambda/2$.

Determine the velocity of sound in air at room temperature by means of Eq. (61.2), using the rated frequency of the fork and the measured value of $\lambda/2$. Determine the error in this value of v, assuming that the error in the frequency of the fork is negligible.

Using Eq. (61.5), determine the velocity of sound at 0°C from your experimentally determined velocity of sound at room temperature. Also, determine the error in this velocity.

Compute the theoretical value of the velocity of sound in air at 0°C, using Eq. (61.3), and compare this value with the experimental value at 0°C. The two values should agree within the limits of the experimental error in the measured value.

Repeat this experiment, using a different set of equipment with a different tuning fork.

Record: Tabulate your data and results.

QUESTIONS

1. If it is assumed that the velocity v of a compressional wave in a medium depends only upon E and ρ for that medium, and that this relationship has the form

$$v = E^x \rho^y$$

show by dimensional argument that $x = \frac{1}{2}$ and $y = -\frac{1}{2}$.

2. Does the velocity of sound in air vary with the barometric pressure? Explain.

3. Using Eq. (61.5), show that the velocity of sound in air in meters per second at a temperature of $t\,°C$ is given approximately by the equation

$$v_t = 332 + 0.61t$$

provided that t is not large.

62

Velocity of
Sound in
Metal Rods

Object: To determine the velocity of sound in two different metal rods by use of a Kundt's tube. To compute the value of Young's modulus for these rods.

Apparatus: Metal rods, Kundt's tube, clamps, and meter stick.

Theory: Stationary longitudinal waves may be generated in a metal rod by clamping it at its mid-point and stroking it lengthwise with a cloth moistened with alcohol or sprinkled with rosin. A series of disturbances (compressions and rarefactions) pass along the rod and are reflected from the free ends of the rod. Incident and reflected waves combine to form stationary waves. The rod will then vibrate strongly, emitting a shrill note. The fundamental mode of vibration of the rod will correspond to a stationary wave train in the rod that has an antinode at each of the free ends of the rod and a node at the clamped point, i.e., the mid-point, (see Fig. 62.1). In this case the length of the rod is just equal to a half wave length of the wave train, i.e., the distance between two successive antinodes.

In order to make an experimental determination of the velocity of sound in this thin metal rod, a light metal-felt piston A is attached to one end and inserted into a glass tube, as shown in Fig. 62.2. At the other end of the tube is a fixed plug B. Along

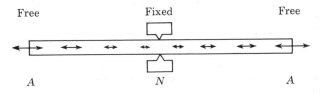

FIGURE 62.1 Fundamental mode of vibration of a metal rod fixed at its midpoint.

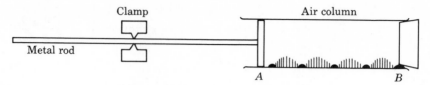

FIGURE 62.2 Kundt's tube apparatus.

the bottom of the tube is sprinkled some light powder, such as cork dust or lycopodium powder.

The metal rod is set into vibration in the manner suggested above. The glass tube is adjusted in position until the air column AB is of such length as to resonate with the note emitted by the rod. Nodes and antinodes are formed in the air column that cause the powder at the bottom of the tube to collect in lateral ridges at the points of maximum disturbance, i.e., at the antinodes. Thus it is an easy matter to determine the wave length of the sound in air. The velocity of sound in the metal rod, v_m, is given by the expression

$$v_m = n\lambda_m \tag{62.1}$$

whereas the velocity of sound in air, v_a (at the same temperature), is given by

$$v_a = n\lambda_a \tag{62.2}$$

where n = frequency of the vibrating rod and also that of the vibrating air column,
  λ_m = wave length of the sound in the metal rod,
  λ_a = wave length of the sound in the air column.

If we divide Eq. (62.1) by Eq. (62.2), we get

$$\frac{v_m}{v_a} = \frac{\lambda_m}{\lambda_a} \tag{62.3}$$

Using Eq. (62.3), the velocity of sound in the metal rod may be determined in terms of the velocity of sound in air (regarded as known) and the ratio of the two wave lengths.

It is also possible to compute the velocity of sound in a thin metal rod by means of the equation

$$v_m = \sqrt{\frac{Y}{\rho}} \tag{62.4}$$

where Y is Young's modulus for the rod, and ρ is the density of the rod. One might suppose from the analysis given in the Theory of Exp. 61 that the above expression for v_m should involve the volume modulus of elasticity E rather than Young's modulus Y. This supposition would not be correct, since in that analysis it was assumed that the medium suffered no lateral change in dimension when a compressional pulse passed through it. This condition is realized when a disturbance originates in an elastic medium of great extent, or when the medium is confined to a tube with rigid walls. However, in this case, when a wave pulse travels along a thin rod, there is a small lateral expansion of that section of the rod undergoing compression. Because of this fact, it may be shown that Young's modulus should be used rather than the volume modulus of elasticity. Equation (62.4) may, of course, be used to compute Y for a rod in terms of measured values of v_m and ρ.

Method: Set up the apparatus as shown in Fig. 62.2. Be sure that the metal rod is firmly clamped exactly in the center. Also, the powder should be distributed evenly along the bottom of the tube between A and B.

Start with the piston A near the end of the tube. Stroke the metal rod with a cloth moistened in alcohol by grasping the rod near the clamp and then pulling away from the clamp. Slowly shorten the length of the air column during this process. When the air column between A and B is of the proper length, it will resonate with the vibrating rod and cause the powder on the bottom of the tube to collect in ridges across the tube, clearly defining the positions of the nodes and antinodes. The inter-nodal distance will be of the order of magnitude 10 cm. Determine this inter-nodal distance accurately by measuring the distance between B (it must be at a node) and the node furthest away from it, and then dividing by the number of antinodes between these two nodes. Make an estimate of the error involved in this measurement of $\lambda_a/2$.

Repeat this process by shortening the air column still further until the next resonance point is reached.

Measure the length of the rod. This is $\lambda_m/2$. Estimate the error in this measurement. Determine the temperature of the room. Determine the velocity of sound in air at this temperature. The velocity of sound in air at 0°C may be taken as 33,170 cm/sec. The velocity of sound at any other temperature may be computed by using the relation that the velocity of sound in air is directly proportional to the square root of the absolute temperature of the air (see Eq. (61.5)). Using these data and results, determine the value of v_m for the metal rod. Also, determine the error in this value.

Finally, determine the value of Young's modulus Y for the rod, using Eq. (61.4). The density of the material of which the rod is made (iron, brass, copper, or aluminum) may be found in Table C, Appendix III. Compare this value of Y with the accepted value. See Table M, Appendix III.

Repeat this experiment, using a rod of different material.

Record: Record your data and results for this experiment in tabulated form.

QUESTIONS

1. Is there a node or an antinode of the vibrating air column at the piston A in this experiment? Explain.

2. The value of Young's modulus for a metal obtained in this experiment is likely to be somewhat larger than its value given in Table M, Appendix III. The latter value is an isothermal value, whereas the former is an adiabatic value. How would this fact account for the difference in the value of Y?

OPTIONAL EXPERIMENT

1. If the glass tube in this experiment is provided with inlet and outlet stopcocks, it may be filled with a different kind of gas, such as CO_2. Therefore, the velocity of sound in a different gas under different conditions may be determined and thus the value of γ for the gas may also be determined.

PART V

LIGHT

Reflection

and

Refraction

Object: To study the reflection of light in a plane mirror; to study refraction of light in glass.

Apparatus: Plane mirror mounted in blocks, rectangular piece of plate glass, plate-glass prism, pins, rule, protractor, and cork-topped stand. Clear plastic may be used in place of glass.

Theory: The image formed by a plane mirror of an object in front of it may be located by tracing the light rays that seem to emanate from the image. Since light travels in straight lines, sighting the image from two different directions, then projecting these two lines of sight until they intersect, will locate the image at the point of intersection. Other lines of sight must then pass through this same point of intersection if the image occupies a fixed position, regardless of the angle of view.

The same technique may be applied to the study of refraction of light in glass. In this case, however, what is seen is not a reflected image, but a refracted image. Sighting toward this image will enable the observer to determine the actual path the rays of light take in traveling from the object, through the glass, to the eye.

Method: *Part I. Reflection of Light by a Plane Mirror.*

a. *Location of a point image by use of sight lines.* Draw a line *AB* about 20 cm long across the center of a record sheet, place this sheet on the cork-topped stand, and stand the mirror with the edge of its reflecting surface on this line. Stick a pin vertically about 6 or 8 cm in front of the mirror to act as the object, *O* (see Fig. 70.1). Be careful that the pin is accurately vertical. The position of the image of this pin as seen in the mirror is to be located carefully.

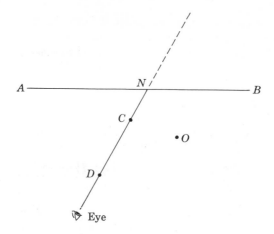

FIGURE 70.1   Ray tracing, plane mirror.

Stick a second pin near the mirror and a few centimeters to one side of the first pin (at a point such as *C*), and another pin (such as at *D*) 6 or 8 cm from *C* and exactly on the line that passes through *C* and the *image* of the original pin. When this has been accomplished, the third pin will hide the second pin and the image of the first pin. The pins should be as nearly vertical as possible, and the eye in sighting should be placed on a level with the paper so that the line of sight is along the points of the pins. Label the points *C* and *D*. The image will lie somewhere on a line through *C* and *D*.

Without disturbing the position of the mirror, determine in the same way two other lines directed toward the image at different angles with *AB*, one on each side of *O*. If the image has the same position when viewed from different directions, these lines (and all others similarly drawn) should intersect in the point that is the position of the image. Has the image a fixed position?

Let *I* denote the position of the image. Draw a line connecting *I* and *O*. What angle does this line make with *AB*? (Measure with the protractor.) How is it divided by *AB*? Describe definitely the position of the image with reference to the mirror and the position of the point object.

It is evident that when sighting along the line *DC* at the image, the light by which it was seen came to the eye along that line, having been reflected by the mirror at the point where *DC* meets it. Call this point *N*. Draw the incident ray *ON* and the perpendicular at *N*. Measure with the protractor and record in your figure the angles of incidence and reflection. Repeat this construction and measurement for the other two reflected rays.

Within what limits (expressed as a percent) do your results agree with the law of reflection?

What is meant by the plane of the angle of incidence and of the angle of reflection? What is the plane of these angles in this experiment?

b. *The position of a point image in a plane mirror* will now be found by parallax. Draw a line *AB* on another record sheet as before, and this time stand the mirror so that its reflecting surface is in the vertical plane through AB, but with its bottom edge about $\frac{1}{4}$ in. above the paper. Adjust it accurately. Stick a pin vertically 6 or 8 in. in front of the mirror. Hold a second pin behind the mirror, but do not stick it into the paper. Move it about until it is in such a position (using parallax) that the portion of it that is seen under the mirror (the eyes again being nearly on a level with the paper) lines up accurately with the portion of the image of the first pin, which is seen at the same time *in* the mirror. Use both eyes and move the head from side to side. When the correct position has been found, the section of the second pin and the image of the first pin will "fit" from all points of view. Let O_1 denote the position of the pin in front and I_1 the position of its image, as indicated by the second pin.

Place the first pin at a different distance from the mirror and again locate the image. Let O_2 and I_2 denote the position of this object and this image, respectively.

Draw lines connecting I and O in each case. What angle does each line make with the line AB? (Protractor.) How is it divided by AB? How do the results compare with those of Sec. a?

Part II. Refraction of Light in Glass.

a. *Apparent displacement of objects seen through plate glass.* Place a sheet of graph paper on the platform. Place the rectangular glass plate flat on the paper near the center in such a way that its long edge is parallel to the horizontal lines of the graph paper. With a *sharp* pencil, draw a line around the glass plate. Now, place a pin near the top of the page and to one side of the center. Place a second pin a few centimeters further down the page and to one side so that the line connecting them contacts the glass plate at a sharp angle.

Looking *through* the glass from the lower side of the page, you can see both pins. As you move your head from side to side, you can find a line of sight along which the pins appear to be directly in line. This line of sight may be defined by placing a third pin between your eye and the glass plate where it appears to fall on the same line as the first two pins. This step must be repeated by placing a fourth pin near the bottom of the page so that it too appears to fall in this line of sight.

Using a straightedge, draw a line through the first two pin points to the edge of the glass plate. This line traces the path of the incident beam of light. Repeat this procedure by drawing a line from the lower edge of the glass plate through the third and fourth pin points. This line traces the path of the beam after it has passed through the glass.

The path of the light beam in the glass may now be traced by drawing a straight line between the point at which the beam enters the glass and the point at which it leaves the glass.

1. Find the relation betwen the incoming beam and the outgoing beam.
2. In what direction is the beam deviated in the glass?
3. Where is the angle of incidence? Where is the angle of refraction? Measure these in degrees with your protractor.

b. *The refraction of light in a prism.* Stick two pins near the center of a fourth record sheet, about two-thirds as far apart as the length of one side of the plate-glass prism. Lay the prism flat on the paper between the two pins so that each side of angle *A* (the angle at which the identifying number is scribed) is touching one pin. With your eyes nearly at the level of the paper, view the setup from such an angle that the portion of the pin *P* on the far side of the prism, which is seen *through* the prism, is just hidden by the corresponding portion of the pin *Q* on the near side (see Fig. 70.2).

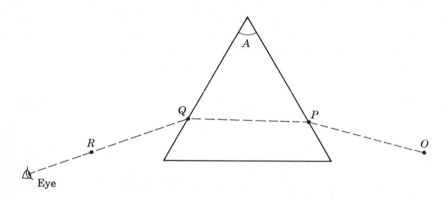

FIGURE 70.2   Ray determination; glass prism.

Stick a third pin, *O*, in the paper to the far side of the prism, several centimeters away from pin *P*, so that it appears to be in the direct line of sight through pins *Q* and *P*, as seen through the prism. Stick a fourth pin, *R*, nearer your eye in the line of sight to pin *Q*. When properly arranged, and with the two far-side pins seen *through* the prism, all four pins must appear to be in a straight line so that only the pin nearest the observer can be seen.

Draw perpendiculars to the refracting surfaces at the points of entrance and emergence of the ray, and indicate the angles of incidence and refraction of the rays, plus the angle of deviation (the angle between lines *OP* and *QR*) caused by the prism. State the direction of deviation of the ray (toward or away from the perpendicular) in passing into and out of the prism. Measure and record the angle of the prism *A* and the angle of deviation.

QUESTIONS

1. In Part IIa, Question 3, of this experiment you measured the angle of incidence *i* and the angle of refraction *r* of a ray of light passing through a glass plate. Compute the index of refraction *n* of the glass by means of the equation

$$n = \frac{\sin i}{\sin r}$$

2. In Part IIb of this experiment you measured the angle of deviation *D* of a ray of light passing through the glass prism. If this ray passes through the prism from *P* to *Q* in Fig.

70.2 in such a way that it is parallel to the base of the prism, then it may be shown that the angle of deviation is a minimum. Under these conditions, the index of refraction n of the glass prism is given by the equation

$$n = \frac{\sin\frac{1}{2}(D + A)}{\sin\left(\frac{1}{2}A\right)}$$

Compute the index of refraction of the glass prism, using this equation.

Index of Refraction
by Apparent
Elevation

Object: To determine the index of refraction of glass and various liquids by apparent elevation.

Apparatus: Microscope and stand, glass dish, liquids, piece of plate glass, and a vernier caliper.

Theory: When a small object that is covered by a layer of some transparent medium, such as glass or water, is viewed from directly above the surface, it appears to be nearer the surface than it actually is. The amount of this apparent elevation depends upon the thickness of the layer and the index of refraction of the layer of transparent material. The relation between these quantities is given by the equation

$$n = \frac{t}{t - e} \tag{71.1}$$

where n = index of refraction of the medium,
$\quad t$ = thickness of layer, and
$\quad e$ = apparent elevation.

Figure 71.1 illustrates a pencil of rays diverging from the source at O directly toward the upper surface of the medium of thickness t. At this upper surface, these rays undergo refraction as they pass into the air and thus appear to come from point O'. It may be shown that the position of O' relative to O obeys Eq. (71.1), provided that the light rays involved make practically normal incidence upon the upper surface (see any good general physics textbook for the development of this relation).

In order to determine n for the layer of material, it is only necessary to measure e and t. This may be done by using a traveling microscope (Appendix II, Note C) and a vernier caliper. The microscope is mounted directly above the layer of material with its axis pointing toward O. It may be focused on O, on O', or on the top of the layer by means of a rack and pinion that moves the entire microscope up or down in a fixed metal sleeve. The position of the microscope relative to the metal sleeve may be measured with a vernier caliper. Therefore, the distances t and e may be determined.

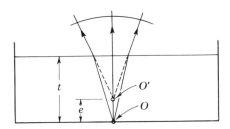

FIGURE 71.1   Ray diagram; apparent-depth apparatus.

Method: *Part I. Glass.* Make a small ink spot upon a piece of stiff white paper. Place it on the microscope stand immediately under the microscope and focus the microscope on it with the rack and pinion. Be sure that the paper is flat against the microscope stand. With the vernier caliper, measure the distance from the top of the metal sleeve to the top of the eyepiece of the microscope. In this process use the protruding shaft of the vernier caliper (see Fig. A.4, Appendix II). Call this reading r_1. Essentially, it gives the position of O in Fig. 71.1. Take three readings of r_1, refocusing the microscope each time.

Then, place the piece of plate glass over the spot and refocus the microscope until the image at O' is in sharp focus. Measure the new distance from the top of the metal sleeve to the top of the eyepiece three times with the vernier caliper. Call this reading r_2. It gives the position of O'.

Finally, focus the microscope on the top surface of the piece of plate glass. In doing this it may be necessary to put a small ink spot on this top surface. Again, measure the position of the microscope relative to the sleeve three times. Call this reading r_3. It is evident that $t = r_3 - r_1$ and that $e = r_2 - r_1$; therefore

$$n = \frac{r_3 - r_1}{r_3 - r_2} \tag{71.2}$$

Calculate, by means of Eq. (71.2), the index of refraction of the glass plate. From the estimated errors in r_1, r_2, r_3, determine the error in n.

Part II. Liquids. In the case of liquids use a flat-bottomed glass dish with a small scratch on the bottom of it (inside surface). This scratch will serve as the point O. Focus the microscope upon the scratch and record the vernier-caliper reading, as in Part I. Be sure to make three trials. Then pour in the liquid, whose index of refraction is to be determined, to a depth of about 3 cm. Focus on the image of O at O' and again record the vernier-caliper readings. Finally, sift a little powder (lycopodium) on to the surface of the liquid and focus the microscope on this powder. Record the vernier-caliper readings for this position of the microscope. From these data, compute the index of refraction of the liquid. Determine the error in this value.

Repeat Part II for a second liquid.

Record: Tabulate your data and results.

QUESTIONS

1. Develop Eq. (71.1), using the wave front or curvature method (law of sagitta).

2. Show that the determinate-error equation corresponding to Eq. (71.2) is

$$\frac{\Delta n}{n} = \frac{-e}{t(t-e)}\Delta r_3 - \frac{1}{t}\Delta r_1 + \frac{1}{t-e}\Delta r_2$$

3. Is the image at O' in Fig. 71.1 real or virtual? Explain.

Lenses

Object: To determine the focal lengths of converging and diverging lenses by different methods. To investigate the images formed by the converging lens.

Apparatus: Optical bench, convex lens, concave lens, telescope, object which may be illuminated, plane mirror, vertical wires, and lamp.

Theory: The locations of object and image in reference to a lens are given by the basic lens formula (for a thin lens)

$$\frac{1}{p} + \frac{1}{q} = \frac{1}{f} \tag{72.1}$$

where p = distance from the object to the center of the lens,
$\quad q$ = distance from the image to the center of the lens,
$\quad f$ = focal length of the lens.

For positive values of q, the image is *real* and may be projected on a screen. For negative values of q, the image is *virtual;* that is to say, the rays of light coming from the object through the lens *diverge* and the eye in viewing these rays sees them as though they came from an image located on the other side of the lens. Such a virtual image cannot be projected onto a screen.

There are two general types of lenses: converging and diverging lenses. The former type is thicker in the middle than at the edges, at least one side of such a lens is *convex*, and rays of light, in passing through such a lens, converge more (or diverge less) on the far side. The diverging lens is thinner in the middle than at the edges, and at least one side is concave; light rays diverge more (or converge less) after passing through such a lens.

The *principal focus* of a lens is the point (in the case of a converging lens) on the axis of the lens through which all rays of light parallel to the principal axis pass, or (in the case of a diverging lens) *appear* to pass, when refracted by the lens. The distance from the center of the lens to this point is called the *focal length* of the lens. A convergent lens has a positive focal length, and is therefore often called a positive lens. A divergent lens has a negative focal length, and is correspondingly called a negative lens. Thus it is clear that in Eq. (72.1) algebraic signs must be carefully observed.

From Eq. (72.1), it is clear that an object distance p, which results in a positive image distance q, implies that the object may be placed at a distance q and form an image at the distance p from the lens. (By convention, a real object distance is always positive; an image distance is positive if the image is on the opposite side of the lens from the object.) Two points, such that an object at one produces an image at the other, are called *conjugate foci*.

Method: *Part I. Converging Lens.* Make three determinations of the focal length by each of the following methods.

1. *Focal length by parallel rays.* Rays of light from distant objects are essentially parallel. For the purposes of this experiment, a building or other object 50 yards or more away may be considered a distant object. Focus the image of such an object upon a screen. Since the incident rays are essentially parallel, the image formed will lie in the principal focal plane of the lens. *Why?* The best focus is obtainable if no extraneous light strikes the screen; it will be helpful if the room is darkened during this phase of the experiment, with window shades up only high enough for a view of a distant object. Measure the distance from the center of the lens to the screen when the best focus has been obtained.

Focus a telescope on some distant object. The telescope is now adjusted to focus parallel rays entering it. Without changing the adjustment of the telescope, place it over the optical bench and aim it toward the lens. On the other side of the lens place an illuminated object. A piece of transparent ruler mounted across a hole in a card and illuminated from behind will do. Move the lens toward and away from this object until the object appears clearly in focus in the telescope. Measure the distance from the center of the lens to the object. *Why is this equal to the focal length?*

2. *Focal length by parallax.* (*Coincidence method.*) Set a plane mirror, whose face is vertical, close behind the lens. Set an object such as a short vertical wire that can be illuminated from the side before it. Move the wire until, with the eye in front of the wire and the lens, an inverted image can be seen in the air just above the tip of the wire. When this has been accomplished, the wire and its image are a focal length away from the lens, since only then will rays of light from the wire be made parallel by the lens and be reflected back on themselves by the mirror. Adjust the position of the wire until there is no parallax between it and its image (see Appendix II, Note E). Since the image is in the focal plane of the lens, the eye must be focused not on the lens, but on the image before it.

3. *Focal length by conjugate foci.* Place the transparent-scale object near one end of the optical bench and illuminate it from behind. Place the lens before it, and adjust

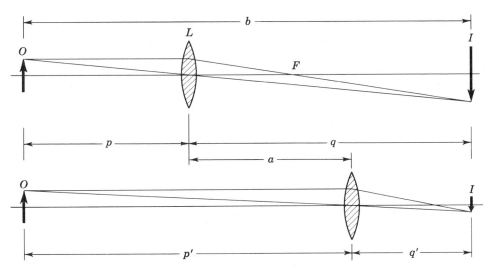

FIGURE 72.1 Ray tracing; conjugate foci.

the screen beyond this object in a position to receive the image of the transparent scale. When a good focus has been achieved, the object and image are at conjugate foci of the lens (Fig. 72.1). To obtain a sharper focus, it will usually be helpful to cover all of the lens, except for a small aperture at the center. This covering reduces spherical aberration, a defect of all lenses made of a single piece of glass and ground with spherical faces. Measure the distances from the object to the center of the lens, and from the center of the lens to the image (distances p and q). Find the focal length from these data.

Without moving either the object or the image after adjustment in the previous step, move the lens until another sharp image is cast on the screen. Again, measure object and image distances (distance p' and q'). How do these new distances compare with the previous measurements? Find the focal length from these data.

Denoting by b the distance between the object and the image and by a the distance between the two positions of the lens, it may be shown that the focal length is given by

$$f = \frac{b^2 - a^2}{4b} \tag{72.2}$$

Use Eq. (72.2) to find the focal length. Substituting an expression for a in terms of b and p in Eq. (72.2) and differentiating, it is found that the longest focal-length lens that will form an image a distance b from an object is one for which $f = \frac{1}{4}b$. In this case $q = p = 2f$. Equation (72.2) can be derived from Eq. (72.1) by substituting expressions for p and q in terms of a and b. Perform this derivation.

Part II. Diverging Lens. Make three determinations of the focal length, using each of the following methods.

1. *Focal length by virtual object.* Set up an illuminated object, convex lens, and screen, as in Part I, Sec. 3, and obtain a sharp image on the screen. Now,

place the diverging lens between the first lens and the screen. Its diverging action will cause the image to lie farther from the convex lens than it did originally (see Fig. 72.2). Move the screen out until the image is again in focus, first recording the original position of the screen. Determine the distance between the center of the concave lens and the two positions of the screen. The position of the original image may be thought

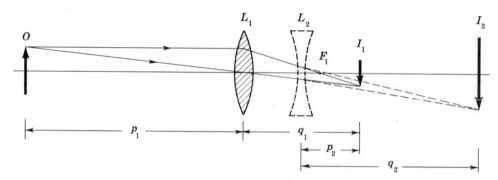

FIGURE 72.2 Ray tracing; measurement of focal length of negative lens.

of as a virtual object (object distance negative) for the diverging lens. This position results in a real image if the converging lens is "stronger" than the diverging lens. Thus the two positions of the screen are in fact conjugate foci of the diverging lens. Determine the focal length of the concave lens, using Eq. (72.1).

2. *Focal length by parallax.* Mount a negative lens on the optical bench, and on the side opposite the viewer place a vertical wire. The vertical wire is fitted into a dowel which, in turn, can be mounted in the optical bench support.

Looking through the lens, determine the general location of the image. At the site (approximate) of the image mount a special wire (see Fig. 72.3). The curved wire is so constructed that an image of it will not be formed by the lens, rather that one can view it outside the lens so as to determine the image position. Use the tip of the curved wire to locate the image of the vertical wire formed by the lens. Adjust the position of

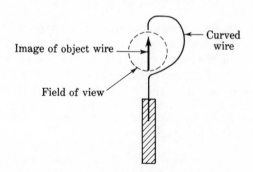

FIGURE 72.3 Special image locater; curved wire.

the curved wire viewed above the lens until its position coincides with the image of the vertical wire.

Measure the distances from the center of the lens to the position of each of the wires. Calculate the focal length of the lens.

Record: Tabulate the six sets of value for the focal length of the converging lens and the two sets for that of the diverging lens, each set consisting of the three trial values and their mean. Calculate the indeterminate error in each case.

Optical

Instruments

Object: To investigate the magnifying properties of a simple magnifying lens. To construct a simple telescope and to investigate its magnifying properties.

Apparatus: Optical bench and accessories, converging lens of short focal length (about 10 cm), converging lens of long focal length (about 50 cm) and wide aperture (about 10 cm), vertical wires, small translucent scale, meter stick, large scale, lamp, and a reading telescope with cross hairs.

Theory:[1] The simple one-lens magnifier and a reading telescope with cross hairs are optical instruments of constant use in the physics laboratory. We use them, in general, because we can see certain things better with them than without them, e.g., the vernier scale on a spectrometer, the deflection of a galvanometer mirror, a distant object which cannot be approached for closer inspection, and so on.

The primary reason why our vision is aided by these and other optical instruments of a similar nature is because they have *magnifying power*. This means that when we use any of these optical instruments for viewing an object, the image formed on the retina of the eye is larger than the corresponding retinal image when we do not use the instrument. The size of the retinal image is directly proportional to the angle subtended at the eye by the object or space image being viewed, since the distance from the crystalline lens of the eye to the retina is fixed for a given individual (2–3 cm). Therefore, a fairly general definition for the *magnifying power* (M. P.) *of an optical instrument is the ratio of the angle subtended at the eye by the image produced out in space by the instrument to the angle subtended at the eye by the object, when viewed in its most favorable position by the unaided eye*. It is necessary to make a sharp distinction between magnifying power and ordinary linear magnification (ratio of image size to

---

[1]For an excellent discussion on the telescope and microscope see J. K. Robertson, *Introduction to Physical Optics* (New York: Van Nostrand Reinhold Company, 1942).

object size) produced by a lens. The distinction between the two will be apparent in the following discussion on the simple magnifying lens.

Part I. The Magnifying Lens. Suppose a small object OO' is placed just inside the principal focus F of a converging lens LL', as shown in Fig. 73.1(a). A virtual image, erect and enlarged, will be formed at II'. The relation between object distance p, image distance q, and focal length f is

$$\frac{1}{p} - \frac{1}{q} = \frac{1}{f} \qquad (q \text{ is negative because image is virtual})$$

If the eye is place immediately behind the lens, it will "see" the image II', provided that it lies somewhere in the interval defined by the relation, $25 \text{ cm} \leqq q \leqq \infty$. This assumes a normal eye with a "near point" of 25 cm and a "far point" of infinity, i.e., an eye with normal accommodation, able to focus on any object or image that lies at least 25 cm away from it. For $q < 25$ cm, this eye would not be able to focus on the image (without considerable eye strain) because of poor accommodation. The angle β in the figure is essentially equal to the angle subtended at the eye by the image II', since the eye is placed immediately behind the lens.

If the same object OO' is to be examined without the visual aid of the lens LL', it must be moved out to a distance of 25 cm from the eye for distinct vision. It then subtends an angle α at the eye, as shown in Fig. 73.1(b).

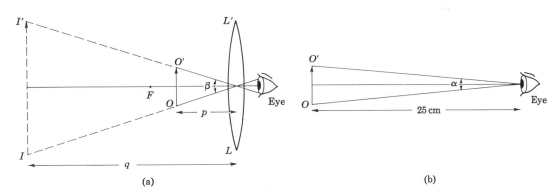

(a)   (b)

FIGURE 73.1 Magnifying power of a single lens.

The magnifying power of the lens is then given by the relation

$$\text{M.P.} = \frac{\beta}{\alpha} \qquad (73.1)$$

Since the angles α and β are small, the following relations are valid:

$$\beta = \frac{II'}{q} = \frac{OO'}{p} \qquad (73.2)$$

$$\alpha = \frac{OO'}{25} \qquad (73.3)$$

If we substitute these values of β and α into Eq. (73.1) and simplify, we get

$$\text{M.P.} = \frac{25}{p} = 25\frac{f+q}{fq} \qquad (73.4)$$

Since q must lie on the interval $\infty \geq q \geq 25$ cm, it is clear that the M.P. may vary from its maximum value

$$\text{M.P. (max)} = \frac{25}{f} + 1 \qquad \text{(image at 25 cm)} \qquad (73.5)$$

to its minimum value

$$\text{M.P. (min)} = \frac{25}{f} \qquad \text{(image at infinity)} \qquad (73.6)$$

If f is small compared to 25 cm, as it generally is, it will be seen that the two values of M.P. differ very little.

Several conclusions can be drawn from this discussion on the magnifying lens.

1. In order for the lens to have a magnifying power greater than 1, its focal length must be less than 25 cm.

2. If the lens is focused on the object so that the image II' appears 25 cm away, then, and only then, is the linear magnification equal to the magnifying power.

3. Ordinarily, the observer does not know where the image II' is located. It may be anywhere between 25 cm and infinity, depending on just how the lens was focused on the object OO'.

4. Since for most people the eye is most relaxed when looking at very distant objects, it is better to focus the lens on the object from a position too far away from it, rather than from a position too near to it. In the former focusing the image comes in at infinity, far from the eye, whereas in the latter focusing the image comes in at 25 cm, close to the eye. The small gain in magnifying power is seldom worth the added eye-strain that usually accompanies the latter type of focusing.

Part. II. The Telescope. The telescope is an optical instrument used to view distant objects. Its range is extremely great. For example, in the laboratory the distance may be only 100 cm, but for astronomical observations, it may be many millions of miles.

In its simplest form the refracting telescope consists of two converging lenses: (1) an objective lens of long focal length f_o and large aperture, and (2) an eyepiece lens of short focal length f_e. Light rays from the distant object located near or on the principal axis of the objective lens enter the lens. These rays are brought to a focus in the principal focal plane of the objective lens, forming a real image there. This assumes that the object distance p is infinite, but for practical purposes it is only necessary that $p \gg f_o$. The real image will be inverted and considerably smaller than object. It may be seen by eye if the eye is placed on the principal axis 25 cm in back of the image, looking toward the image and the objective lens. It is interesting to note that the magnifying power in this case is $f_o/25$ and therefore is only greater than one provided that $f_o > 25$ cm. Can you show this?

The eyepiece in its simplest form is essentially a magnifying lens used to examine the real image formed by the objective lens. The function of the magnifying lens is the same as when it is used to examine a small object, except that the "small object" is now the small real image produced by the objective lens of the telescope. It is advisable to focus the eyepiece from a position too far away from the real image so that the enlarged virtual image formed by the eyepiece comes in at infinity, rather than at 25 cm. Thus the eye looking at this virtual image will suffer a minimum of strain. Some of these details will be clarified in the following discussion.

Figure 73.2 shows the optical paths of two sets of rays through a telescope having an objective lens LL' and an eyepiece EE'. The size of the eyepiece is greatly exag-

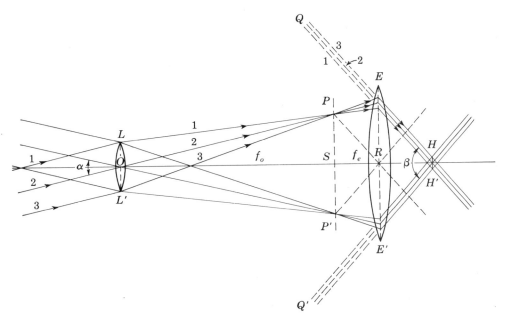

FIGURE 73.2   Telescope; optical paths of two sets of rays.

gerated in this diagram, because the angles that the rays form with the principal axis OR are greatly exaggerated. Rays 1, 2, 3, are a set of rays, practically parallel, coming from a point on the lower edge of the distant object (not shown). These rays pass through the objective lens LOL' and are brought to a focus at P in the focal plane of the objective lens. Corresponding parallel rays from the uppermost edge of the distant object are brought to a focus at the point P'. Thus a *real inverted image* of the distant object is formed in the focal plane of the objective lens. $OS = f_o$, the focal length of the objective lens.

This real image PP' is examined by means of the eyepiece EE', which is so adjusted in position that it forms a virtual image QQ' of the real image PP' at a very great distance from EE'. Thus the distance $SR = f_e$, the focal length of the eyepiece. Also, the rays 1, 2, 3 emerging from the eyepiece are, for all practical purposes, parallel rays. When the eye of the observer is placed behind the eyepiece EE', these parallel rays enter the eye and the eye sees the final virtual image QQ' essentially at infinity.

It is evident that this final virtual image will be inverted with respect to the distant object. This is a characteristic of the astronomical telescope as distinguished from the terrestrial telescope.

The magnifying power of the telescope is simply the angular size of the final virtual image divided by the angular size of the distant object. The distant object obviously cannot be brought to the position of most distinct vision, i.e., 25 cm. Therefore, one may write

$$\text{M.P.} = \frac{\beta}{\alpha} \tag{73.7}$$

where β and α are now the angles shown in Fig. 73.2. If it is remembered that these angles are very small in the case of the telescope (greatly exaggerated in the figure), it is an easy matter to show that $\beta/\alpha = f_o/f_e$; thus

$$\text{M.P.} = \frac{f_o}{f_e} \tag{73.8}$$

This is left as an exercise for the student.

An examination of Fig. 73.2 shows that the rays coming from the opposite edges of the distant object and finally emerging from the eyepiece pass through a small circle at HH' called the *exit pupil*, or *eye ring* of the telescope. In fact, all rays coming from the object via the telescope pass through the eye ring, since the eye ring is just the real image of the objective lens LL' formed by the eyepiece EE'. Therefore, in using the telescope the eye is placed at the eye ring in order to see the maximum range of the object. Unfortunately, a single eyepiece lens gives an eye ring that is too far away from the lens for convenient observing. In order to avoid this and other defects, an eyepiece is usually made up of a combination of two lenses, e.g., a Ramsden eyepiece. In this experiment we use only a single lens eyepiece and make no attempt to place the eye at the eye ring for observation.

The position and size of the eye ring may be computed by treating the lens LL' as an object in front of the eyepiece, and by finding the position and size of the real image of this object. Show that the diameter of the eye ring in Fig. 73.2 is

$$HH' = LL'\left(\frac{f_e}{f_o}\right) \tag{73.9}$$

and therefore, using Eq. (73.8), that

$$\text{M.P.} = \frac{f_o}{f_e} = \frac{LL'}{HH'} \tag{73.10}$$

Thus the M.P. of the telescope described is equal to the diameter of the objective lens divided by the diameter of the eye ring. By properly illuminating the objective lens, its real image (the eye ring) may be projected on a screen and measured.

For convenient observation, the eye ring should be approximately the same size as the pupil of the eye. Hence, Eq. (73.10) clearly indicates the need of an objective lens of large aperture and long focal length if any considerable magnifying power is demanded in a telescope. Also, a study of diffraction effects in telescopes shows the desirability of having an objective lens of large aperture in order to increase the resolving power (ability to show detail) of the instrument.

Method: In this experiment several precautions must be taken. It is necessary that the lenses and other items used be carefully aligned on the optical bench. Poor alignment generally produces images that are distorted and out of position. This distortion is especially true for an optical system consisting of more than one lens, e.g., the telescope. Also, in the case of lenses with very short focal lengths (high magnifying power), the consequences of poor alignment are usually disastrous. For this reason we make no attempt to use such lenses in the experiment.

In using the parallax method it should be realized that lens defects and poor alignment may cause a false movement if the eye is moved too far. Adjustments of this sort should be made with small head motions.

Part I. Magnifying Lens. Determine the focal length of the short focus lens. Use it as a magnifier to examine a small section of a horizontal scale that is properly illuminated. Set the lens at a convenient mark near the end of the optical bench and set the object (the scale) beyond it, near the principal focus of the lens. Place the eye just behind the lens and focus the system, moving the object rather than the lens. This movement is merely a matter of convenience in this experiment. Normally, the lens is moved rather than the object. Note that there is a considerable range of focusing for which one obtains a sharp image of the scale. This corresponds to the possible range of the image distance; i.e., q may range from 25 cm to infinity.

In this experiment we will focus for just two values of q, namely $q = 25$ cm and $q = 100$ cm (the approximate length of the optical bench). Furthermore, in the case where $q = 25$ cm, we will make a direct determination of the magnifying power of the lens by matching the virtual image of the scale with an actual scale set at this distance.

In order to focus the system so that $q = 25$ cm, set a long vertical wire at $q = 25$ cm on the optical bench. Then, focus the system by moving the object (scale) until the virtual image, as seen through the lens, coincides in position with the wire, as seen above the lens. Use the method of parallax for this determination. Remove the wire and replace it with a section of a well-marked meter stick mounted in a horizontal position. Match the virtual image of the small scale, as seen through the lens, with the actual scale, as seen above the lens. The linear magnification in this case is simply the magnifying power of the lens. Why? Determine the magnification by direct observation and compare it with the value $25/p$, as given by Eq. (73.4) for the magnifying power. The value of p may be read directly on the optical bench scale.

For $q = 100$ cm, set the wire at this position and focus the system so that the virtual image now coincides with the position of the wire. Note the value of p for this case. Is it larger or smaller than the value obtained in the first case? Why is it impossible in this case to obtain the magnifying power of the lens directly, as we did in the first case?

Part II. Telescope. Determine the focal length of the objective lens, then remove it from the optical bench. Set the short focus lens, which you have just used as a magnifier and will now use as an eyepiece for the telescope, approximately 20 cm from one end of the optical bench. On the other side of the lens, place a short vertical wire (a substitute for cross hairs) at the principal focus of the lens. Point the optical bench toward an object several meters away in the laboratory. It is convenient to use the large scale as the distant object, but two parallel lines on a blackboard or a large piece of paper will suffice. Focus the eyepiece on the wire by moving the wire toward the eyepiece until the virtual image of the wire, as seen through the lens, coincides in position with the distant scale. Use the method of parallax for this adjustment. In the use of a real telescope this procedure is equivalent to focusing the eyepiece on the cross hairs of the telescope.

Place the objective lens on the optical bench in front of the wire, at a distance equal to its focal length. Then, adjust the position of the objective lens so that you get a sharp image of the distant scale when looking through the eyepiece. If there is any parallax between the image of the vertical wire and the image of the distant scale, remove it by continued adjustment of the position of the objective lens. In the case of a real telescope this adjustment amounts to focusing the eyepiece and cross hairs on a distant object.

If the preceding adjustments have been properly made, then (1) the real image of the distant scale formed by the objective lens alone coincides in position with the vertical wire, and (2) the virtual image of the vertical wire, along with that of the real image as seen through the eyepiece, is located at the position of the distant scale that is being viewed. Furthermore, if the telescope that we have just constructed and focused has a magnifying power greater than unity, then the virtual image of the scale as seen through the telescope will be larger than the scale when viewed directly. Finally, since the telescope has been focused so that the virtual image of the scale is as far away from the eye as is the actual scale, the linear magnification of the scale produced by the telescope is just equal to β/α, the magnifying power of the telescope.

In order to determine experimentally the magnifying power of the telescope, it is only necessary to compare the distance between two scale marks, as seen through the telescope and as seen directly. This comparison may be done in the following manner. With both eyes open, one seeing the scale directly and the other seeing it through the telescope, align the telescope so that the images on the two retinas of the eyes merge into one, i.e., are superimposed. This requires some practice, since your eyes are not normally required to do this sort of thing. Therefore, one of your eyes may initially fail to function and you will see only one of the images. If this happens, close the eye that is accepting the image so that the other image appears, then open the eye again. Eventually, you will obtain the dual image and make a direct comparison of scale lengths, i.e., one scale division on the virtual image scale equals X scale divisions on the actual scale. Then, $X = $ M.P. You may be so far away from the scale that you cannot read off the value of X. In this case your partner should be stationed at the scale with a pointer of some kind, which he moves across the scale under your directions. He can read the value of X for you.

Compare this observed value of the magnifying power of the telescope with that given by Eq. (73.8). Explain any discrepancies.

Illuminate the objective lens of your telescope from the side and project its real image, formed by the eyepiece lens, on a white screen placed behind the eyepiece. This real image is the eye ring of the telescope. Note that in this case the real image is much too large and far away from the eyepiece to serve any good purpose. This is not the case for a good telescope. Measure the diameters of the eye ring and the objective lens. Then, compute the magnifying power by use of Eq. (73.10) and compare it with the observed M.P. Explain any discrepancies.

As a final step in this experiment, focus a reading telescope with cross hairs on the distant scale so that a sharp image of the scale is obtained and there is no parallax between the markings on the scale and the cross hairs of the telescope. Have the instructor check your focused telescope.

Record: Tabulate all data and results.

QUESTIONS

1. Using Fig. 73.2 and Eq. (73.7), develop Eq. (73.8).
2. Develop Eq. (73.9).

Index of
Refraction with
Spectrometer

Object: To determine the index of refraction of a glass prism for the mercury green line by spectrometer measurements of the prism angle and angle of minimum deviation.

Apparatus: Spectrometer, mercury arc, glass prism, and level.

Theory: It is possible to make a very accurate determination of the index of refraction, n, of glass for a given wave length of light by using refraction through a glass prism.

Suppose that a ray of monochromatic light (light of a single wave length) is incident upon one face of a glass prism at point M, as shown in Fig. 74.1. This ray will be refracted as it passes from air into the glass and will proceed through the glass along some line MN. At point N, it will again be refracted as it emerges from the glass into the air on the other side of the prism. The angle D in the figure represents the total deviation of the ray resulting from its passage through the prism; i.e., D is the angle between the incident ray and the emergent ray. It may be shown from the geometry of the figure that

$$D = (i_1 - r_1) + (i_2 - r_2)$$

where i_1 = angle of incidence,
$\quad r_1$ = corresponding angle of refraction,
$\quad i_2$ = angle of emergence,
$\quad r_2$ = corresponding angle of refraction.

It may also be shown by geometry that $r_1 + r_2 = A$, where A is the angle of the prism as shown. Thus

$$D = i_1 + i_2 - A \tag{74.1}$$

The complete proof of these formulas is left as an exercise for the student.

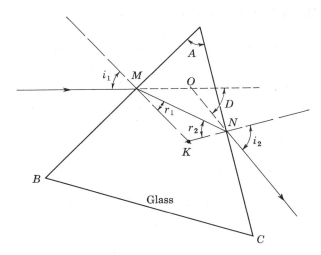

FIGURE 74.1   Refraction of light in prism.

For a given prism, the magnitude of the angle D depends only upon the incident angle i_1. This fact is evident since, for a given prism of fixed n and A, the angles r_1, r_2, and i_2 may be calculated once i_1 is given. D may therefore be regarded as a function of i_1.

By actual trial, it is found that there is only one value of i_1 that makes D a minimum. For this value, the angles i_1 and i_2 must be equal, for, if they were not equal, there would exist two different angles giving minimum deviation, i.e., i_1 and i_2. This condition follows from the fact that it is always possible to reverse the direction of the ray through the prism without changing the value of D.

In this experiment the condition of minimum deviation is secured by trial. Under this condition

$$\left.\begin{array}{l} i_1 = i_2 = i \\[2mm] r_1 = r_2 = r = \dfrac{A}{2} \end{array}\right\} \tag{74.2}$$

It follows that Eq. (74.1) may be written

$$D_m = 2i - A \tag{74.3}$$

where D_m = angle of minimum deviation,
$\quad\quad i$ = angle of incidence or emergence,
$\quad\quad A$ = prism angle.

The index of refraction n of the glass prism is given by the defining relation

$$n = \frac{\sin i}{\sin r} \tag{74.4}$$

Using Eqs. (74.2) and (74.3), the value of n may then be written in the form

$$n = \frac{\sin \frac{1}{2}(D_m + A)}{\sin (A/2)} \tag{74.5}$$

Relation (74.5) provides us with one of the standard methods of determining the index of refraction of a substance, such as glass, when it is in the form of a prism. Using a spectrometer, the angles A and D_m may be measured with great accuracy, and therefore a very accurate value of n may be obtained.

Method: *Part. I. Adjustment of Spectrometer.* Details concerning the construction and the approximate adjustment of the spectrometer are given in Note J, Appendix II. The student should read this section in the Appendix and make the adjustments called for before proceeding further in this experiment.

Part II. Measurement of Prism Angle. Place the prism on the spectrometer table with its refracting edge near the center of the table and with its base approximately perpendicular to the axis of the collimator, as shown in Fig. 74.2.

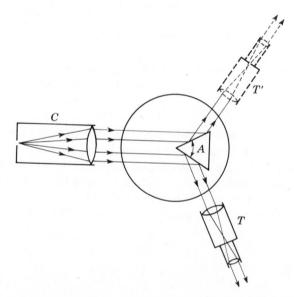

FIGURE 74.2   Prism spectrometer; angle of prism.

Illuminate the slit of the collimator C with light from a lamp bulb or a mercury arc. Make the slit fairly wide to begin with so that there will be no difficulty in seeing it. Parallel rays of light coming from the collimator will then fall upon both faces of the prism and be reflected. Before trying to find these reflected rays with the telescope, try finding them with your eyes. You should be able to see a clear image of the collimator slit by reflection from either face of the prism. If you cannot find these images with your eyes, there is little chance that you will be able to do so with the telescope. Some adjustment of the position of the prism on the table may be necessary during this process.

Clamp the prism table so that it cannot turn with the telescope. Seek the two images of the collimator slit by using the telescope. These should be found with the telescope before any readings are attempted. Then, set the cross hairs of the telescope on the center of the image of the collimator slit in one of the faces of the prism, e.g., in position T. In order to get an accurate setting, it is advisable to make the collimator slit very narrow. Read both verniers to the nearest minute of arc and record these readings. Distinguish between the two vernier readings by labeling them. Move the cross hairs off the collimator image, using the slow-motion screw of the telescope; then, move them back on to the image. Again read and record the position of the telescope.

Next, unclamp the telescope and move it to the position T', setting the cross hairs on the other image of the collimator slit. Repeat the process described in the foregoing paragraph. In this part of the experiment it is essential not to move the prism, prism table, or collimator. Why?

By subtraction, determine the angle through which the telescope has been turned. This is twice the angle of the prism. Why? Make an estimate of the error involved in measuring the angle of the prism. (NOTE: All prisms used in the laboratory are very nearly equiangular; thus your prism angle should be very nearly 60°. Use as angle A the angle in which the identifying number is scribed.)

Part III. Measurement of Angle of Minimum Deviation. Set the prism on the spectrometer table, with its refracting edge near the center of the table and its base making an angle of about 30° with the collimator axis, as shown in Fig. 74.3.

Illuminate the slit of the collimator with light from the mercury arc. Some of the parallel rays from the collimator will be incident upon the first face of the prism and be deviated in their passage through the prism toward the telescope in position T. Others will miss the face and pass directly toward the telescope in position T_0, provided that the first prism face does not extend too far across the spectrometer table.

With the collimator slit fairly wide, try to find the image of the slit with your eye in

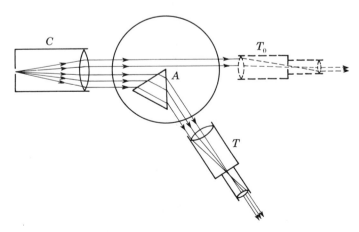

FIGURE 74.3 Prism spectrometer; minimum angle of deviation.

position T_0 and then in position T. There will be only one image at position T_0, but there should be a whole series of slit images at position T because of dispersion by the glass prism; that is, one gets the line spectrum of the mercury arc in this position.

Set the telescope in the position T_0 to find the direct image of the slit. When this image has been found, make the slit of the collimator very narrow and set the cross hairs at the center of the slit, using the slow-motion screw of the telescope. Read and record both verniers for this position of the telescope. Then move the cross hairs away from, and then back to, the center of this image. Again read and record the verniers. These readings give the zero position of the telescope.

Now, swing the telescope around to position T and search for the green line of the mercury arc. This line is quite intense and can hardly be missed, unless you happen to be color blind. Set the cross hairs of the telescope on this line. Then, slowly rotate the prism table and prism in the direction that causes the green line to move toward the position of the direct or central image at T_0. Follow this line with the telescope until the line has moved as far toward T_0 as possible. At this point, further rotation of the prism table in the *same* direction will cause the line to reverse its motion and move away from T_0. Set the cross hairs of the telescope on the green line when it is nearest the position T_0. Read and record both verniers for this position of the telescope, i.e., the position of minimum deviation. Repeat this process two more times. In each trial rotate the prism table slightly so that the green line moves away and then back to its position of minimum deviation.

The angle of minimum deviation D_m is the smallest angle obtainable between the two angular positions T_0 and T of the telescope and may be computed by subtraction of vernier readings (*the same vernier*) for the two positions. Compute this angle and estimate the error in its determination.

Using Eq. (74.5), compute the index of refraction for the green mercury line ($\lambda = 5460.74\ A$). Use a five-place log table for this calculation. Compute the error in the index of refraction, using the method given in the sample record.

Record: (Sample.)
    App. Nos. Prism No. 8
             Spectrometer No. 6
             Mercury arc No. 5
             Wave length (green line) = 5460.74 A

Part II. Prism Angle A.

	Vernier		Vernier	
	A	*B*	*A*	*B*
Left T	246° 13′	66° 3′	246° 15′	66° 2′
Right T'	126° 17′	306° 9′	126° 18′	306° 12′
Diff	119° 56′	119° 54′	119° 57′	119° 50′

Ave Diff = 119° 54′ ± 2′ A = 59° 57′ ± 1′

Part III. Angle of Minimum Deviation D_m.

ZERO POSITION

Vernier	Trial I	Trial II	Average
A	193° 48′	193° 45′	193° 46′
B	13° 39′	13° 41′	13° 40′

DEVIATED POSITION

Vernier	Trial I	Trial II	Trial III	Average
A	245° 34′	245° 38′	245° 32′	245° 37′
B	65° 25′	65° 21′	65° 28′	65° 27′

DEVIATION

Vernier	Dev pos	Zero pos	D_m
A	245° 37′	193° 46′	51° 51′
B	65° 27′	13° 40′	51° 47′
		Ave 51° 49′ $\pm$ 2′	

$$D_m = 51° \ 49′ \pm 2′$$

$$n = \frac{\sin \frac{1}{2}(D_m + A)}{\sin \frac{1}{2}A} = \frac{\sin 55° \ 53′}{\sin 29° \ 58′} \quad \begin{array}{c} \bar{1}.91798 \\ \bar{1}.69853 \\ \hline 0.21945 \end{array}$$

$$n = 1.6575$$

In order to compute the error in n, we may proceed in the ordinary manner by developing the error equation. In this case the error equation is complicated by the presence of trigonometric functions. Those students who are familiar with the calculus may find it worthwhile developing the error equation to fit this case. For the sake of variety, we shall proceed in a different manner.

The errors in angles D_m and A are given as $\pm 1′$ and $\pm 2′$, respectively. In order to get the error in n, we have only to recalculate the value of n, using for D_m its average value $\pm 2′$, and for A its average value $\pm 1′$. This new value $n'(= n + \Delta n)$ minus the original value will then represent the error in n. The signs to be chosen for ΔD_m and ΔA should be such as to make n as large as possible. A little thought indicates that D_m should be increased by 2′ and A decreased by 1′. Therefore

$$n + \Delta n = \frac{\sin 55° \ 53′ \ 30″}{\sin 29° \ 57′ \ 30″} \quad \begin{array}{c} \bar{1}.91802 \\ \bar{1}.69842 \\ \hline 0.21960 \end{array}$$

$$= 1.6581$$

$$\therefore \Delta n = \pm 0.0006$$

In calculating the error in the result by this method it is essential that the calculations be made very carefully without arithmetical mistakes, because the error is the difference between two relatively large numbers that are very nearly equal.

Finally, it should be pointed out that the error obtained in this experiment is probably too small because no account is taken of the fact that the spectrometer was only approximately in adjustment. Lack of complete adjustment will probably increase the error, but it is difficult to say by how much.

QUESTIONS

1. Develop Eq. (74.1), using Fig. 74.1.

2. Using the calculus, show that D is a minimum in Eq. (74.1) when, and only when, $i_1 = i_2$.

3. In Part II (measurement of prism angle) it frequently happens that the observer is able to see with his eyes a clear image of the collimator slit by reflection from either face of the prism, but is unable to "see" the image by use of the telescope. Using Fig. 74.2, explain this anomaly. (HINT: Suppose the prism lies nearer collimator than shown.)

4. In Part III (measurement of angle of minimum deviation) the refracting edge of the prism is supposed to be placed near the center of the prism table. Why is this necessary?

5. Using the calculus, develop the error equation corresponding to Eq. (74.5).

6. The angle between T and T' in Fig. 74.2 is twice the angle of the prism. Prove this by using a diagram and the laws of reflection.

OPTIONAL EXPERIMENT

1. The prism and spectrometer in Exp. 74 may be used as a spectroscope since the prism disperses any light beam sent through it into a spectrum. This condition results from the fact that the index of refraction of glass is a function of the wave length, usually decreasing with increasing wave length. Therefore, the deviation of a ray of monochromatic light by the prism is a function of the wave length of the light. Thus the prism spectroscope may be used to determine wave lengths if it is first calibrated by sending rays of known wave lengths through it and observing their deviations. A graph showing the relation between these wave lengths and their corresponding deviations can then be used to determine an unknown wave length in terms of its deviation. The spectrum of known wave lengths is called the comparison spectrum.

Calibrate the prism spectroscope, using the prominent lines in the mercury-arc spectrum as a comparison spectrum. See Table H, Appendix III, for values of these wave lengths. Plot a calibration curve. Then, determine the wave lengths of the prominent lines in the spectra from some different sources furnished by the instructor.

Wave Length
of Light:
Diffraction
Grating

Object: To measure the wave lengths of some of the prominent spectral lines in the mercury-arc spectrum with the diffraction grating.

Apparatus: Spectrometer, transmission diffraction grating, level, and mercury arc.

Theory: The diffraction grating is essentially a multiple-slit device used in optics for producing interference fringes that are exceedingly sharp and widely spaced. This diffraction is in contrast to single and double slit diffraction, where the fringes are quite broad and closely spaced. As a consequence, a much more accurate determination of wave length can be made with the former device than with the latter.

Original gratings are made by ruling a large number of fine lines on a carefully prepared piece of glass or metal with a diamond point. In the case of glass the transparent openings between the opaque lines constitute the parallel slit system. The gratings used in this experiment are replicas of an original ruled grating that has several thousand lines per inch.

The fundamental action of the grating in diffracting light is discussed in practically all general physics textbooks. The analysis gives the angular positions of the diffracted rays as a function of wave length and grating space by means of the equation

$$n\lambda = a \sin \theta_n \qquad (75.1)$$

where λ is the wave length of the diffracted light, θ_n is the diffraction angle, i.e., the angle that the diffracted ray makes with the incident ray, a is the grating space, and n is the order of diffraction, i.e., $n = 1$ for the first order diffraction, $n = 2$ for second order diffraction, and so on. In deriving Eq. (75.1), only parallel rays (or plane waves) are considered (Fraunhofer diffraction). Furthermore, it is assumed that the incident rays from the source fall perpendicularly upon the plane of the grating, i.e., the angle of incidence is zero.

Fraunhofer diffraction is realized experimentally by focusing both the collimator and the telescope of the spectrometer for parallel rays. This is standard procedure. However, if Eq. (75.1) is to be used for calculating wave lengths, it is necessary to set the plane of the grating so that its normal is in alignment with the axis of the collimator. The accuracy of wave-length determinations in the experiment depends upon the accuracy of this alignment. In this experiment we wish to determine the wave lengths of some of the prominent lines in the mercury-arc spectrum to within 2 parts in 1000, i.e., 0.2% error. This fact should enable us to set limits on the position of the grating relative to the collimator axis.

In order to set these limits it is necessary to determine the diffraction equations for a grating when the angle of incidence $i \neq 0$. By an analysis similar, but necessarily more complex than that given for $i = 0$, we obtain the two equations

$$n\lambda = a[\sin (\theta_n - i) + \sin i] \tag{75.2}$$

and
$$n\lambda = a[\sin (\theta'_n + i) - \sin i] \tag{75.3}$$

The angles θ_n, θ'_n, and i are shown in Fig. 75.1, where SO is an incident ray from the collimator, OC is an undeviated ray giving the central image, ON is the normal to

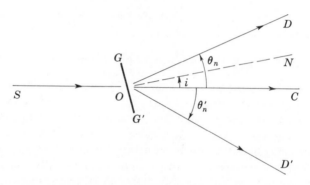

FIGURE 75.1   Transmission grating; $i \neq 0$.

the grating GG', OD is a diffracted ray giving one spectral line corresponding to $n\lambda$, and OD' is a diffracted ray giving the other. Note that the angular positions of the two spectral lines for a given $n\lambda$ are no longer symmetrical with respect to the position of the central image, i.e., θ_n is not equal to θ'_n, but in fact is always smaller for the case shown. When $i = 0$, Eqs. (75.2) and (75.3) reduce to Eq. (75.1), and $\theta_n = \theta'_n$.

Eliminating $n\lambda/a$ from Eqs. (75.2) and (75.3), we get

$$\tan i = \frac{\sin \theta'_n - \sin\theta_n}{2 - (\cos \theta'_n + \cos \theta_n)} \tag{75.4}$$

For small values of i (the case in which we are interested), Eq. (75.4) reduces to

$$i = \frac{\cos \theta_n}{2 - 2 \cos \theta_n}\delta\theta_n \tag{75.5}$$

where $\delta\theta_n = \theta'_n - \theta_n$, and the angles i and $\delta\theta_n$ are expressed in radians.

By adding Eqs. (75.2) and (75.3) and making use of Eq. (75.5), we get

$$2n\lambda = a[\sin(\theta_n - i) + \sin(\theta'_n + i)]$$
$$= 2a \sin \tfrac{1}{2}(\theta_n + \theta'_n) \cos \tfrac{1}{2}(\theta'_n - \theta_n + 2i) \tag{75.6}$$
$$= 2a \sin \tfrac{1}{2}(\theta_n + \theta'_n) \cos \left(\frac{i}{\cos \theta_n}\right)$$

Equation (75.6) clearly shows the effect of not having $i = 0$ in this experiment. A correction factor $\cos(i/\cos \theta_n)$ must be introduced in all wave length determinations, or i must be sufficiently small so that this correction differs from 1 by less than 2 parts in 1000 for this experiment. Fortunately, the correction represents a second-order effect since the cosine of a small angle is only slightly less than unity. Thus in this experiment it will be sufficient to set the grating so that i is less than $2°$.

There are several methods of making this adjustment. The standard method is to first align the axis of the telescope with that of the collimator. Then, using the grating as a plane mirror and illuminating the cross hairs of the telescope, the grating is turned until the reflected image of the cross hairs coincides with the actual cross hairs. This is a very precise method that will reduce i to 1 or 2 min of arc, but it requires complete adjustment of the spectrometer and a telescope with a gaussian eyepiece. A simpler, but less precise method, is to proceed as follows. Align the axis of the telescope with the collimator. Read the position of the telescope. Then, turn the telescope through $90°$ so that its axis is now at right angles to that of the collimator. Clamp it in this position. Turn the table holding the grating until the grating, acting as a mirror, reflects light from the collimator into the telescope. The plane of the grating will then make an angle of $45°$ with the collimator axis. If the grating is now turned back through $45°$, it will be perpendicular to the collimator axis. Obviously, the adjustments of the clamps on the telescope, table, and collimator must be made in the right order for this method to succeed. If done correctly, the grating may be set perpendicular to the collimator axis, certainly to within $1°$. The setting of the grating may be checked by measuring θ_1 and θ'_1 for a given spectral line. If these angles differ by more than 5 or 6 min of arc, either the spectrometer is not properly adjusted, or the angle i is much larger than $1°$.

After proper setting of the grating, the wave lengths of the spectral lines in the mercury-arc spectrum are determined by measuring the diffraction angles θ_n and θ'_n with the spectrometer and using either Eq. (75.1) or Eq. (75.6), where the correction factor is set equal to unity.

Method: Before starting this experiment, adjust the spectrometer by the method outlined in Note J, Appendix II. Then, place the diffraction grating in its mounting at the center of the spectrometer table. Using the simple method outlined in the theory, take the necessary steps to set the plane of the grating perpendicular to the axis of the collimator.

Illuminate the slit of the collimator with light from a mercury arc. Parallel rays of light will emerge from the collimator lens and fall perpendicularly upon the grating G, as shown in Fig. 75.2. These rays will be diffracted by the grating, forming bright-line spectra of the mercury arc on both sides of the central image at T_0. Look for

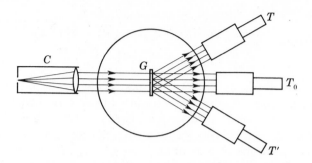

FIGURE 75.2 Diffration grating; telescope positions.

these spectra with the eyes before trying to find them with the telescope. It should be possible to see two or three orders of spectra. There are several rather prominent lines in the mercury spectrum: a bright yellow line (doublet), an intense green line, and several blue and violet lines. One of the violet lines is fairly intense (see Table H, Appendix III).

Check the setting of the grating, using the first-order green line of the mercury arc. Make the slit of the collimator quite narrow. Set the cross hairs of the telescope, first upon the central image (read the angular position), second upon the first-order green line to the right of the central image (read the angular position), and finally upon the first-order green line to the left of the central position (read the angular position). Compute angles θ_1 and θ'_1 for this line. If they differ by more than 5 or 6 minutes of arc, repeat the procedure for the proper setting of the grating. The angle i is probably greater than $1°$.

Once the grating has been properly set, it is no longer necessary to compute θ_n and θ'_n separately. It will be sufficient to compute only $\frac{1}{2}(\theta_n + \theta'_n)$ for each line, i.e., one half the angle between T and T' in Fig. 75.2.

Swing the telescope into the position T so that it is directed toward the *first-order* spectrum. Make the slit of the collimator quite narrow. Then, set the cross hairs of the telescope in turn upon the brightest violet line, the bright green line, each of the yellow lines (these are very close together and may not be resolved unless the collimator slit is very narrow). In each case, record both vernier readings to the nearest minute of arc. Then, rotate the telescope into the corresponding position at T' on the other side of T_0. Take a similar set of readings. Be sure to set the cross hairs on the same lines to the right and left of T_0.

Repeat the above process for these same lines in the second-order spectrum.

Compute the wave length in angstroms for each of the mercury lines measured, using the number of lines per inch stamped on the grating mounting. In each case determine the amount of error in the wave length, using estimated errors in the diffraction angles. It is advisable to use a five-place log table to make the calculations. Errors in the wave lengths may be calculated as they were in Exp. 74.

Record:

Spectrometer No._____

Grating No._____

No. of lines per inch_____

Grating space in angstrom units $= \dfrac{2.540 \times 10^8}{\text{lines per inch}} =$ _____

YELLOW LINE NO. 1—FIRST ORDER

Vernier	Pos T	Pos T'	θ_1	$\Delta\theta_1$
A				
B				
			Ave	

Make similar tables for the other lines and other orders.

	FIRST ORDER				SECOND ORDER				
Mercury lines	θ_1	$\Delta\theta_1$	λ	$\Delta\lambda$	θ_2	$\Delta\theta_2$	λ	$\Delta\lambda$	Ave λ
Yellow, No. 1									
Yellow, No. 2									
Green									
Violet									

Percent difference between measured λ and standard value: _____

_____

_____

_____

QUESTIONS

1. Show that the error equation corresponding to Eq. (75.1) is

$$\frac{\Delta\lambda}{\lambda} = \frac{\Delta a}{a} + \cot\theta_n\Delta\theta_n$$

where $\Delta\theta_n$ is expressed in radians. It is assumed that there is no error in n, the order of the spectrum.

2. Using the error equation given in Question 1, show that the wave length of a given spectral line can be determined most accurately by making observations on the image of highest order that appears in the spectrum.

3. Discuss the differences between the grating spectrum and the prism spectrum. Which color is deviated the most in each case?

4. Develop Eqs. (75.2) and (75.3) and then show that $\theta'_n > \theta_n$ for $i > 0$.

5. Using Eq. (75.2), show that θ_n is a minimum for varying i when $\theta_n - i = i$. Also, show that no such minimum exists for θ'_n in Eq. (75.3).

OPTIONAL EXPERIMENT

1. Determine the wave lengths in the mercury-arc spectrum, using the method of minimum deviation as suggested by Problem 5.

Laser:
Diffraction
Phenomenon

Object: To study the diffraction patterns formed by single and double slits when illuminated by laser light.

Apparatus: Laser ($\lambda = 6328$A° or similar), wedge-shaped single slit, double slits (Cornell plates), optical bench, and metric rule (meter stick or tape with adhesive backing).

Theory: The word LASER is an acronym meaning, "Light Amplification by Stimulated Emission of Radiation." Many different types have been developed in the last decade. In general, all lasers—solid, liquid, or gas—are basically the same.

The laser is a unique light source in that the emitted light is self-collimated, monochromatic, and coherent. Because of these characteristics, it is most suitable for observing certain interference phenomena.

The laser consists of an optical cavity filled with a suitable medium such as He-Ne gas mixture. Under normal conditions most atoms would be found in the ground state. However, it is possible to supply the medium with radiation of appropriate frequency and sufficient intensity to raise the majority of the atoms to energy levels above the ground state. This process is called population inversion. The method of accomplishing the inversion, however, may be thermal, electrical, mechanical, or by electromagnetic radiation.

Up to this point we have considered the process of excitation and emission of light by an excited atom. This theory was first advanced by Bohr in his explanation of the characteristic spectral lines of hydrogen.

Figure 76.1 is an energy level diagram showing the ground state E_1 and two excited states E_2 and E_3. If, therefore, an electron of an atom is raised to the level of E_3, it might return to the ground state in two consecutive jumps or a single jump.

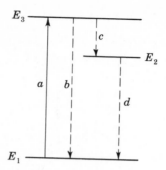

FIGURE 76.1 Energy-level diagram.

Each jump represents a release of a discrete amount of energy. Also E_2 might represent a metastable state where the electron would remain for a relatively long period of time.

In the laser an external source causes the population inversion, the first condition needed for light amplification. The second prerequisite is the optical cavity (Fig. 76.2). Light of proper wave length travels between the parallel mirrors with little loss of transmission. As the light travels between the mirrors, some of the excited atoms are stimulated to return to the ground state, and consequently the beam is amplified. Also, as the two photons join, the resultant light is in phase (coherent) and of the same wave length as the stimulating photon.

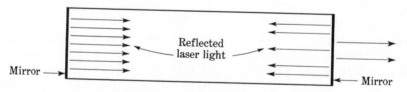

FIGURE 76.2 Optical cavity of laser.

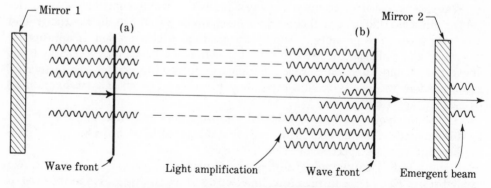

FIGURE 76.3 Laser action; light amplification.

Figure 76.3 shows in a crude way the amplification of light as the beam travels from mirror 1 to mirror 2. At (a) a plane wave front is drawn. This is possible, since the inherent characteristic of the laser beam is that the waves are in phase no matter how many atoms have returned to the ground state. At (b) you will note that the beam is more intense and that another plane may be drawn to represent the phase relationship of the separate waves.

Typical diffraction patterns are shown in Fig. 76.4. In the single-slit diffraction pattern (a) the central bright image is easily recognized because it is broad and rela-

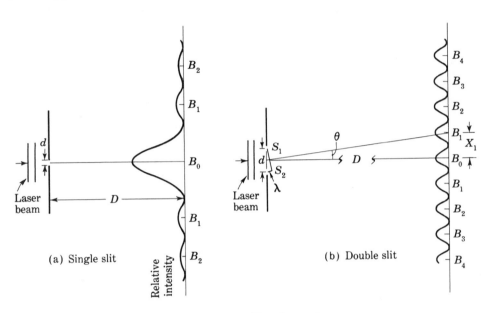

FIGURE 76.4   Diffraction patterns.

tively very intense. Several weak secondary diffraction images can be observed in this experiment. In addition, theory predicts that for a given wave length the diffraction pattern will expand as the slit width is reduced.

In the double-slit diffraction pattern (b) distances between consecutive dark or bright images are equal and the bright images are very nearly equal in intensity.

For constructive interference to occur, such as at B_1, the wavelets emanating from slits S_1 and S_2 must be in phase at point B_1 on the screen, hence the optical paths must differ by one wave length. We can, therefore, write $\lambda/d = X_1/D$ for small angles of θ. Hence we may write the general equation for bright images, as

$$\frac{n\lambda}{d} = \frac{X_n}{D}, \qquad n = 0, 1, 2, \cdots, N \qquad (76.1)$$

Similarly for the dark images, we may write

$$\frac{(n + \frac{1}{2})\lambda}{d} = \frac{X_n}{D}, \qquad n = 0, 1, 2, 3, \cdots, N \qquad (76.2)$$

From Eq. (76.1) we can write

$$X_n = \frac{n\lambda D}{d}$$

and

$$X_{n+1} = \frac{(n+1)\lambda D}{d} \qquad (76.3)$$

Hence

$$X_{n+1} - X_n = \Delta X = \frac{\lambda D}{d} \qquad (76.4)$$

where ΔX is the separation between consecutive bright images. This is also true for dark images.

Method: CAUTION: DO NOT LOOK DIRECTLY INTO THE LASER BEAM OR INTO THE BEAM REFLECTED FROM ANY MIRROR SURFACE BECAUSE OF PROBABLE EYE DAMAGE.

For both single- and double-slit diffraction, mount the slits (Cornell or scribed exposed photographic plate) approximately 10 cm in front of the laser and the screen several meters (2 to 6 m) on the other side of the slits. If the slit is mounted on an optical bench set transverse to the laser beam, one can adjust the diffraction plate laterally. Make certain that the plate is transverse to the laser beam. Usually one observes several bright spots due to multiple reflection. The plate is adjusted for perpendicularity by rotating the plate to cause coincidence of secondary and primary spots.

Single-slit diffraction. Note the single-slit pattern which has a bright and broad central image flanked by images of lesser intensities. Use either a tapered single slit (Cornell plate) or several single slits of different widths to make a comparison study of the diffraction patterns.

This part of the experiment is qualitative and your results might best be summarized by using graphs such as Fig. 76.4(a).

Step 2. Double-slit diffraction. Obtain a clear diffraction pattern by proper adjustments of the laser, double-slit, and screen. There should be ten or more well-defined images in the pattern. Determine the value of ΔX by using the scale values of the successive positions of either the bright or the dark images in the diffraction pattern. The procedure for determining the value of ΔX is essentially the same as that used in Exp. 3A for determining the value of $v\tau$. Record the scale positions of the images and the computed value of ΔX.

Calculate and record the slit separation d for the double slit by use of Eq. (76.4) and the given value of λ for the laser. Compare this value of d with the given value, or that obtained with a traveling microscope.

Photoelectric

Tube

Object: To determine the work function of cesium-antimony. To determine the ratio h/e.

Apparatus: Light-tight box containing phototube RCA #929, 40 w incandescent lamp housed in suitable lamp holder, sharp cut-off Kodak Wratten filters #22, 24, 25, and 29 (or similar), potentiometer (~100 ohms), 2 dry cells, and 1 nano-ammeter (10^{-9} amp).

Theory: Certain metals emit electrons when light falls on their surfaces. This effect was first observed by Hertz in 1887 in his experiments on radio waves. He noted that a spark occurred across a spark gap in his receiver when ultraviolet light from his transmitter fell on the metal balls forming the gap. A year later, Hallwachs was able to show that the spark occurred because the ultraviolet light caused the metals to emit negative electricity, aiding the spark to start. If such a sensitive metal surface is placed in an evacuated tube with a collector plate nearby, then completion of the circuit between emitter and collector outside the tube would permit the establishment of a current. In the conventional sense, this current is from collector to emitter inside the tube, but the electron current, of course, is from the emitter to the collector. If the collector is made positive with respect to the emitter, the current becomes larger with larger potential until a point is reached beyond which no further increase of potential causes an increased current. The collector is now receiving all the electrons emitted by the light-sensitive surface, the effect of space-charge having been overcome.

The amount of electron current emitted by the light-sensitive surface varies directly with the intensity of illumination falling on it. This relation is exact over a very wide range of light intensities, from approximately 0.0001 to 10,000 foot-candles.

Apparent deviation from this linear relationship may be caused by proximity of the sensitive surface to the walls of the tube and by space charges.

The electron current also depends on the wavelength of the light falling on the emitter, and the dependency is not in general a simple one. Different metals have different spectral distribution of preferred frequencies and the emission rises to peaks at these frequencies of light. For each metal, however, there is a maximum wavelength beyond which no emission occurs, regardless of intensity. This *photoelectric threshold* depends essentially on the work function of the material of the emitter. According to Einstein, the energy ϵ of a photon of light is directly proportional to the frequency f. This relation is given by the fundamental equation

$$\epsilon = hf \tag{77.1}$$

where h is Planck's constant. An electron, to escape from a metal surface, must obtain a certain minimum amount of energy, the work function of the metal. In order for light, therefore, to cause the ejection of any electrons from the surface, the frequency must be such that a photon of light has more energy to give up to an electron than the value of the work function of the metal. Any excess energy appears as kinetic energy of the ejected electron, i.e.,

$$hf = w_0 + \tfrac{1}{2}mv_{\max}^2 \tag{77.2}$$

where w_0 is the work function of the metal. Only light above a certain minimum frequency can do this, thus the existence of a photoelectric threshold. The threshold frequency is that which is just sufficient to eject an electron with no kinetic energy. Thus the threshold frequency f_0 is given by the relation

$$hf_0 = w_0 \tag{77.3}$$

The peaks of emission current occurring at different light frequencies come about because of selective absorption; i.e., the sensitive surface is not black, and thus does not absorb all wavelengths of light with equal ease. Therefore, some phototubes have peak outputs for light in the infrared region of the spectrum, some in the ultraviolet region, and some in the visible region.

Method: The housing of the photoelectric tube is made to lie flat on a table with a window on the top side. A light source is then mounted above the photoelectric tube. The lamp and distance are not critical but a 40 w lamp in a goose-neck support at a distance of approximately 1 ft is quite reasonable.

Connect the components as shown in the circuit diagram (Fig. 77.1), and vary the potentiometer until the anode voltage reads zero. With V set at 0 v and the phototube in total darkness, adjust the nano-ammeter to zero current. If now the phototube is illuminated the nano-ammeter will register a current.

If a mechanical shutter is not used a black cloth will serve to shut out the light. Either turn off the lamp or cover the window of the photoelectric tube when changing filters.

With a filter in place and the lamp turned *ON* vary the anode potential until the anode current is returned to zero. When the anode potential is sufficiently negative the most energetic photoelectrons are stopped from reaching the anode. For this

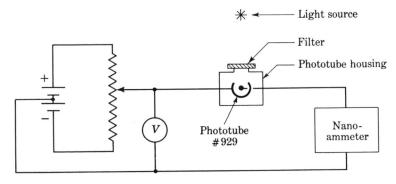

FIGURE 77.1 Phototube circuit diagram.

situation we may set $\frac{1}{2}mv^2_{\max}$ in Eq. (77.2) equal to eV_s and write

$$eV_s = hf - w_0 \qquad \text{or} \qquad V_s = \frac{h}{e}f - \frac{w_0}{e} \qquad (77.3)$$

where V_s is the stopping potential.

Measure the stopping potential for each of four filters and plot a graph of V_s versus f. From the graph determine

1. The work function of the cathode in electron volts.
2. The slope of the curve h/e (see Fig. 77.2) and the value of f_0. Compare with accepted value.

Tables and graphs in Appendix III list accepted values for constants, work functions of various metals, and cut-off wavelengths for the Wratten filters.

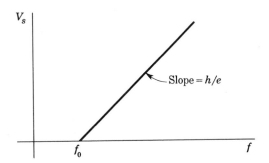

FIGURE 77.2 Graph of V_s versus f.

PART VI

ATOMIC
PHYSICS

Geiger Counter:
Half Life of
Radioactive Indium

Object: To determine (1) the plateau curve for a Geiger counter, and (2) the half life of radioactive indium.

Apparatus: Scaler (Nuclear-Chicago #8770), detector (D-34 end window) plus mount, radioactive source, indium, and neutron source.

Theory: Radioactive decay is a statistical phenomenon. It is impossible to predict when a single radioactive nucleus will decay. However, if there is a very large number N of such radioactive nuclei, then the time rate of decay dN/dt is directly proportional to the number of nuclei. Thus we may write

$$\frac{dN}{dt} = -\lambda N \tag{80.1}$$

where the proportionality constant λ is called the *decay constant* (also the transformation or disintegration constant). λ is a *positive* constant characteristic of the kind of radioactive nuclei under consideration. The negative sign in the equation is necessary since N is always a decreasing function of the time. Integration of Eq. (80.1) gives

$$N = N_0 e^{-\lambda t} \tag{80.2}$$

where N_0 is the number of radioactive nuclei at zero time.

The *half life* T of a large number N_0 of the same kind of radioactive nuclei is defined as the time that it takes for half of this number to decay or disintegrate. Its relation to λ may be found by setting $N = N_0/2$ and $t = T$ in Eq. (80.2). This relation gives

$$T = \frac{\ln 2}{\lambda} = \frac{0.693}{\lambda} \tag{80.3}$$

where ln 2 means $\log_e 2$. The spread in the values of T for different natural radioactive nuclei is enormous, ranging from about 10^{10} yr to 10^{-7} sec. Obviously, nuclei with a long half life are fairly stable, whereas those with a short half life are quite unstable.

In observing radioactive decay, it is more feasible to measure the *rate of decay* (or something proportional to it) than to measure N. In order to get a working equation that involves rates of decay, substitute the value of N from Eq. (80.2) in the right side of Eq. (80.1), obtaining

$$\frac{dN}{dt} = -\lambda N_0 e^{-\lambda t} \tag{80.4}$$

Now, at zero time ($t = 0$), $dN/dt = -\lambda N_0$, thus Eq. (80.4) may be written in the form

$$R = R_0 e^{-\lambda t} \tag{80.5}$$

where R ($= dN/dt$) is the rate of decay at any time and R_0 is that rate at zero time. Equation (80.5) may now be written in logarithmic form

$$\ln R = \ln R_0 - \lambda t \tag{80.6}$$

where $\ln R$ means $\log_e R$. Note that in Eq. (80.6) the $\ln R$ is a *linear* function of t. Therefore, a graph of $\ln R$ plotted against t yields a straight line whose slope is $-\lambda$. Since R may be measured directly by observing the counting rate of a Geiger counter or the discharge rate of an electroscope as a function of time, Eq. (80.6) enables us to determine the value of λ, and thus the half life of the radioactive nuclei under consideration.

In this experiment a Geiger counter is used to determine the rate of decay, and thus the half life of radioactive In^{116} nuclei. Normal indium is a metallic element of atomic number 49. It consists of two isotopes, one of mass number 115 (95.8 %), and the other of mass number 113 (4.2 %). When In^{115} nuclei are bombarded with thermal neutrons, they are converted into In^{116} nuclei, according to the following nuclear reaction

$$In_{49}^{115} + n_0^1 \longrightarrow In_{49}^{116} \tag{80.7}$$

The resulting In_{49}^{116} is radioactive and decays to Sn_{50}^{116} with the emission of β particles. Actually, the reaction given in relation (80.7) is more complex than is indicated, since it turns out that two different forms of In_{49}^{116} are produced. These forms are called nuclear isomers. These isomers have quite different half lives, and reach the end product, Sn_{50}^{116}, by different paths. One of the isomers decays directly to the ground state of Sn_{50}^{116} with the emission of a β particle, according to the reaction

$$In_{49}^{116} \longrightarrow Sn_{50}^{116} + \beta_-^0 \tag{80.8}$$

The half life for this isomer is only 13 sec. This isomer will be of no concern in this experiment, since most of it will have vanished before the experiment gets underway.

The other isomer of In_{50}^{116} decays first to an excited state of Sn_{50}^{116*}, with the emission of a β particle. Then, the excited Sn_{50}^{116*} nucleus decays to the ground state, with the emission of a γ particle. The reaction is as follows

$$In_{49}^{116} \longrightarrow Sn_{50}^{116*} + \beta_-^0$$
$$\phantom{In_{49}^{116} \longrightarrow} \longrightarrow Sn_{50}^{116} + \gamma \tag{80.9}$$

The half life of this isomer is 54.20 min, a convenient time interval to determine during the course of the laboratory period. It is this isomer whose half life is to be determined in the experiment.

Geiger Counter and Scaler. The Geiger counter tube in its usual form is a thin cylindrical conducting shell with a coaxial wire suspended along its axis. The region between the wire and the shell is filled with a gas, such as argon. The tube is connected to a high voltage source, as shown in Fig. 80.1, in order to make the metal tube negative

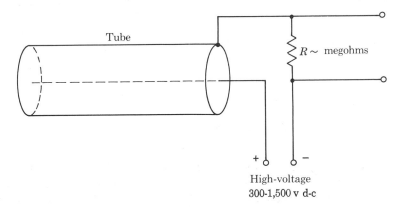

FIGURE 80.1   Geiger tube connections.

with respect to the wire. The voltage across the tube must be greater than a certain threshold value, but less than the voltage that would cause a continuous discharge of electricity through the gas. Under this condition, a slight ionization in the gas produced by the entrance of a single ionizing particle will start a process of cumulative ionization (avalanche) between the electrodes. The avalanche in turn produces a current pulse in the external resistance R. The resulting drop in potential across R reduces the potential across the tube below the threshold value and the current ceases. This process, requiring about 200 μsec, is called the "dead time" of the tube, since a second ionizing particle entering the tube during this time will not produce a separate pulse. Thus the "dead time" of the tube sets an upper limit on the counting rate of the tube. The succession of current pulses in R corresponding to the number of ionizing particles passing into the tube may be picked up with a suitable detector and counted.

The wire-cylinder potential difference V plotted against the counting rate CR, owing to a radioactive source of fixed intensity placed near the tube, is called a plateau curve. A typical plateau curve is shown in Fig. 80.2. At voltage A, pulses begin to appear. As the voltage across the tube is increased, the counting rate increases rapidly to point B and then levels off to an almost constant value until voltage C is reached. Above C, a continuous discharge takes place. Obviously, the operating voltage used in the experiment should lie somewhere in the region BC, since in this region the counting rate is independent of the voltage. The voltage interval BC is commonly 200–400 v wide.

Statistical Errors. Since radioactive decay is a purely statistical process, it should obey statistical laws.[1] Given N radioactive nuclei of the same species, we can observe

[1]See Appendix I, Note B, for a discussion of the Poisson distribution law that applies here.

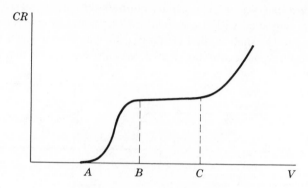

FIGURE 80.2 Pleateau curve.

the number n of these nuclei, those which decay in any fixed time interval τ. But re-peated observations of n, using the same values of N and τ, will not yield the same value of n. There will be a spread about some mean value in the values of n, which appears to be entirely resultant from the statistical nature of the process of decay. It is customary to indicate this spread of values by specifying the standard deviation σ. This is defined as the *square root of the mean square deviation*, in short the root mean square deviation. For example, if a sequence of m observations gives values of n equal to n_1, n_2, n_3, . . . , n_m, then the mean value of n, designated by $\bar{n}$, is defined by the equation

$$\bar{n} = \frac{1}{m}[n_1 + n_2 + \cdots + n_m] = \frac{1}{m}\sum_{i=1}^{m} n_i \qquad (80.10)$$

In a similar manner the mean value of n^2, designated by $\overline{n^2}$, is defined by the equation

$$\overline{n^2} = \frac{1}{m}\sum_{i=1}^{m} n_i^2 \qquad (80.11)$$

Finally, the square of the standard deviation, designated by σ^2, is defined by the equation

$$\sigma^2 = \frac{1}{m}\sum_{i=1}^{m} (n_i - \bar{n})^2 \qquad (80.12)$$

Using the defining equations (80.10), (80.11), and (80.12), it follows that

$$\sigma^2 = \overline{n^2} - \bar{n}^2 \qquad (80.13)$$

The proof of this is left as an exercise for the student.

When the time interval τ is small in comparison with the half life T of the nuclei, it turns out that σ^2 is approximately equal to $\bar{n}$. Therefore, it is customary in decay experiments to assign an error of $\pm n^{1/2}$ to any single observed value of n. Thus if n were observed to be 100, then the error associated with this count would be ± 10, i.e., a 10% error. However, if n were 10,000, then the assigned error would be ± 100, i.e., an error of only 1%. Obviously, it is advantageous in this and similar experiments to make n as large as circumstances will permit in order to reduce the statistical errors.

This may be done by increasing N, or τ, or both. However, experimental conditions usually place an upper limit on the value of N, and the value of τ is limited by the requirement that it must be small compared to T.

Method: *Part I.*

1. Make sure that the leads between the Geiger counter and the scaler unit that contains the power supply are securely connected. [CAUTION: Be sure that the power to the scaler (110 a-c) is off when the connection is made, since voltages as high as 1000 v are used in this experiment.] The Geiger counter tube is housed in a massive iron box, which shields the tube from stray radioactivity. The tube itself is made of very thin metal and is very fragile. It should not be touched or handled in any way by the student without specific instructions from the instructor.

2. Turn the voltage-control knob to its lowest position (counter clockwise) before turning the power on. Also, be sure that the count switch is on.

3. Place the uranium glass (used as radioactive source) on the shelf just below the counter tube and increase the high voltage slowly until the scaler begins to indicate pulses. Pulses should occur at approximately 800 v or less.

4. Determine the counting rate at 50 v intervals from the threshold voltage up to 1000 v, or to a point where the counting rate has increased 10% above the voltage along the plateau, whichever is lower. When raising the high voltage, be sure that the count switch is always turned on so that excessive counting rates may be avoided. Never exceed the high voltage value that raises the counting rate 10% above the plateau value. *Excessive voltage applied to the counter tube will ruin it.*

5. For the remainder of the experiment reduce the high voltage to a value approximately 50 v above the lower end of the plateau region.

6. Plot the data obtained in the previous item, with the counting rate as the ordinate and the high voltage as the abscissa. Indicate the error in counting rate ($n^{1/2}$) by means of a vertical line of appropriate length through each point. Determine the slope of the "plateau" in percentage count-rate per volt.

Part II. (Two hours necessary).

1. Set up with the high voltage about 50 v above the lower knee of the plateau region.

2. Determine the background counting rate for 5 min.

3. The instructor will give you a sample of indium which has been irradiated by slow neutrons. Place the sample on a shelf just below the window of the tube. Take a 3 min count. Wait 3 min and repeat the process. Repeat the above steps for at least $1\frac{1}{2}$ hr. Keep a continuous time record of the experiment.

4. After making the correction for background count, plot the logarithm of the count rate against the time lapsed, assigning the center of the count interval to the corresponding point on the time record. These points should lie on a straight line whose slope is the decay constant [see Eq. (80.6)]. Determine the decay constant and then the half life of the radioactive indium. Compare with the accepted value of 54.20 min. On your plot of the decay curve, indicate the error in counting rate at each point by means of a vertical line. Note that the statistical error in each 3 min count

should be the square root of the total number of counts observed uncorrected for background count.

OPTIONAL EXPERIMENT

1. Check the validity of the equation $\sigma = \bar{n}^{1/2}$ for statistical fluctuations. Use a source with a very long half life. Choose τ so that the number of counts in time τ is rather large compared to the background count. However, the value of τ must be small compared to the length of the laboratory period.

Cloud
Chamber

Object: To become familiar with the operation of a continuously active cloud chamber; to observe alpha, beta, and cosmic ray tracks in the chamber; to estimate the range of the alpha ray tracks from the source provided in this experiment; to compare the absorption of alpha and beta rays.

Apparatus: Raymaster[1] cloud chamber and accessories, 45 v battery, and radioactive sources of beta rays and alpha rays.

Theory: When ionizing particles such as alpha particles (helium nuclei) or beta particles (high-speed electrons) pass through a medium such as a gas, they produce ions along their paths. If this region contains a supersaturated vapor such as water or alcohol, condensation of the vapor takes place on the ions, thus forming a set of small droplets along the tracks. It is then possible to observe these tracks visually under proper conditions of illumination. Note that only the tracks of the alpha or beta particles become visible, not the particles themselves.

The cloud chamber is a closed vessel in which a supersaturated vapor is maintained (usually water vapor or alcohol vapor). This vapor permits visual observation of the track of any ionizing particle passing through the chamber.

The cloud chamber, invented by C. T. R. Wilson in 1911, has played an extremely important role in the field of modern physics. In the cloud chamber one can observe nuclear transmutations of all kinds, along with the birth and death of many new and strange particles.

There are many different types of cloud chambers in use. The one used in this

[1]With the kind permission of Atomic Laboratories, the majority of the written material in this experiment has been taken from the *Manual of Experiments* furnished by the manufacturer of the cloud chamber, Atomic Laboratories, Berkeley, California.

experiment is a *continuously* operating chamber, using methyl alcohol as a vapor and generally called a "diffusion" cloud chamber. This type of cloud chamber was invented in 1939 by Langsdorf.

Part I. Operation of the Raymaster Cloud Chamber. The cloud chamber consists of a cylindrical vessel about 6 in. in diameter and 4 in. high, in which the vapor of methyl alcohol is maintained. A cross section of the chamber is shown in Fig. 81.1.

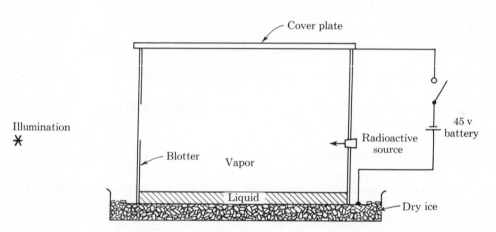

FIGURE 81.1 Cross section of cloud chamber.

In operation, the chamber is viewed from above through the cover glass. Tracks appear in the lower part of the chamber.

Instructions for putting the cloud chamber into operation are as follows:

1. *Carefully* remove, wash, and dry the cover glass. Do not scratch the fused silver electrode around the edge of the cover glass. Unless the cover glass is perfectly clean, tracks in the chamber are difficult to see.

2. Before replacing the clean cover glass, fill the bottom of the chamber with methyl alcohol to a depth of approximately $\frac{1}{8}$ in. Be sure that the level of the alcohol in the chamber is high enough to touch the bottom edge of the black blotter that lines the wall of the cloud chamber. The blotter will then become saturated with alcohol. It must remain saturated during the course of the experiments. Note that there is a black mat on the bottom of the chamber, thus no black dye need be added to the alcohol to produce a black background for viewing the white tracks.

3. After the alcohol is at the proper level in the chamber, replace the cover glass (with the silver electrode on the bottom side) and fit the aluminum top ring upon it, so that there is a snug fit.

4. Powder a sufficient amount of Dry Ice to fill the tray (which holds the chamber) to a depth of about 1 in. Then, place the chamber on the layer of Dry Ice in the tray.

5. Connect the 45 v battery to the chamber in such a way that the *positive* terminal of the battery is connected to the *bottom* of the chamber and the negative terminal is

connected to the top of the chamber. This method of connection enables the operator to switch an electric field, known as the clearing field, on and off across the chamber. This field sweeps out the stray ions in the chamber, which, if left in, would produce fog in the chamber, thus obscuring the tracks. Since the stray negative ions are usually electrons that have a high mobility (velocity per unit field), the lower part of the chamber (the sensitive region) is quickly cleared of stray ions. This clearing of stray ions would not occur so quickly if the connections to the battery were reversed because of the low mobility of positive ions. It is not necessary to keep the clearing field on at all times. In fact, some of the best viewing occurs just after the field is turned off.

6. The source of illumination is placed about 6 in. from the clear space in the cloud chamber side wall and directed so that it skims the top of the alcohol surface in the cloud chamber. With a good light source, the tracks may be viewed directly from above the chamber.

7. The radioactive alpha and beta sources are identified by name and mounted on the heads of needles embedded in rubber stoppers. A plain rubber stopper is also provided. A hole in the side of the chamber allows any one of the stoppers to be inserted tightly so that there are no air leaks. Otherwise, convection currents are set up that are very detrimental to cloud-chamber operation. Sources may be interchanged at any time. The plain rubber stopper may be used for observing cosmic ray tracks. (CAUTION: The alpha source contains radium; the beta source contains strontium 90. Both sources are relatively safe since they contain very small amounts of radioactive material. Handle the sources only by contact with the rubber stoppers. If by chance your hands come into direct contact with the radioactive material on the needle points, wash your hands thoroughly.)

8. After the cloud chamber has been placed upon the Dry Ice, it takes about 5 min for it to cool sufficiently for proper operation. However, a slight precipitation of droplets will be observed almost at the start. Gradually, this precipitation will subside and cosmic ray tracks will begin to appear in the bottom region of the chamber. If radioactive alpha and beta sources are now inserted, alpha and beta tracks will appear. From time to time, it may be necessary to switch on the clearing field. After the chamber has been in operation for about half an hour, the cover glass may get so cold that water vapor from the air may condense on it. Remove this moisture by wiping it off with a clean handkerchief. If this does not suffice, remove the cover glass and rinse it in warm tap water. Dry carefully, replace, and use as before. The latter operation will interrupt the operation of the chamber for 5 or 10 min. Why? When you are through with the cloud chamber, disassemble it and rinse it with tap water. Be sure to save the alcohol by pouring it into a specified beaker. Never leave alcohol in in the chamber overnight. Also, clean up any alcohol that has spilled on the table.

Part II. Properties of Alpha, Beta, and Cosmic Ray Tracks. The purpose of this part of the experiment is to become familiar with the qualitative differences of tracks in the cloud chamber and to learn to identify the various types of tracks by visual observation. Tracks can vary with respect to range (length), density of ionization, which can be inferred from the heaviness of the tracks, and scattering. The last refers to whether the track is quite straight (very little scattering) or changes direction frequently (scatters) because of collisions. In this experiment we shall study the last two properties only. Range will be the subject of Part III.

In order to observe *cosmic ray tracks* in the chamber, operate the chamber with the blank rubber stopper inserted. After becoming familiar with the appearance of these tracks, make a drawing of what they look like. Cosmic ray tracks are for the most part very faint and strongly scattered, although some very straight tracks occasionally occur. The horizontal tracks are usually made by electrons, whereas the vertical tracks are mostly due to μ mesons.

Next, insert the *beta-ray source* into the chamber, observe the appearance of the tracks, and make a drawing of them. The beta source is the radioactive isotope, strontium 90. It emits beta particles (electrons) with an energy of 0.53 mev (million electron volts). The tracks made by these particles are similar to the cosmic ray tracks, but are usually much longer and straighter.

Finally, insert the alpha ray source. Observe and record, by using drawings, the nature of these tracks. This alpha source consists of a Ra^{226} compound. Note that each of the alpha tracks is very dense and straight.

In general, the density of ionization produced by a particle is proportional to the square of its charge and inversely proportional to the square of its speed. Thus it is clear why alpha tracks are so much denser than beta tracks.

Part III. Range of Alpha Particles. The purpose of this part of the experiment is to measure the maximum range of the alpha particles and then to infer the energy of the alpha particles and the half life of the radioactive emitter of these particles.

Measure the range of about 20 alpha particles that are traveling nearly horizontally. To measure these ranges, make a wax pencil mark on the cover glass directly above the source, and a similar mark at the end of each track. The distance between the two marks will be the approximate range of the particle, provided that the track is horizontal. Ranges will be found to vary from very short lengths up to a maximum of about 3.3 cm. Actually, all the alpha particles emitted by Ra^{226} have the same energy (therefore, the same virtual range). However, some of the particles are emitted from under the surface of the emitter and must expend some of their energy in order to reach the surface. This expended energy results in a diminished range in the cloud chamber. Thus it is only the maximum range of the alpha particles that gives information about the initial energy of the particles. Determine this maximum range for the 20 tracks.

The following table gives the relation between the range of alpha particles in air (15°C and 76 cm Hg) and their energies:

Energy (mev)	Range (cm)
0	0
1	0.51
2	1.01
3	1.66
4	2.47
5	3.57
6	4.66
7	5.93

Temperature and pressure corrections need not be made, since they are small. Using this table, determine the initial energy of the alpha particles emitted by Ra^{226}.

The Geiger-Nuttall law may be used to infer the half life of Ra^{226} from the maximum range (in air) of the alpha particles emitted. This law, first discovered empirically and later put on a sound theoretical basis, gives a relation between the range of alpha particles and the half life of the emitter. It may be written

$$\log_{10} R = 0.707 + 0.0174 \log_{10} \lambda \qquad (81.1)$$

where

$$\lambda = \frac{0.693}{T} \text{ (decay constant)},$$

$T = $ half life in sec,
$R = $ range in cm in air.

The constants in Eq. (81.1) are only valid for the Uranium Series. It has been pointed out that the source of alpha particles in this experiment is a compound of Ra^{226}, which has a half life of about 1620 yr, corresponding to a range of emitted alpha particles of about 3.3 cm in air. Occasionally, the source is contaminated with a small amount of other radioactive material, which may emit alpha particles having a quite different range. If such particles are observed, estimate and record the maximum range of these particles.

Part IV. Absorption of Alpha, Beta, and Cosmic Rays. The purpose of this part of the experiment is to compare the penetrating power of the different kinds of rays. To make this comparison, cover the hole in the cloud chamber with various materials, such as paper, cellophane, aluminum foil, and lead foil. In succession, place each radioactive source just outside the hole and note the rays that penetrate the material. Record the results.

Franck-Hertz
Experiment

Object: To study quantized transitions caused by electron collisions with mercury atoms.

Apparatus: Franck-Hertz tube mounted in thermostatically controlled oven (Klinger Scientific Apparatus Company), power supply (fil-6 v; plate 0–60 v), voltmeter (0–50 v), current amplifier (Leybold or similar), potentiometers (1000 ohms and 50,000 ohms), rheostat (2–3 ohms, 6 amp), and battery (3 v).

Theory: The Franck-Hertz tube is essentially a four-electrode tube containing a small amount of mercury. It is used to demonstrate inelastic collisions of slow electrons with mercury atoms. Depending on the type of tube being used, the cathode may be coated with barium oxide, thereby aiding the release of electrons.

At approximately 180°C the mercury within the tube is highly vaporized, so the chance for collisions between electrons and mercury atoms is very good. In the collision process the mercury atom may absorb all or part of the energy of the incident electron. When the kinetic energy of the electron has reached 4.9 ev, mercury resonance has been achieved; i.e., an electron in the outermost orbit of the mercury atom can, by collision, receive the appropriate energy to raise it to its next higher energy state. When this occurs, the current that is registered by the collector plate will show a decided drop. At higher voltages similar results are obtained, with peaks or minima separated by energy differences of 4.9 ev (see Fig. 82.1). The dips in the plate current, therefore, occur where electrons lose energy to the mercury atoms, enough so that they can no longer travel "up hill" from electrode A to the plate P. The first maximum, however, appears at approximately 7 v, which is due to the difference in contact potential between the barium of the cathode and the iron of the collector plate.

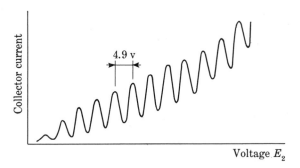

FIGURE 82.1 Collector current versus voltage.

Method: For proper operation the tube is mounted in an oven which is thermostatically controlled to hold the temperature at approximately 180°C. NOTE: Because it takes about 20 min to stabilize the tube at the proper operating temperature, it would be advisable to apply power to the oven at the beginning of the lab period or even earlier.

Figure 82.2 shows a schematic view of the apparatus. E_1 and E_2 are usually common to many portable power supplies. E_3 can be a pair of dry cells.

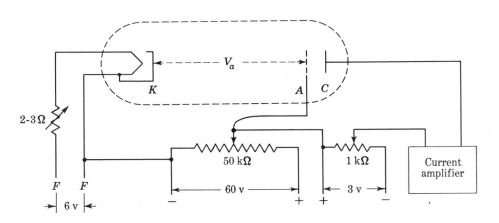

FIGURE 82.2 Circuit diagram.

The amplifier detector is a very sensitive current-detecting device whose output may be ammeter, scope, or x-y plotter. In each case adjust the retarding potential, filament voltage, and amplifier gain, such that several peaks may be observed for a given excursion of the accelerating potential.

Step 1. Plot I_c (collector current) versus V_A (accelerating potential).

Step 2. Determine the first excitation potential of mercury ($^1S_0 - {}^3P_1$). See Fig. 82.3. Is metastable $^3P_0$ also excited? (Energy = 4.66 ev.) Explain.

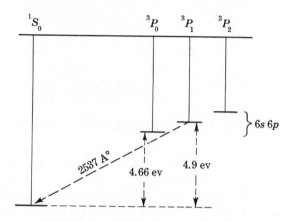

FIGURE 82.3 Energy-level diagram.

A. Method of
Least Squares;
B. Poisson's
Distribution

A. METHOD OF LEAST SQUARES

In 1806 Legendre proposed the principle of least squares as a solution to problems of measurement where the observational data is more than sufficient to determine the unknowns. The simplest example of this is the determination of a physical quantity by repeated direct measurement. We have already discussed this case in some detail in the Introduction. There we described the procedures necessary to obtain from a set of n direct measurements of a quantity the best estimate of its value and standard error. The principle of least squares, closely related to the normal error law, is basic in the theory of measurements.

1. Direct Measurements. We return to our example in the Introduction of n measurements of a quantity having observed values of $X_1, X_2, \ldots, X_n$, where n is greater than one. It is assumed that all of the values are equally reliable, i.e., observed under the same conditions, but that not all of them are the same because of random errors.

We first write the n independent equations of condition in the simple form,

$$X \cong X_i, \qquad i = 1, 2, \ldots, n \tag{A.1}$$

where X is the unknown. Obviously these equations are incompatible, since there is no single value of X that will satisfy all of the equations. What we need is some principle that tells us how to obtain from Eqs. (A.1) the best value of X, i.e., the one that will most nearly satisfy the equations of condition. The principle of least squares tells us how to do this.

We first make Eqs. (A.1) compatible by putting into the equations quantities called deviations and writing the modified equations in the form,

$$X_i - X = d_i, \qquad i = 1, 2, \ldots, n \tag{A.2}$$

where the deviations d_i are completely defined by these equations for any arbitrary value of X. Eqs. (A.2) are now compatible for any value of X. The principle of least squares states that the best (most probable) value of X is that value that reduces the sum of the squares of the deviation to a minimum.

In order to apply this principle, let the sum of the squares of the deviations for any value of X be denoted by $U(X)$. Then

$$U(X) = \sum_{i=1}^{n} (d_i)^2 = \sum_{i=1}^{n} (X_i - X)^2 \tag{A.3}$$

The function $U(X)$ has the following properties:

(a) $U(X)$ is positive for all real values of X, since it is the sum of squares.
(b) $U(X)$ is a quadratic function of X and, therefore, can be represented geometrically by a parabola.
(c) Thus if we plot values of X along the horizontal axis and corresponding values of U along the vertical axis, we obtain a parabola whose axis is vertical and whose vertex lies above the X axis, as shown in Fig. A.1.

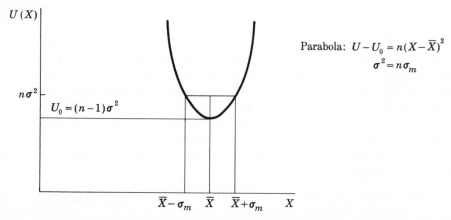

FIGURE A.1   Geometric properties of the parabolic function $U(X)$.

(d) The vertex of the parabola represents the unique minimum of $U(X)$. Therefore the coordinates of this vertex give respectively the minimum value of U and hence, by the principle of least squares, the best value of X. Both coordinates of the vertex are determined from Eq. (A.3) by setting $dU/dX = 0$, the condition for a minimum. When we do this, we find that the X coordinate of the vertex is just the mean of the measured values and that the U coordinate is just the sum of the squares of the deviations from the mean. Thus $U_0 \cong (n - 1)\sigma^2$, where U_0 is the minimum value of $U(X)$ and σ is the standard deviation.
(e) Finally, the equation of the parabola can be written in the compact form,

$$U(X) - U_0 = n(X - \bar{X})^2 \tag{A.4}$$

which is precisely equivalent to Eq. (I.4) when expression (I.4a) is used.

Thus we see that the deviation function $U(X)$ given by Eq. (A.3) and graphically represented in Fig. A.1 is of prime importance in the method of least squares. The general form of the function requires a parabolic representation with a definite orientation. The position of the vertex of the parabola is determined by the values of $\bar{X}$ and U_0. The intrinsic shape of the parabola is determined solely by the value of n, the total number of independent measurements. As n increases, the corresponding parabola gets thinner, since its latus rectum is just $1/n$. Finally it follows from Eq. (A.4) and the relation $U_0 \cong (n - 1)\sigma^2$ that the half-width of the parabola at $U = n\sigma^2$ is just σ_m, the standard error.

It is important to realize that in any specific case we can get all of the information needed from a table of values of U as a function of X. For any given value of X we can always compute the corresponding value of U by Eq. (A.3). If the values of X are carefully chosen for the computation of U values, the table of values reveals by direct inspection approximate values of $\bar{X}$ and U_0 and, hence, of σ and σ_m. In order to illustrate this procedure, we give in Table A.1 the computed values of U for a set of X values in the region of $\bar{X}$, where the X_i are just those taken from Table I.1 in the Introduction.

Table A.1

Values of $U = \sum_{i=1}^{n} (X_i - X)^2$ as a function of X (in the region of $\bar{X}$) from data in Table I.1: $n = 10$, X is in units of 10^{-4} cm, U is in units of 10^{-8} cm^2.

X	4760	4761	4762	4763	4764	4765
U	191	136	102	88	94	120

By direct inspection of Table A.1 we see that U has a minimum close to the value $X = 4763$ and slightly to the right. Good first approximations to the values of $\bar{X}$ and U_0 are respectively 4763 and 88. The corresponding value of σ^2 is $\frac{88}{9}$ or about 10. Now if we add σ^2 to U_0, we get $U = 98$ at which the corresponding X value is $\bar{X} + \sigma_m$. From the table we see that $\sigma_m \cong 1$ unit or 10^{-4} cm. These values of $\bar{X}$ and U_0 may be obtained from Table I.1.

More exact values of $\bar{X}$ and U_0 may be obtained from Table A.1 by use of the interpolation formula obtained from Eq. (A.4) by taking its differential, i.e.,

$$\Delta U = 2n(X - \bar{X})\Delta X \qquad (A.5)$$

where X is to be taken at the center of the interval ΔX. By use of Eq. (A.5) we get for the exact values of $\bar{X}$ and U_0 respectively 4763.2 and 87.6.[1]

The principle of least squares that we have just applied to the direct measurement problem can be extended to more complex situations where the equations of condition come from repeated measurements of one or more functions of the values of several physical quantities. Although the method is very powerful and very general, it has the one disadvantage of requiring an enormous amount of computation compared with

[1] Take $X = 4763$, $\Delta X = (4764 - 4762)$, $\Delta U = (94 - 102)$, and $n = 10$ in Eq. (A.5). Then solve for $\bar{X}$.

other simpler but less accurate methods. With the advent of modern computer facilities now available at many educational institutions there is good reason for making more use of the method of least squares.

2. Indirect Measurements. Very often the physical quantity that we wish to determine (the unknown) appears as a parameter in some functional relation between two or more variables of observation. Frequently, the dependent variable is just a linear function of a single independent variable with the unknown parameters as coefficients. In this case, the determination of the parameters by the method of least squares is especially simple and direct. Geometrically the problem reduces to that of fitting a straight line to the plotted points of the observed values of the variables in order to determine its slope and intercept.

Linear Function: Best Straight Line. Suppose the linear function has the form

$$Y = A + B(X - X_0) \tag{A.6}$$

where the constants A and B are to be determined from a set of observed values of X, Y, say X_j, Y_j, taken in the neighborhood of X_0. We assume that the X_j values are uniformly spaced and ordered so that the running index j takes the integral values $-m, -m+1, \ldots, -1, 0, 1, \ldots, m-1, m$. Thus there are $2m+1$ observed values of X in the sequence with X_0 at its center. We further assume that the X_j values are measured with considerably more precision than are the corresponding Y_j values, so that the random errors in the X_j values are negligible in comparison with the Y_j values. When the observed values of X_j and Y_j are substituted into Eq. (A.6) we get a set of *approximate* equations

$$Y_j \cong A + B(X_j - X_0), \quad j = -m, \ldots, 0, \ldots, m \tag{A.7}$$

that are called the *equations of condition* (or observation).

Equations (A.7) may be simplified by shifting the origin of the X, Y coordinates to the point X_0, Y_0. To do this let $X_j = Y_0 + j\Delta X$, where ΔX is the constant interval $(X_{j+1} - X_j)$; and let $Y_j = Y_0 + y_j$, where y_j are simply the measures of Y_j with respect to Y_0. When we substitute these values of X_j and Y_j into Eq. (A.7), we get

$$Y_0 + y_j \cong A + B j\Delta X, \quad j = -m, \ldots, 0, \ldots, m \tag{A.8}$$

Note that the equation for $j = 0$ is simply $Y_0 \cong A$ which means that $A - Y_0$ is a *small* constant ϵ, called the *zero point correction*. This correction occurs in each of Eqs. (A.8), so the equations may now be written in the compact form

$$y_j \cong \epsilon + bj, \quad j = -m, \ldots, 0, \ldots, m \tag{A.9}$$

where $b = B\Delta X$.

Equations (A.9) are the equations of condition reduced to their simplest form. They are independent and equally reliable equations, since the y_j are *equally reliable measured* quantities. But these equations of condition are also incompatible when their number is greater than the number of unknowns (the usual case). This means that there exists no values of ϵ and b that will satisfy all of the equations. Under these circumstances the best that we can do is to seek a pair of values of ϵ and b that will most nearly satisfy the equations of condition. The principle of least squares tells us how to do this.

We modify Eqs. (A.9) so that they are exact (compatible) by writing

$$y_j - (\epsilon + jb) = d_j, \qquad j = -m, \ldots, 0, \ldots, m \qquad \text{(A.10)}$$

where the d_j, called deviations, are defined by Eqs. (A.10) for any real values of ϵ and b. The principle of least squares states that the most probable values of ϵ and b are those that make the sum of the squares of the deviations an absolute minimum. Thus the principle of least squares as applied to indirectly measured quantities is simply a generalization of its application to directly measured quantities. We therefore proceed in the same manner.

Let the deviation function $U(\epsilon, b)$ be defined by the equation

$$U(\epsilon, b) = \sum_{-m}^{m} (d_j)^2$$
$$= \sum_{-m}^{m} [y_j - (\epsilon + jb)]^2 \qquad \text{(A.11)}$$

for real values of ϵ and b. Then this function U has properties similar to the deviation function used in the previous section except that here U is a function of two independent variables, ϵ and b, rather than just one. In this case $U(\epsilon, b)$ is positive for all real values of ϵ and b and is a quadratic function of these variables. Therefore the function can be represented geometrically in three dimensions (U, ϵ, b) by an elliptic paraboloid as shown in Fig. A.2. We note that the coordinates U_0, ϵ_0, and b_0 of the

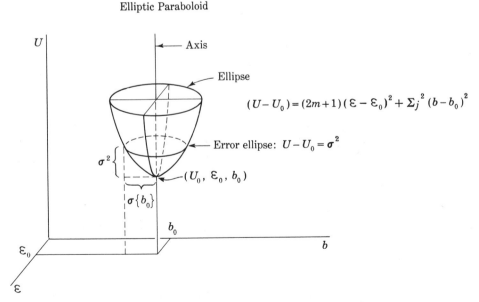

Elliptic Paraboloid

$$(U - U_0) = (2m+1)(\epsilon - \epsilon_0)^2 + \Sigma j^2 (b - b_0)^2$$

Error ellipse: $U - U_0 = \sigma^2$

FIGURE A.2 Geometric properties of the elliptic paraboloidal function $U(\epsilon, b)$.

vertex of the paraboloid give the minimum value of U and the least squares values of ϵ and b.

In order to determine ϵ_0 and b_0, we could, of course, use a "cut and try" method by computing the value of U for various values of ϵ and b until we found the pair that gave the smallest possible value of U. This can actually be done by using a computer. But the usual method (calculus) is to set the partial derivatives of U, with respect to ϵ and with respect to b, equal to zero. When we do this, we obtain two equations, called *normal equations*, whose solutions give the values of ϵ_0 and b_0. The normal equations in this case are

(I) $$\sum_{-m}^{m} [y_j - (\epsilon_0 + jb_0)] = 0 \qquad (A.12)$$

(II) $$\sum_{-m}^{m} [jy_j - (j\epsilon_0 + j^2 b_0)] = 0 \qquad (A.13)$$

In carrying out the summations indicated in Eqs. (A.12) and (A.13), we note that the sum over jb_0 vanishes in the first normal equation and that over $j\epsilon_0$ vanishes in the second. This is a direct consequence of choosing X_0 at the center of the sequence of X_j values rather than somewhere else. See footnotes 1, 2 in Exp. 4. As a consequence of this choice, the first normal equation gives the value of ϵ_0 and the second that of b_0. Thus

$$\epsilon_0 = \frac{\sum_{-m}^{m} y_j}{2m+1} = \frac{\sum_{-m}^{m} Y_j}{2m+1} - Y_0 \qquad (A.14)$$

and $$b_0 = \frac{\sum_{-m}^{m} jy_j}{\sum_{-m}^{m} j^2} = \frac{\sum_{-m}^{m} jY_j}{\frac{1}{3}(m)(m+1)(2m+1)} \qquad (A.15)$$

The corresponding least squares values of A and B in our initial equation (A.6) are therefore

$$A_0 = Y_0 + \epsilon_0 \qquad (A.14a)$$

and $$B_0 = \frac{b_0}{\Delta X} \qquad (A.15a)$$

The method of least squares enables us to compute not only the most probable values of ϵ and b but also their respective standard errors. To do this, however, we must first determine the standard deviation σ in the measured Y_j values. The best estimate of σ for a finite set of observations is given by the familiar relation,

$$\sigma \cong \sqrt{\frac{U_0}{2m-1}} \qquad (A.16)$$

where $2m-1$ is the number of observations minus the number of unknowns, and where

$$U_0 = \sum_{-m}^{m} [y_j - (\epsilon_0 + jb_0)]^2 \qquad (A.17)$$

For large values of m the computations required are extensive and a computer is a necessity.

Once the value of σ has been determined, the standard errors in ϵ_0 and b_0 are respectively

$$\sigma\{\epsilon_0\} = \frac{\sigma}{\sqrt{2m+1}} \tag{A.16a}$$

and
$$\sigma\{b_0\} = \frac{\sigma}{\sqrt{\tfrac{1}{3}(m)(m+1)(2m+1)}} \tag{A.16b}$$

Equations (A. 16a) and (A.16b) are obtained from Eqs. (A.14) and (A.15) respectively by a direct application of the variance theorem.[2]

All of the results that we have obtained in this section are inherent in the deviation function $U(\epsilon, b)$ defined by Eq. (A.11) and represented geometrically by the surface of an elliptic paraboloid shown in Fig. A.2. The measured values of y_j essentially fix the position of the axis of the paraboloid while the number of observations determine the intrinsic shape of the paraboloid and the height of its vertex above the ϵ, b plane; i.e., as the number of observations increases, the value of U_0 increases, and the elliptical cross sections are reduced in size. The elliptical cross section for which $U - U_0 = \sigma^2$ may be called the error ellipse, since its ϵ and b semi-axes may be shown to equal the standard errors in ϵ_0 and b_0 respectively.

3. Quadratic Function: Best Parabola. In the previous section we considered the dependent measurable variable Y to be a linear function of the independent measurable variable X. In this section we consider Y to be a quadratic function of X and write it in the form

$$Y = A + B(X - X_0) + C(X - X_0)^2 \tag{A.18}$$

where again X_0 is an arbitrary reference value of X, and A, B, and C are the unknown parameters to be determined by the method of least squares. We proceed in the same manner and with the same notation as for the linear case. The results are somewhat more complex than in the linear case for we now have three unknown parameters instead of two, but no new difficulties are encountered. For this reason we present the significant features of the least squares analysis in this case but omit the details.

Let the sequence of $2m + 1$ precisely measured values of X with uniform spacing ΔX be represented by X_j, where $j = -m, \ldots, -1, 0, 1, \ldots, m$. The corresponding measured values of Y are Y_j. Then $X_j = X_0 + j\Delta X$ and $Y_j = Y_0 + y_j$. The equations of condition (or observation) are then

$$Y_0 + y_j = A + Bj\Delta X + Cj^2(\Delta X)^2, \qquad j = -m, \ldots, m \tag{A.19}$$

These equations may be simplified by setting $A - Y_0 = \epsilon$, $B\Delta X = b$, and $C(\Delta X)^2 = c$. We get the $2m + 1$ equations of condition (all of equal weight)

$$y_j \cong \epsilon + jb + j^2 c, \qquad j = -m, \ldots, m \tag{A.20}$$

which are to be solved for the least square values of ϵ, b, and c assuming that the number of equations is greater than the number of unknowns and hence that the equations are incompatible.

The principle of least squares states that the most probable values of ϵ, b, and c are

[2]See footnote 6 in the Introduction.

those that make the sum of the squares of the deviations an absolute minimum. These deviations for arbitrary values of the unknowns are defined by the equations

$$y_j - (\epsilon + jb + j^2 c) = d_j, \qquad j = -m, \ldots, m \qquad \text{(A.21)}$$

We define the deviation function $U(\epsilon, b, c)$ by the identity

$$U(\epsilon, b, c) = \sum_{-m}^{m} d_j^2$$

$$= \sum_{-m}^{m} [y_j - (\epsilon + jb + j^2 c)]^2 \qquad \text{(A.22)}$$

where $U(\epsilon, b, c)$ is now a quadratic function of the three parameters ϵ, b, and c.

The necessary conditions that U be a minimum, say U_0, are that $\partial U/\partial \epsilon = \partial U/\partial b = \partial U/\partial c = 0$. These in turn require that $\sum_{-m}^{m} d_j = \sum_{-m}^{m} jd_j = \sum_{-m}^{m} j^2 d_j = 0$. By imposing these conditions for minimum U on Eqs. (A.22) we obtain three normal equations. Upon solving these normal equations explicitly for each of the unknowns, we get

$$\epsilon_0 = \frac{J_4 \sum y_j - J_2 \sum j^2 y_j}{J_0 J_4 - J_2^2} \qquad \text{(A.23)}$$

$$b_0 = \frac{\sum j y_j}{J_2} = \frac{\sum j Y_j}{J_2} \qquad \text{(A.24)}$$

and

$$c_0 = \frac{J_0 \sum j^2 y_j - J_2 \sum y_j}{J_0 J_4 - J_2^2} \qquad \text{(A.25)}$$

where $J_n = \sum j^n$ and *where we assume* that the *summation limits for j are* $-m$ to $+m$ *unless otherwise specified.*

The Eq. (A.24) giving the value of b_0 is especially simple. But the value of c_0 given in Eq. (A.25) is complicated by the presence of ϵ_0, the zero point correction, in the normal equations. If this correction is assumed to be negligible, then $J_4 \sum y_j$ is approximately equal to $J_2 \sum j^2 y_j$ and we get two approximate simple expressions for c_0, namely,

$$c_0 \cong \frac{\sum y_j}{J_2} \qquad \text{(A.26)}$$

and

$$c_0 \cong \frac{\sum j^2 y_j}{J_4} \qquad \text{(A.27)}$$

It may be shown that the second approximation is better than the first and that any significant difference in the values of c_0 given by these two expressions is a clear indication of the significance of the zero point correction. See, for example, the use of these approximate values in Exp. 4 where the quadratic relation (A.18) is used to represent the position ($= Y$) of a freely falling body as a function of the time ($= X$).

Once the least squares values ϵ_0, b_0, and c_0 have been determined, there remains the problem of determining the standard error in each of these values. To do this we must first determine the standard deviation σ in the measured Y_j values. The best estimate of σ is given by the relation

$$\sigma = \sqrt{\frac{U_0}{2m - 2}} \qquad \text{(A.28)}$$

where U_0 is the minimum value of U and $2m - 2$ is the number of observations minus the number of unknowns. In order to compute U_0 it is necessary to evaluate all of the unknowns even though we may be interested in only one. Thus the evaluation of U_0, and thence σ, becomes exceedingly tedious without the use of some computer device.

The relation between σ and the standard error of any of the quantities ϵ_0, b_0, or c_0 is obtained by applying the variance theorem to the corresponding least squares expression for that quantity. Let us take c_0 given by Eq. (A.25) as an example. We note that this equation gives c_0 as a linear function of the y_j values. Therefore it may be written in the form

$$c_0 = \frac{1}{D} \sum \lambda_j y_j \qquad (A.29)$$

where $D = J_0 J_4 - J_2^2$ and where $\lambda_j = J_0 j^2 - J_2$. Therefore

$$\sigma^2\{c_0\} = \left(\frac{\sigma^2}{D^2}\right) \sum \lambda_j^2 = \frac{J_0}{D}\sigma^2 \qquad (A.30)$$

In a similar manner

$$\sigma^2\{\epsilon_0\} = \frac{J_4}{D}\sigma^2 \qquad (A.31)$$

and

$$\sigma^2\{b_0\} = \frac{\sigma^2}{J_2} \qquad (A.32)$$

For purposes of computation we have

$$J_0 = 2m + 1$$
$$J_2 = \tfrac{1}{3}(m)(m+1)(2m+1) \qquad (A.33)$$
$$J_4 = \frac{2m^5}{5} + m^4 + \frac{2m^3}{3} - \frac{m}{15}$$

Further explications of the method of least squares may be found in footnotes (2), (3), and (6) cited in the Introduction. Especially thorough discussions may be found in *The Calculus of Observations* by E. L. Whittaker and G. Robinson (London: Blackie and Son, Ltd., 1929) and *Data Reduction and Error Analysis for the Physical Sciences* by Philip R. Bevington (New York: McGraw-Hill Book Company, 1969).

For an example illustrating the method of least squares when Y is a quadratic function of X, see Exp. 4. There we determine the acceleration of gravity by a least squares analysis of the position-time data of a freely falling body. The equation of motion of a freely falling body given in that experiment by Eqs. (4.4) and (4.5) can always be put into the form of Eqs. (A.18) and (A.19) by setting $S = Y$ and $t = X$. Then $\tau = \Delta X$, $\epsilon = \epsilon$, $v_0\tau = b$ and $\tfrac{1}{2}g\tau^2 = c$.

We have pointed out before that there is an alternative "cut and try" method of determining the least squares value of an unknown and its standard error. A simple example of this method is given in Table A.1 where values of the deviation function U are tabulated for a properly chosen sequence of values of the one unknown parameter X. As shown there, the least squares value of X and its standard error can be obtained from this tabulation practically by inspection.

In theory this method can be extended to any number of unknown parameters.

But the extension is only feasible when computer facilities are available, since the amount of computation required by the method is extensive.

We have used this "cut and try" method in the analysis of the data taken from a tape obtained in Exp. 4. In fact the sample data shown in Exp. 4 is part of these data. In Table A.2 we show the computed values of the deviation function $U(\epsilon, b, g)$ for the tabulated values of ϵ and g and for $b = b_0$. All numerical values in this table and in the following discussion are in cgs units. Inspection of this table clearly indicates a minimum value of U, i.e., $U_0 = 0.021$ cm^2, to two significant figures. The best position of the true minimum seems to be dubious, but of the three U values of 0.021 only the one at the center of the table has the symmetrical distribution of U values about it that is required for a true minimum. Hence the least squares values of ϵ and g are respectively -0.01 cm and 979.1 cm/sec^2.

Table A.2*

Values of $U(\epsilon, b_0, g)$ in cm^2.

g(cm/sec^2) \\ ϵ(cm)	-0.02	-0.01	0.00	
977.9	0.037	0.028	0.024	
978.3	0.031	0.024	0.022	$g_0 - \sigma\{g_0\}$
978.7	0.027	0.021	0.022	
979.1	0.024	0.021	0.024	g_0
979.5	0.022	0.021	0.027	
979.9	0.022	0.024	0.031	$g_0 + \sigma\{g_0\}$
980.3	0.024	0.028	0.037	
	$\epsilon_0 - \sigma\{\epsilon_0\}$	ϵ_0	$\epsilon_0 + \sigma\{\epsilon_0\}$	

*We are indebted to Leo Lake of the School of Physics and Astronomy, University of Minnesota, for his invaluable aid in preparing the master table of U values from which this table is taken.

The function $U = U(\epsilon, b_0, g)$ can be represented by an elliptic paraboloid similar to that shown in Fig. A.2. In that figure the curves for $U = $ constant $> U_0$ are ellipses. If we project these ellipses on to the ϵb plane, they form a family of ellipses with common axes and a common center at $\epsilon = \epsilon_0$ and $b = b_0$.

We can do the same sort of thing with the function $U(\epsilon, b_0, g)$. Curves for $U = $ constant $> U_0$ in the ϵg plane are also a family of ellipses with common axes and a common center at $\epsilon = \epsilon_0$ and $g = g_0$. This feature appears in Table A.2 when we note, for example, the pattern of tabular positions for which $U = 0.024$. The locus of these positions is an ellipse with its center at ϵ_0, g_0 but with tilted axes in this case.

Not only does Table A.2 give us the least squares value of ϵ and g but it gives us, in addition, a simple method of estimating the standard errors of these values. From the value of $U_0 = 0.021$ we can at once get σ^2 by dividing U_0 by the number of observations (29) minus the number of unknowns (3). Thus $\sigma^2 = 0.00081$. The value of U giving the error ellipse is thus $U_0 + \sigma^2 = 0.022$, to two significant figures. By inspection of Table A.2 we see that this error ellipse extends through an interval of g values from

$g_0 - 0.8$ to $g_0 + 0.8$. Now on the basis of the normal law of errors, there is a $\frac{2}{3}$ chance that any repetition of this experiment under the same conditions would give values of g_0 and ϵ_0 that lie somewhere in the error ellipse. Therefore there is the same chance that the new value of g_0 would lie somewhere in the interval of g values just given. This means that the standard error in our value of g_0 is $0.8 = \sigma\{g_0\}$. In a similar manner we see that $\sigma\{\epsilon_0\} = 0.01$. If we take three times the standard error in g_0 as a reasonable upper bound on the random errors in the value of g_0, then the value of the acceleration of gravity given in this experiment could be written (979 ± 2) cm/sec^2. This is the same result as obtained in Exp. 4 by the usual method of least squares.

4. Computer Program for *Least Squares Computations.* As indicated in Note A.1 of Appendix I, computations in least squares determinations are quite tedious. For this reason, the digital computer is of great assistance in such computations. The following program, written in FORTRAN IV, Subset D, is designed to accept data and determine the least squares values of functions up to the quadratic. This program is designed particularly for use with Exp. 4, and the formal statements provide for the proper number of significant figures for each of the values expected in that experiment and use the word "cm" to designate the units of the results. However, the terminology, other than this, is general and can be used with any reasonable data set. Furthermore, the program can be used as is (although this makes much of the computation unnecessary) for linear functions. The constant "c" will drop out (be found nearly or equal to zero) in this case.

The final portion of the program, which computes and prints a table of residuals in the neighborhood of the least squares values for c and ϵ, may be dispensed with by inputting a *zero* in the third line of data. If the table is desired, the entry should be a *one*. The tabulation is designed so that it is centered on the rounded off least squares values for c and ϵ. The intervals in the table depend on the standard errors in these quantities. Thus, for large standard errors, the intervals are automatically selected larger, so that the error ellipses described in Exp. 4 will be included within the area of the table. The *minimum* sizes of the intervals are arbitrarily set, so that even "perfect" data will be shown in a table with some "spread."

The program is liberally sprinkled with "comment" statements, so that the steps may be followed easily. This will assist in converting to other computer languages or FORTRAN subsets used in particular laboratories.

```
C
C     FORTRAN PROGRAM FOR LEAST-SQUARES DETERMINATION OF QUADRATIC FUNCTION
C
      DIMENSION JAY(16), SNEG(16), S(16), SMDSQ(9), EPVALU(9)
      REAL JAY, JAYO
      IG=5
      IP=6
C
C     READ AND WRITE DATA 'OWNERSHIP' IDENTIFICATION
C
      READ(IG,31) XNAME
      WRITE(IP,32) XNAME
```

```
C
C       READ IN NUMBER OF DATA POINTS - M NOT GREATER THAN 16
C
        READ(IG,33) M
C
C       READ INSTRUCTION WHETHER TABLE OF RESIDUALS IS DESIRED (1=YES,0=NO)
C
        READ(IG,33) K
C
C       READ IN DATA SET - SEE EXPERIMENT 4 FOR FORMAT
C
        DO 2 J=1,M
2       READ(IG,34) JAY(J), S(J), SNEG(J)
C
C       COMPUTE SUMS OF DISTANCES AND SUMS OF POWERS OF J
C
        SUMJ=0
        SUMS=0
        SUMSNG=0
        SUMJSQ=0
        SUMJFO=0
        SUMJ2S=0

        DO 10 J=1,M
        SUMJ=SUMJ+JAY(J)
        SUMS=SUMS+S(J)
        SUMSNG=SUMSNG+SNEG(J)
        SUMJSQ=SUMJSQ+JAY(J)**2
        SUMJFO=SUMJFO+JAY(J)**4
        SUMJ2S=SUMJ2S+(JAY(J)**2)*(S(J)+SNEG(J))
10      CONTINUE

C
C       COMPUTE VALUES OF CONSTANTS FROM NORMAL EQUATIONS
C
        WRITE(IP,35)
        C1=(SUMS+SUMSNG)/(2.0*SUMJSQ)
        C2=SUMJ2S/(2.0*SUMJFO)
        WRITE(IP,36) C1
        WRITE(IP,37) C2
        XM=M
        EPSLN=-0.75*(XM**2)*(C2-C1)
        WRITE(IP,38) EPSLN
        B=(SUMS-SUMSNG)/(2.0*SUMJ)
        WRITE(IP,39) B
        C=C2+1.25*(C2-C1)
        WRITE(IP,40) C
        EPSLNN=-1.0*EPSLN
        JAYO=0.0
C
C       COMPUTE LEAST SQUARES DEVIATIONS OF ALL MEASURED VALUES
C
        WRITE(IP,41)
        SUMDSQ=EPSLN**2
        WRITE(IP,42)
        WRITE(IP,43) JAYO,EPSLNN

        DO 20 J=1,M
        D=S(J)-EPSLN-B*JAY(J)-C*JAY(J)**2
        DNEG=SNEG(J)-EPSLN+B*JAY(J)-C*JAY(J)**2
        WRITE(IP,43) JAY(J),D,DNEG
        SUMDSQ=SUMDSQ+D**2+DNEG**2
20      CONTINUE
```

```
C
C       CALCULATE ERRORS IN COMPUTED VALUES
C
        WRITE(IP,44)
        WRITE(IP,45) SUMDSQ
        SIGMA=SQRT(SUMDSQ/(2.0*XM-2.0))
        WRITE(IP,46) SIGMA
        SIGMAC=SQRT(((2.0*XM+1.0)*SIGMA**2)/((2.0*XM+1.0)*2.0*SUMJFO
       1-4.0*SUMJSQ**2))
        WRITE(IP,47) SIGMAC
        SIGMAB=SQRT(SIGMA**2/(2.0*SUMJSQ))
        WRITE(IP,48) SIGMAB
        SIGMEP=SQRT((SUMJFO*SIGMA**2)/((2.0*XM+1.0)*SUMJFO
       1-2.0*SUMJSQ**2))
        WRITE(IP,49) SIGMEP
C
C       CHECK WHETHER CHART OF RESIDUALS IS DESIRED.  IF NOT, CALL EXIT.
C
        IF(K) 100,100,54
C
C       PREPARE CHART OF RESIDUALS
C
   54   WRITE(IP,50)
C
C       ROUND OFF C AND EPSILON FOR USE AS CHART CENTER-VALUES
C
        CX=10000.*C+.5
        ICX=CX
        CX=ICX
        CRND=CX/10000.

        EPX=1000.*EPSLN+.5
        IEPX=EPX
        EPX=IEPX
        EPRND=EPX/1000.
C
C       SET SIZES OF CHART INTERVALS - ROUND OFF AND SET MINIMUM INTERVALS
C
        SIGC=100000.*SIGMAC+.5
        ISIGC=SIGC
        SIGC=ISIGC
        CINT=SIGC/200000.
        IF(CINT-.00003) 55,55,56
   55   CINT=.00003
   56   CONTINUE

        SIGEP=1000.*SIGMEP+.5
        ISIGEP=SIGEP
        SIGEP=ISIGEP
        EPINT=SIGEP/1000.
        IF(EPINT-.005) 57,57,58
   57   EPINT=.005
   58   CONTINUE
C
C       COMPUTE THE RESIDUALS FOR VALUES OF C AND EPSILON SURROUNDING THE
C       LEAST SQUARES VALUES
C
C       COMPUTE INTERVALS AND LABEL THE HORIZONTAL AXIS OF THE CHART
C
        XN=-4.*EPINT
        DO 60 N=1,9
        EPVALU(N)=EPRND+XN
   60   XN=XN+EPINT
```

```
          WRITE(IP,51) (EPVALU(N), N=1,9)
          WRITE(IP,52)
          CVAL=CRND-9.*CINT
C
C         SET UP 19 ROWS OF 'C' VALUES, 'CINT' APART AND CENTERED
C         ON LEAST SQUARES VALUE
C
          DO 90 JC=1,19
          EPVAL=EPRND-4.*EPINT
C
C         SET UP 9 COLUMNS OF 'EPSILON' VALUES, 'EPINT' APART AND
C         CENTERED ON LEAST SQUARES VALUE
C
          DO 80 JEP=1,9


          SMDSQ(JEP)=EPVAL**2
C
C         CALCULATE A RESIDUAL SUM OF SQUARES FOR A PARTICULAR
C         PAIR OF C AND EPSILON VALUES
C
          DO 70 J=1,M
          D=S(J)-EPVAL-B*JAY(J)-CVAL*JAY(J)**2
          DNEG=SNEG(J)-EPVAL+B*JAY(J)-CVAL*JAY(J)**2
          SMDSQ(JEP)=SMDSQ(JEP)+D**2+DNEG**2
70        CONTINUE

          EPVAL=EPVAL+EPINT
80        CONTINUE

          WRITE(IP,53) CVAL,(SMDSQ(JEP), JEP=1,9)

          CVAL=CVAL+CINT
90        CONTINUE

100       CALL EXIT

31        FORMAT (1X,A4)
32        FORMAT (/////44H FALLING BODY LEAST SQUARES COMPUTATIONS FOR,7X,
          1A4)
33        FORMAT (I3)
34        FORMAT (F4.0,2F8.2)
35        FORMAT (/////44H VALUES OF CONSTANTS FROM NORMAL EQUATIONS -)
36        FORMAT (//5H C1 =,F8.5,3H CM)
37        FORMAT (/5H C2 =,F8.5,3H CM)
38        FORMAT (/10H EPSILON =,F8.4,3H CM)
39        FORMAT (/4H B =,F6.3,3H CM)
40        FORMAT (/4H C =,F8.5,3H CM)
41        FORMAT (/////44H LEAST SQUARES DEVIATIONS OF MEASURED VALUES)
42        FORMAT (/15X,1HJ,5X,7HD(J) CM,5X,8HD(-J) CM/)
43        FORMAT (14X,F3.0,4X,F7.5,6X,F7.5)
44        FORMAT (/////18H ERROR ESTIMATES -)
45        FORMAT (/31H LEAST SUM SQUARES DEVIATIONS =,F8.5,6H CM SQ,
          110H (= U-MIN))
46        FORMAT (/43H STANDARD DEVIATION OF MEASURED DISTANCES =,
          1F6.4,3H CM)
47        FORMAT (/22H STANDARD ERROR IN C =,F9.7,3H CM)
48        FORMAT (/22H STANDARD ERROR IN B =,F9.7,3H CM)
49        FORMAT (/28H STANDARD ERROR IN EPSILON =,F9.7,3H CM)
50        FORMAT (/////30H SUM OF SQUARES OF RESIDUALS -)
51        FORMAT (//9H EPSILON ,F6.3,8F8.3)
52        FORMAT (/9H C VALUES)
53        FORMAT (/F9.5,F7.4,8F8.4)

          END
```

```
/DATA
WALL
14
1
 1.    4.95    -4.70
 2.   10.25    -9.15
 3.   15.75   -13.35
 4.   21.50   -17.20
 5.   27.60   -20.80
 6.   33.95   -24.20
 7.   40.55   -27.20
 8.   47.40   -30.05
 9.   54.55   -32.60
10.   62.00   -34.85
11.   69.65   -36.75
12.   77.70   -38.50
13.   85.85   -39.95
14.   94.40   -41.10
/END RUNM.0073 ACTION IN PROGRESS.
                    SIZE OF COMMON  00000    PROGRAM  04476
END OF COMPILATION  MAIN
```

FALLING BODY LEAST SQUARES COMPUTATIONS FOR   WALL

VALUES OF CONSTANTS FROM NORMAL EQUATIONS -

C_1 = 0.13581 CM

C_2 = 0.13589 CM

EPSILON = -0.0116 CM

B = 4.840 CM

C = 0.13599 CM

LEAST SQUARES DEVIATIONS OF MEASURED VALUES

J	D(J) CM	D(-J) CM
0.	0.01157	
1.	0.01489	0.01606
2.	0.03667	0.00144
3.	0.01625	0.04092
4.	0.02615	0.00237
5.	0.00947	0.01420
6.	0.02312	0.04123
7.	0.01477	0.03137
8.	0.01556	0.01798
9.	0.01786	0.03934
10.	0.00788	0.03268
11.	0.03838	0.05200
12.	0.04340	0.01471
13.	0.04683	0.00543
14.	0.00902	0.02419

ERROR ESTIMATES -

LEAST SUM SQUARES DEVIATIONS = 0.02104 CM SQ (= U-MIN)

STANDARD DEVIATION OF MEASURED DISTANCES =0.0284 CM

STANDARD ERROR IN C =0.0000845 CM

STANDARD ERROR IN B =0.0006314 CM

STANDARD ERROR IN EPSILON =0.0079322 CM

SUM OF SQUARES OF RESIDUALS -

EPSILON	-0.043	-0.035	-0.027	-0.019	-0.011	-0.003	0.005	0.013	0.021
C VALUES									
0.13564	0.1230	0.0992	0.0793	0.0630	0.0504	0.0415	0.0364	0.0349	0.0372
0.13568	0.1113	0.0888	0.0701	0.0552	0.0439	0.0363	0.0325	0.0323	0.0359
0.13572	0.1004	0.0792	0.0618	0.0482	0.0382	0.0319	0.0294	0.0305	0.0354
0.13576	0.0903	0.0705	0.0544	0.0420	0.0333	0.0283	0.0271	0.0295	0.0357
0.13580	0.0810	0.0625	0.0477	0.0366	0.0292	0.0256	0.0256	0.0294	0.0368
0.13584	0.0726	0.0554	0.0419	0.0321	0.0260	0.0236	0.0250	0.0300	0.0388
0.13588	0.0649	0.0490	0.0368	0.0283	0.0236	0.0225	0.0251	0.0315	0.0416
0.13592	0.0581	0.0435	0.0326	0.0254	0.0219	0.0222	0.0261	0.0338	0.0451
0.13596	0.0521	0.0388	0.0292	0.0233	0.0211	0.0227	0.0279	0.0369	0.0495
0.13600	0.0470	0.0349	0.0266	0.0220	0.0212	0.0240	0.0305	0.0408	0.0548
0.13604	0.0426	0.0319	0.0249	0.0216	0.0220	0.0261	0.0340	0.0455	0.0608
0.13608	0.0391	0.0296	0.0239	0.0219	0.0237	0.0291	0.0382	0.0511	0.0676
0.13612	0.0363	0.0282	0.0238	0.0231	0.0261	0.0329	0.0433	0.0574	0.0753
0.13616	0.0344	0.0276	0.0245	0.0251	0.0294	0.0374	0.0492	0.0646	0.0838
0.13620	0.0333	0.0278	0.0260	0.0279	0.0335	0.0428	0.0559	0.0726	0.0931
0.13624	0.0331	0.0288	0.0283	0.0315	0.0384	0.0491	0.0634	0.0814	0.1032
0.13628	0.0336	0.0307	0.0315	0.0360	0.0442	0.0561	0.0717	0.0911	0.1141
0.13632	0.0350	0.0333	0.0354	0.0412	0.0507	0.0640	0.0809	0.1015	0.1259
0.13636	0.0371	0.0368	0.0402	0.0473	0.0581	0.0726	0.0909	0.1128	0.1385

M.0070 ACTION COMPLETE.

B. POISSON'S DISTRIBUTION

The random fluctuations that occur in nuclear disintegration come from the statistical nature of nuclear decay. But, as indicated in Exp. 2, this randomness does not obey the normal distribution law but rather one known as *Poisson's distribution.* In contrast with the former, this latter distribution is both discontinuous and asymmetrical. The Poisson distribution is especially useful in particle counting experiments of many different types.

Since Poisson's distribution is a limiting case of the binomial distribution (as is the normal distribution), we start with the binomial distribution and its relation to the probability theory of radioactive decay.

A single radioactive nucleus observed during some time interval may persist or may decay. Let p equal the probability that it persists and q equal the probability that it decays in time interval τ. Then $p + q = 1$. If now a large number N of such nuclei are observed in the same time interval τ, then there is a definite probability $w_N(n)$ that n of these will decay in this time. The value of n may take any integral value from 0 to N. The corresponding probability function $w_N(n)$ is called the *probability density* or *distribution function* and must always satisfy the fundamental relation

$$\sum_{n=0}^{N} w_N(n) = 1 \qquad \text{(B.1)}$$

It may be shown that the values of $w_N(n)$, $n = 0, 1, 2, \ldots, N$, are just the terms in the binomial expansion:

$$(p + q)^N = \sum_{n=0}^{N} \frac{N!}{n!(N-n)!} p^{N-n}q^n$$
$$= \sum_{n=0}^{N} w_N(n) = 1 \qquad \text{(B.2)}$$

This expansion is of prime importance in the theory of probability and is associated with the names of Newton and Bernoulli, each of whom made notable contributions in this field of analysis.

In accordance with Eq. (B.2), we write

$$w_N(n) = \frac{N!}{n!(N-n)!} p^{N-n}q^n, \qquad n = 0, 1, 2, \ldots, N \qquad \text{(B.3)}$$

It is interesting to observe that the binomial coefficient $N!/n!(N-n)!$ is just the number of ways in which n indistinguishable labels (decay labels) can be put on N objects (the nuclei).

Once the probability distribution function is known, as in Eq. (B.3), it is possible to determine the mean value of any function of n. For example, the mean value of n itself will be given by the relation

$$\bar{n} = \sum_{n=0}^{N} n w_N(n) \qquad \text{(B.4)}$$

since the sum on the right is just the weighted average of all possible values of n; i.e., each term in the series is the product of a specific value of n multiplied by the probability of its occurrence. In a similar manner the mean value of any function of n can be computed by the equation

$$\overline{f(n)} = \sum_{n=0}^{N} f(n)w_N(n) \tag{B.5}$$

Let us first compute the mean value of n by means of Eq. (B.4) after substituting in it the values of $w_N(n)$ given in Eq. (B.3). We have

$$\begin{aligned}
\bar{n} &= \sum_{n=0}^{N} n \frac{N!}{n!(N-n)!} p^{N-n} q^n \\
&= \sum_{n=1}^{N} n \frac{N!}{n!(N-n)!} p^{N-n} q^n \\
&= Nq \sum_{n=1}^{N} \frac{(N-1)!}{(n-1)!(N-n)!} p^{N-n} q^{n-1} \\
&= Nq(p+q)^{N-1} = Nq
\end{aligned} \tag{B.6}$$

Here we observe that the mean value of n equals the total number of nuclei N multiplied by q, the probability of decay in time interval τ for a single nucleus. If $\bar{n}$ and N can be determined by experiment, we are able to compute q by means of this relation. In fact this relation, for most practical purposes, *defines* the meaning of q, the probability of decay in the time interval τ, for sufficiently large values of N.

The next problem is to determine the spread in the individual values of n about the mean value $\bar{n}$. We start with the determination of the mean value of $n(n-1)$ which according to Eq. (B.5) is

$$\begin{aligned}
\overline{n(n-1)} &= \sum_{n=0}^{N} n(n-1)w_N(n) \\
&= \sum_{n=0}^{N} n(n-1) \frac{N!}{n!(N-n)!} p^{N-n} q^n \\
&= \sum_{n=2}^{N} \frac{N!}{(n-2)!(N-n)!} p^{N-n} q^n \\
&= N(N-1)q^2(p+q)^{N-2} = N(N-1)q^2 \tag{B.7}
\end{aligned}$$

Now $\overline{n(n-1)} = \overline{n^2} - \bar{n}$, hence Eq. (B.7) may be written as

$$\begin{aligned}
\overline{n^2} - \bar{n} &= N^2 q^2 - Nq^2 \\
&= \bar{n}^2 - \bar{n}q \tag{B.8}
\end{aligned}$$

by use of Eq. (B.6). Rearranging the terms in Eq. (B.8) and making use of the facts that $1 - q = p$ and that the square of the standard deviation equals $\overline{n^2} - \bar{n}^2$, we get

$$\sigma^2 = \overline{n^2} - \bar{n}^2 = \bar{n}p \tag{B.9}$$

and hence,

$$\sigma = (\bar{n}p)^{1/2} \tag{B.10}$$

a result of considerable importance in the theory of statistical fluctuations.

If we now assume that the chosen time interval τ is sufficiently small compared to the half life of the nuclei so that the probability of persistence p in this time interval is almost equal to 1, we get

$$\sigma \cong \bar{n}^{1/2} \tag{B.11}$$

The Newton-Bernoulli distribution function given by Eq. (B.3) is of fundamental importance in statistical theory, but its form is unsuitable for computational purposes when N is large. Because of this fact several successful attempts have been made to approximate this function by functions more amenable to computation. The range of validity of any of these approximate functions is of course restricted by the assumptions made in obtaining them. Perhaps the best known of the approximate functions is the famous Gaussian distribution function used so extensively in the treatment of random errors. But the distribution function due to Poisson is much more applicable to problems of radioactive decay and similar counting problems, since the assumptions made in obtaining it usually correspond closely with the experimental conditions.

The Poisson distribution function is obtained from the Newton-Bernoulli function Eq. (B.3) by assuming that the product Nq remains finite even when N becomes infinitely large provided, of course, that the time interval τ and the probability q both go to zero. Let $Nq = \nu$, a finite parameter, for this limiting process. Accordingly in Eq. (B.3) we replace q by ν/N and p by $(1 - \nu/N)$ obtaining

$$w_N(n) = \frac{N!}{n!(N-n)!}\left(1 - \frac{\nu}{N}\right)^{N-n}\left(\frac{\nu}{N}\right)^n$$

$$= \frac{\nu^n}{n!}\left(1 - \frac{\nu}{N}\right)^N\left[\frac{N!\left(1 - \frac{\nu}{N}\right)^{N-n}}{N^n(N-n)!}\right] \tag{B.12}$$

We then take the limit of Eq. (B.12) as N goes to infinity keeping n and ν fixed. In this limit $(1 - \nu/N)^N$ becomes $e^{-\nu}$, the quantity in the bracket becomes 1, and $w_N(n)$ is replaced by $P(n, \nu)$, the Poisson distribution function. We get

$$P(n, \nu) = \frac{\nu^n}{n!}e^{-\nu} \tag{B.13}$$

where n is any non-negative integer and ν is any non-negative parameter (not necessarily integral).

The function $P(n, \nu)$ is a much simpler function to deal with than is the function $w_N(n)$ for large N and small q, but it has the same significant properties. For example the function $P(n, \nu)$ satisfies the three equations:

$$\sum_{n=0}^{\infty} P(n, \nu) = 1 \tag{B.14}$$

$$\sum_{n=0}^{\infty} nP(n, \nu) = \nu \tag{B.15}$$

$$\sum_{n=0}^{\infty} n(n-1)P(n, \nu) = \nu^2 \tag{B.16}$$

We conclude from these three equations that: (1) the Poisson distribution function is normalized; (2) the parameter ν equals the weighted mean value of n; and (3) the square of the parameter ν equals the weighted mean value of $n(n-1)$ from which we infer the $\nu^{1/2}$ equals the standard deviation σ of the variable n.

In Fig. B.1 we plot the discrete values of $P(n, \nu)$ for $\nu = 2$ and $n = 0, 1, 2, \ldots, 7$. In this case the asymmetry of the distribution is quite marked. If we carry out the same procedure for $\nu = 10$, for example, we find that the pattern of significant P

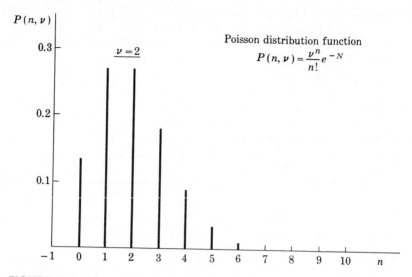

FIGURE B.1 Plot of the Values of $P(n, \nu)$ for $\nu = 2$ and $n = 0, 1, \ldots, 7$.

values shifts to the right and spreads out about a peak value near $n = 10$. The distribution in this latter case is much more symmetrical than it is for $\nu = 2$, and appears to be much like a normal distribution. The student should verify these statements by computing and plotting the values of $P(n, 10)$ for $n = 0, 1, 2, \ldots, 20$.

It is possible to show that, for values of $\nu \geq 30$, the Poisson distribution is practically equivalent to a normal distribution in which $\sigma = \nu^{1/2}$. The results in Exp. 2 should indicate this fact.

Notes on
Equipment

A. THE VERNIER

1. The vernier is a convenient attachment for accurately determining a fraction of the finest division on the main scale of a measuring instrument. A portion of a typical main scale of an instrument without a vernier is shown in Fig. A.1. Usually, the main scale is fixed in some manner and a sliding index indicates the position on the scale corresponding to the measurement in question.

In Fig. A.1, the finest division on the main scale is $\frac{1}{10}$ cm, and by the position of the index it is known that the measurement in question is between 2.3 and 2.4 cm.

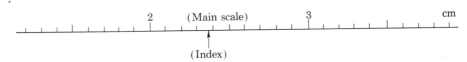

(Index)

FIGURE A.1   Interpretation of main-scale reading (no vernier).

We can *estimate* the fraction of the division to be $\frac{7}{10}$ cm so that the measurement is 2.37 cm, with the first decimal place known accurately and the second one estimated.

The vernier scale is an auxiliary to the main scale and its divisions are different from those of the main scale, but related to them in a simple manner. The zero mark of the vernier scale takes the place of the simple index in the illustration above. However, using the vernier scale, the fraction of the smallest division on the main scale can be read accurately.

Figure A.2 illustrates the same main scale as in Fig. A.1, with the sliding index now replaced by a vernier scale. Note that the same measurement (2.37 cm) is indicated.

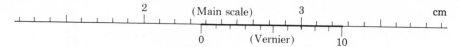

FIGURE A.2   Interpretation of main-scale reading (with vernier).

In the case of a centimeter scale, such as the one shown, the most convenient fraction of the smallest division of the main scale is $\frac{1}{10}$. In order to make such a measurement, the vernier scale is so graduated that ten of its divisions correspond exactly to nine of the main scale divisions. That is to say, each vernier division is only $\frac{9}{10}$ as long as a main scale division. In Fig. A.3 we see the vernier with its num-

FIGURE A.3   Scale reading (with vernier).

ber-one mark exactly corresponding to a mark on the main scale (in this case the 2.8 cm mark). Since the vernier division is only $\frac{9}{10}$ as long as a main division, the vernier's *index* must be $\frac{1}{10}$ of a main division to the right of the 2.7 cm mark on the main scale. Remembering that the vernier index shows the actual measurement, this reading must, therefore, be 2.71 cm.

Now, if the sliding vernier were moved another $\frac{1}{10}$ of a main division to the right, the vernier's number *two* mark would be exactly opposite a mark on the main scale. (*Which* mark on the main scale, of course, does not matter, since it is the position of the vernier *index* which gives the reading.) This means now that the vernier index is exactly $\frac{2}{10}$ of a main scale division to the right of the 2.7 cm mark on the main scale. The actual reading, therefore, is now 2.72 cm.

Now the difference in length between a main division and a vernier division is called the *least count* of the vernier. In the case just studied the least count was 0.01 cm. *It is also clear that the closest measurement that can be made accurately* (coincidence of a vernier line and a main line) *is just the least count.* We have seen that when the number-one mark of the vernier is exactly opposite a mark on the main scale, the vernier *index* is a distance equal to the least count to the right of the next lower main scale mark; when the number-two mark of the vernier coincides with a main scale mark, the index has moved two times the least count; and so on. Thus, by noting which vernier line coincides with a main scale mark, we know immediately what fraction of a main division the vernier index has moved.

Now, looking back at Fig. A.2, which shows the same reading as Fig. A.1, we can read accurately that the measurement is 2.37 cm, where now the last figure is read directly, not estimated.

2. *Vernier Caliper.* In Fig. A.4 an ordinary vernier caliper, measuring the diameter of a cylinder that is placed between the jaws of the caliper, is shown. The left-

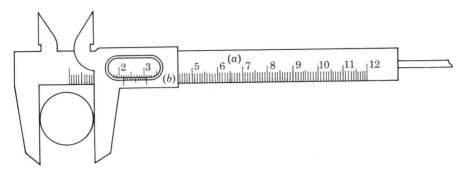

FIGURE A.4 Vernier caliper.

hand jaw is fixed to the main scale (a) and is perpendicular to it. The right-hand jaw is parallel to the left-hand jaw, but can slide along the main scale and carries with it the vernier scale (b).

When the two jaws touch each other, the index of the vernier scale should be exactly coincident with the zero mark on the main scale. The two upper jaws are for measuring *inside* dimensions of a cavity, whereas the protruding shaft at the right is for measuring the depth of holes.

3. *Zero Error.* The actual dimension of an object will be the difference in readings when it is measured and when the calipers are closed. Ideally, the caliper will read exactly zero when closed; if it does not, there is a *zero error* that must be taken into account. This zero error will, of course, be the same for all measurements, and must be subtracted from all readings. If, with the jaws closed, the reading is positive, the zero error is said to be positive.

4. *Angular Measurement.* The verniers used on instruments that measure angles are constructed in the same manner as the one described above. A common main scale on a spectrometer is one that is marked every $\frac{1}{3}$ of a degree. The vernier scale may run from -2 divisions up to 20 divisions, with each one-half division marked. With such an arrangement, the vernier scale from 0 to 20 covers exactly $13°$ on the main scale. That is to say, 40 vernier divisions equal 39 main divisions, so the least count is $\frac{1}{40}$. Since each main scale division is $20'$ of arc, each vernier mark will then indicate $\frac{1}{40}$ of this or $\frac{1}{2}'$. Thus the vernier will read directly in minutes and half minutes if it is numbered from 0 to 20. The negative $2'$ are for convenience. Note that the vernier reading must be added to the reading of its index. That is, if the index shows $65°20'$ plus, and the vernier shows $12\frac{1}{2}'$, the measurement is $65°32\frac{1}{2}'$.

B. THE MICROMETER

1. In its simplest form, the micrometer is simply a screw, very accurately made so as to be uniform along its entire length, moving in a fixed nut. The pitch of the screw is known, and the head of the screw moves forward by the amount of this

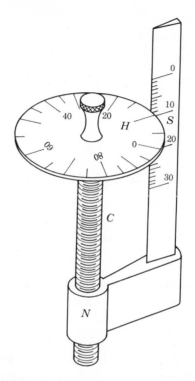

FIGURE B.1   Micrometer screw and dial.

pitch for one complete revolution. A scale is attached to the screw like a wheel to an axle, and by means of the scale, the fraction of a revolution may be measured (see Fig. B.1). For instance, if the scale is divided into 100 parts, then turning the screw and the scale through one such part will advance the screw 0.01 of its pitch. A fixed scale is usually mounted on the nut in which the micrometer screw turns, to count the number of whole revolutions. The moving scale is called the "micrometer head."

2. *Micrometer Caliper* (Fig. B.2). The micrometer caliper is a modification of the simple micrometer principle. The fixed nut and scale are part of the frame F. The knurled knobs K and K' and the moving scale H are mounted on the accurate screw. The pitch of the screw in the better calipers is $\frac{1}{2}$ *mm*. That is, one complete turn of K will advance the point P of the screw $\frac{1}{2}$ mm toward the anvil A, which is a part of the frame and may or may not be adjustable.

On a micrometer with this pitch, the movable scale H is divided into 50 parts so that each part represents $\frac{1}{50}$ of $\frac{1}{2}$ mm, or 0.01 mm. In other words, in *two* complete turns, the moving scale turns past 100 divisions, and advances the screw 1 mm. To indicate $\frac{1}{2}$ mm (one turn), the fixed scale S has alternate marks set below the full millimeter marks (see the sketch Fig. B.3). In this sketch the movable scale has

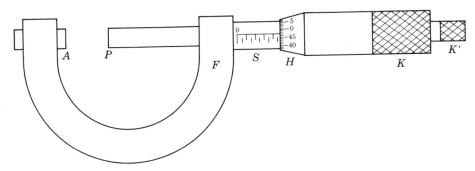

FIGURE B.2   Micrometer.

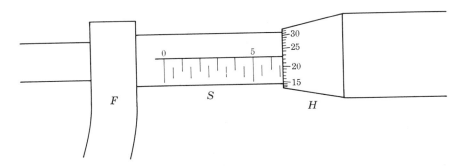

FIGURE B.3   Micrometer scale.

turned beyond a $\frac{1}{2}$ mm mark. The reading (remembering to estimate $\frac{1}{10}$ of the smallest scale division) is 6.727 mm.

The zero error must be subtracted from the reading made with the calipers. This error is determined by closing the caliper—bringing the point P of the screw in contact with the anvil A. If the reading of the scales is positive, the zero error is positive.

In marking any measurements, the knob K' is to be used so as not to force the instrument. This knob will continue to turn after the screw stops, with a little friction or with the clicking of a ratchet. Closing the jaws on the measured object *gently* until this point is reached will ensure the same force being applied at each measurement, and prevent damage to the screw and jaws. The object measured should be held loosely so that it will be able to align itself perpendicularly to the jaws.

C. THE TRAVELING MICROSCOPE

The traveling microscope, or micrometer, is used to measure short distances with accuracy. It consists of a low-power microscope mounted on a moving chassis, which slides along a carefully constructed track. The sliding motion is produced by a micrometer-screw arrangement (see the preceding section, Appendix II, Note B).

In the usual case the pitch of the thread is $\frac{1}{2}$ mm, and the movable scale (the "micrometer head") has 100 divisions. Thus to move the microscope a full millimeter,

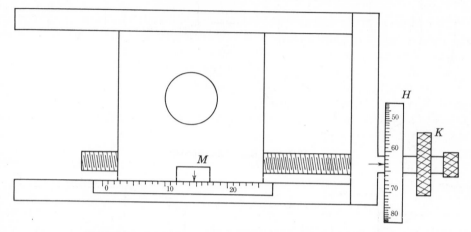

FIGURE C.1 Micrometer screw on traveling microscope.

the head must be turned twice, or past 200 divisions. This movement imposes special care on the user, since each division has a value of $\frac{1}{200}$ mm; it means that the reading on the micrometer head must be divided by 2 to get the motion in hundredths of a millimeter. The index mark M will indicate on a fixed scale the travel in full millimeters, and its position must be noted carefully to find whether the micrometer head is in the first or second turn of the millimeter in question. The reading of the index in Fig. C.1 is 14.5 mm plus. The micrometer head indicates 63.3% of a revolution, or 31.65 hundredths of a millimeter. The index shows the head to be in the second revolution, so our reading, upon adding the two figures, is 14.8165 mm.

The main precaution to be taken in using this instrument concerns *lost motion*. The screw is purposely cut so that there is some free play in the instrument. It will be noticed that the micrometer head may be turned back and forth through several divisions, without causing any motion of the chassis. It is obvious, therefore, that readings must always be taken *with the screw turning in the same direction*. Whether this is to be clockwise or counter-clockwise is decided by noticing which direction gives the closest check on zero. That is to say, when the micrometer head is on zero, the index M should be exactly on a line or halfway between two lines of the fixed scale. In taking a reading, if the screw is accidentally turned too far, back it off approximately one turn and approach the setting from the same side as before.

Do not handle the micrometer head H, but always turn the screw by means of the knurled knob K.

In using the microscope, the eyepiece must first be adjusted so that a sharp image of the cross hairs will be formed. Then, the main tube must be moved up and down (moving the whole microscope) until the point under observation is in good focus. This can be checked by seeing whether there is any parallax, that is, whether the cross hairs seem to move with respect to the image when the eye is moved from side to side. If such parallax is found, it must be eliminated by further adjustment of the eyepiece and refocusing the microscope (see Note E).

D. THE MERCURY BAROMETER

The mercury in the mercury barometer partially fills an evacuated glass tube and a reservoir, which is open to the air. The atmospheric pressure is balanced by the column of mercury, and the height of this column is measured from the surface of the mercury in the reservoir. This surface must be brought to a fixed known height in order for the calibration of the column to be accurate. To establish this height, a small ivory pointer is provided above the mercury surface in the reservoir, and an adjusting screw, located below the reservoir, will raise or lower the mercury in the reservoir.

To use the barometer, first turn this adjusting screw until the surface of the mercury is just at the level of the tip of the ivory pointer. This level has been reached when the point and its image in the mercury appear just to touch each other. Then, standing with the eye just at the level of the top of the mercury column, adjust the movable tube surrounding the column by means of the knob at the right-hand side of the barometer, until the bottom of the tube is just at the level of the meniscus of the mercury; that is, until light from behind the barometer is just shut off by the tube's edges coming to the level of the mercury. A vernier attached to this moving tube will then allow the height to be read to the $\frac{1}{20}$ mm.

For accurate work, the effect of temperature on the density of mercury and the length of the scale must be taken into account. To reduce the readings as obtained by the above directions for standard conditions (the barometer is calibrated at 0°C), use Table F in Appendix III.

E. PARALLAX

This phenomenon is defined as the apparent displacement of one body with respect to another when the position of the *observer* is changed. As an example, two pencils may be held at arm's length in line with one eye, but with one pencil a few inches behind the other. Hold one with the point up, and the other above and behind it with the point down. Without moving the pencils, move the head from side to side. The pencils appear to shift with respect to each other, but as they are brought closer together the shift becomes less pronounced, until one is actually above the other. They will continue to appear so, regardless of the angle from which viewed.

Parallax may be put to use in reading instruments like electric meters, in which a pointer indicates the reading by means of a dial. Since the pointer is a certain distance above the dial, the reading obtained depends upon the angle of observation. The correct reading will be the one obtained when the line of sight passes through the pointer, perpendicular to the plane of the dial. Some dials are equipped with mirrors. The perpendicular line is then easy to find: the line of sight will pass through both the pointer and its image, that is, the pointer appears to be directly above its image and covers it.

In optical instruments parallax may be used to determine when two images are in the same plane, or when a set of cross hairs is accurately placed in the plane of an image. A motion of the head back and forth should not produce any relative motion of the two images.

F. ELECTRICAL CIRCUIT ELEMENTS

1. *Standard symbols* for the various elements encountered in electric circuits are given in the table below:

Connecting wire (negligible resistance) . . . . _____

Resistance (fixed) . . . . R ⌐\/\/\/⌐

(NOTE: Symbol "Ω" stands for "ohms.")

Variable resistance (two terminals) such as a dial box . . . . ⌐\/\/\/⌐

Rheostat or potentiometer (three terminals) . . . . ⌐\/\/\/⌐

Capacitance (fixed) . . . . ⊣⊢

Capacitance (variable) . . . . ⊣⊬

Inductance . . . . ⌐ΟΟΟΟΟ⌐

Inductance (iron core) or choke . . . . ⌐ΟΟΟΟΟ⌐

Transformer (or mutual inductance) . . . .

Transformer (iron core) . . . .

Autotransformer . . . .

Voltaic cell . . . . ⌐ − ⊣⊢ + ⌐

Battery (two or more cells in series) . . . . ⌐ − ⊣|⊢ + ⌐

Fuse . . . . ⌐Ο∿Ο⌐

Switch [single-pole single-throw (SPST)] . . . . ⌐•/ Ο⌐

Switch [double-pole double-throw (DPDT)] . . . .

Tap key (SPST momentary contact) . . . .

Galvanometer . . . . ⌐(G)⌐

Ammeter .... —(A)—

Milliammeter .... —(MA)—

Vacuum tubes:

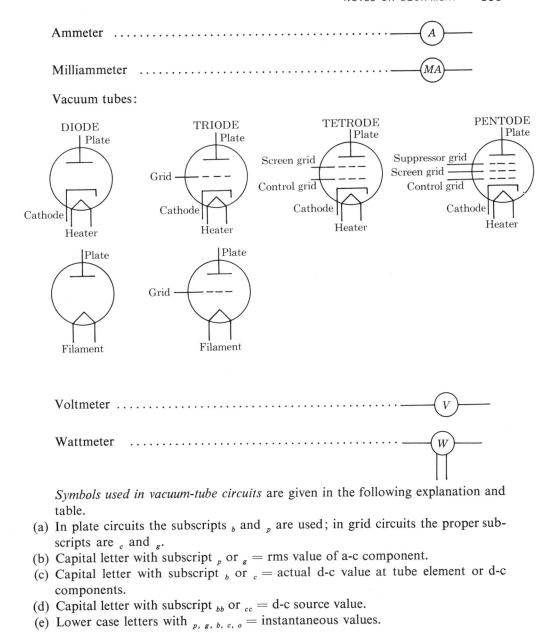

Voltmeter .... —(V)—

Wattmeter .... —(W)—

Symbols used in vacuum-tube circuits are given in the following explanation and table.

(a) In plate circuits the subscripts $_b$ and $_p$ are used; in grid circuits the proper subscripts are $_c$ and $_g$.

(b) Capital letter with subscript $_p$ or $_g$ = rms value of a-c component.

(c) Capital letter with subscript $_b$ or $_c$ = actual d-c value at tube element or d-c components.

(d) Capital letter with subscript $_{bb}$ or $_{cc}$ = d-c source value.

(e) Lower case letters with $_{p, g, b, c, o}$ = instantaneous values.

$$\text{Amplification factor of tube} \left(= -\frac{\partial e_b}{\partial e_c} \right) \quad .... \quad \mu$$

$$\text{Plate resistance of tube} \left(= \frac{\partial e_b}{\partial i_b} \right) \quad .... \quad r_p$$

$$\text{Control-grid, plate transconductance} \left(\frac{\partial i_b}{\partial e_c} \right) \quad .... \quad g_m$$

Plate resistor (load resistor) .... R_L
Net d-c resistance of external plate circuit .... R_b
Net impedance of external plate circuit .... z_b

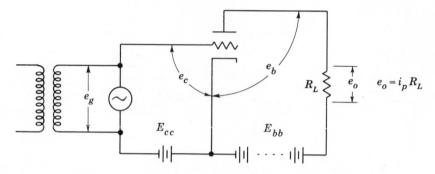

FIGURE F.1 Schematic of simple triode circuit; polarities.

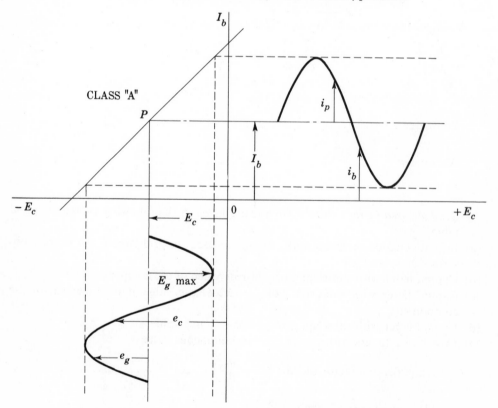

FIGURE F.2 Symbolism; plate current, plate and grid voltages.

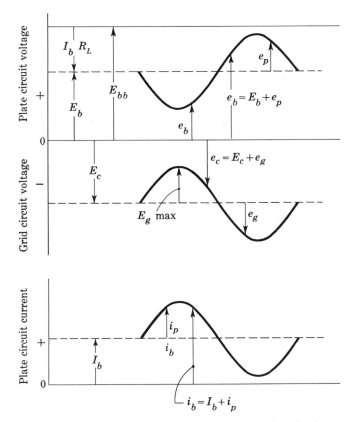

FIGURE F.3 Symbolism; plate current, plate and grid voltages.

2. *Resistance Boxes.* A resistance box is a two-terminal variable resistance so designed that any desired known noninductive resistance may be introduced into the circuit of which it is a part. In the dial-type box several dials are arranged in decades, so that internal switches select the desired resistance. For example, one dial may select any number of thousands of ohms from one to nine. The next, any number of hundreds of ohms from one to nine; the next, any number of tens of ohms, and the fourth, any number of ohms from one to nine. Thus by manipulating all four dials, any value from 1 to 9999 ohms may be set to the nearest ohm.

In the plug-type box, resistance is inserted by *removing* a plug, which, when in place, short-circuits the resistance. The plugs must, therefore, be kept clean, and when removed from the box, placed on a clean sheet of paper. They are inserted with a clockwise motion, with only slight pressure. When a slight resistance to turning is felt, a good contact has been obtained.

In either type, each coil is constructed so that it can dissipate $\frac{1}{4}$ w safely. This rated capacity of 0.25 w must not be exceeded, since overheating a coil may change its resistance permanently.

$$P = I^2R \qquad \text{so} \qquad I = \sqrt{\frac{0.25}{R}}$$

A 10 ohm coil can carry 0.16 amp; a 100 ohm coil, 0.050 amp; a 1000 ohm coil, 0.016 amp.

Most resistance boxes contain coils of manganin resistance wire, the resistivity of which does not change appreciably with small changes of temperature. The accuracy of such a resistance standard may be taken as $\pm 0.25\%$.

3. *Rheostats.* A rheostat usually consists of a solenoid of bare resistance wire wound on an insulating cylinder. A sliding contact introduces more or less of the resistance into the circuit, and thus controls the current of the circuit. Rheostats are able to dissipate a certain amount of power in the form of heat, and thus, like resistance boxes, have maximum current values. Those rheostats in use in the laboratory may be mounted on wooden bases, at the end of which is stamped the total resistance and the maximum allowable current. The rheostat will be too hot to touch at currents considerably under this maximum value.

Potentiometers are rheostats connected so that the entire resistance is in the circuit containing the main current. The slider then is able to tap off any fraction of the voltage drop across the resistance (see also Exp. 37). In small potentiometers the solenoid is often bent into a toroidal shape (doughnut) so that the slider may be mounted on a shaft, tapping off a voltage proportional to the angle of rotation, if the winding is uniform.

4. *Capacitors.* The simplest form of a capacitor is a pair of metal plates facing each other. The capacitance of such a capacitor varies directly with the area of the plates and inversely with the distance separating them. It also varies directly with the dielectric constant of the material separating the plates. The relation may be written in the form

$$C \propto \frac{KA}{d} \tag{F.1}$$

where K = dielectric constant (= 1 for air),
$\quad A$ = area of each plate,
$\quad d$ = distance separating the plates.

The capacitance of such a pair of plates of any convenient size is very small and totally inadequate for most electrical purposes. One way of increasing the capacitance is to form a stack of plates connecting every other plate together, and having a sheet of dielectric between each pair of adjacent plates (Fig. F.4). Another way is to make the sheets of dielectric thinner, decreasing d. This cannot be carried too far, however, since, if the dielectric becomes too thin, it becomes mechanically weak; it also may break down when even reasonable voltages are applied between the plates.

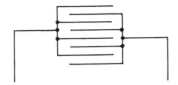

FIGURE F.4 Schematic; capacitor.

For some purposes, air is used as the dielectric. An example of this is the tuning capacitor of a radio, the capacitance of which is smoothly varied by interleafing one set of plates to any arbitrary extent with a fixed set of plates. A dielectric with useful properties of high resistance to voltage breakdown and stable value of K is mica. A stack of plates, such as the plates described in the preceding paragraph, using mica as the dielectric is the usual form of a *standard capacitor*. This type is still too bulky for most purposes, so the rigid plates are replaced by a long ribbon of aluminum foil, and the stiff mica by a strip of oil-impregnated paper. Using two strips of foil and two strips of paper and then rolling the four strips tightly into a cylinder forms a tubular paper capacitor, in which each portion of one foil is adjacent to *two* equal portions of the other foil, one above it and one below it. In the laboratory, such a paper capacitor has been rolled into a somewhat flat form and the whole is encased in a metal container.

One other type that may be mentioned is the electrolytic capacitor. In this capacitor the metal of one plate is coated by a thin film, only a few molecules thick, of its own oxide, and the other "plate" of the capacitor is an electrolyte, which makes intimate contact with the oxide film. Thus the distance d between the "plates" is exceedingly small and filled with a material of high dielectric constant; a very high capacitance is therefore possible in a limited volume. Such a capacitor is polarized. The oxide film will stand considerable voltage in one direction; however, placing even a small voltage of incorrect polarity on the plates will destroy the coating and render the capacitor useless.

5. *Autotransformer.* The voltage drop across an ordinary inductance is uniform from turn to turn of the coil, if the coil winding itself is uniform and the coil has an iron core. A sliding contact will "tap off" a fraction of the total voltage drop proportional to the number of turns the coil is from a reference turn, and thus performs the same function as the sliding contact on a potentiometer. An autotransformer often uses as a reference the first turn of the winding. In addition, the voltage source may be connected at points short of the ends of the winding. By transformer action, the sliding contact may tap off a voltage greater than that of the voltage source. A typical autotransformer connection is shown in Fig. F.5.

Since this instrument is a transformer in every sense of the word (although it differs from the ordinary transformer in that it has only one winding), it will operate only on alternating current. Connecting it to a d-c source, whose emf is higher than a

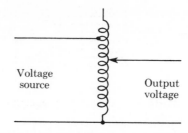

FIGURE F.5 Schematic; autotransformer.

very small fraction of the rated a-c voltage, will cause excessive current, burning out the winding.

6. *Voltaic Cells.* These cells furnish, through chemical action, a source of emf and current. The theory of these cells is covered in lectures and need not be described here. The so-called *dry cell* is used in the laboratory, where only small currents (0.1 amp or less) are needed for short periods of time. For larger current demands, *storage cells* are used. These can furnish currents up to approximately 15 amp for short periods, or 1 or 2 amp for relatively long periods. A third and very important type of cell is the *standard cell.* This type is constructed so as to have a very constant emf. However, it must not be used to deliver any appreciable current—no more than a few microamperes. The standard cells in the laboratory, consequently, may be equipped with internal series resistances to limit the current to a safe value. However, the current that can still be drawn by short-circuiting the terminals, or by having low resistances across the terminals for any length of time exceeding a few seconds, can affect the cell to the extent that its emf changes greatly. Several days of non-use are necessary to restore it to normal. It should not be called upon to furnish even enough current to operate a voltmeter (see Note G, Sec. 2); *never call upon it to do more than deflect a table galvanometer slightly for a few seconds.*

7. *Switches.* An SPST switch is used to interrupt or complete a single branch of a circuit. A DPST switch simultaneously performs the same operation in two branches of the circuit. The DPDT switch may be used to select either of two circuit elements for inclusion in the main circuit. It can also be used as a *reversing switch*, changing the direction of the current in any portion of a circuit. The alternate corner contacts of the switch are connected electrically, as shown in Fig. F.6, so that with the blades

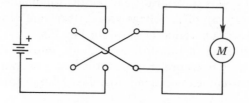

FIGURE F.6 Reversing switch.

to the right, the current through the motor is shown by the arrow. With the blades to the left, the flow is opposite to that indicated.

G. GALVANOMETERS

1. *General Notes.* A galvanometer is an instrument designed to detect small electric currents. The small portable type used in the general physics laboratory will deflect one (small) division with a current of approximately 0.00002 amp (20 μa). The wall galvanometer, or reflecting galvanometer, is considerably more sensitive. One type in common use will deflect one division (1 mm) with a current of about 0.02 μa, and thus, on this basis, is 1000 times as sensitive as the portable or table type of galvanometer. An idea of the magnitude of this current is obtained with the realization that an ordinary 100 w lamp bulb uses about 1 amp—one hundred million times the current detectable on the wall galvanometer. Other galvanometers may be designed for different uses ranging to sensitivities 1000 times greater than that of the wall galvanometer. The amount of current necessary to deflect the galvanometer one scale division is called the *current sensitivity.* Other types of sensitivity may also be defined, such as voltage sensitivity or charge sensitivity.

Most galvanometers consist essentially of a coil of wire suspended in a stationary magnetic field (the d'Arsonval movement). See Fig. G.1. The interaction of this field with the field produced by a current in the wire causes the coil to turn in the stationary field. Since the field produced by the current in the wire is proportional to the current, the stationary field is designed to have a constant value regardless of the angular position of the coil, in order that the deflecting torque be proportional to the current. A cylinder of soft iron (Core of Fig. G.1) is mounted at the center of the coil, and the

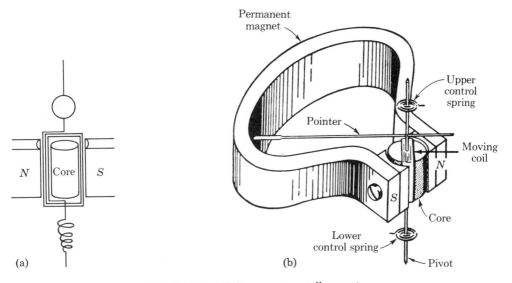

FIGURE G.1 Galvanometer; coil mount.

pole pieces are often curved, to produce a radial field (Fig. G.2), which for relatively large deflections of the coil is constant in value. The coil is mounted in such a way that as it turns it twists a spring or suspension; it will turn, therefore, until the torque caused by the current will just be balanced by the restoring torque of the spring or suspension. The coil may be pivoted in jeweled bearings, or it may be suspended by a fine wire or ribbon, which also serves to conduct the current to the coil. *By the very*

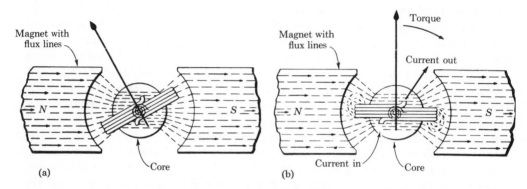

FIGURE G.2 Cross section of coil mounted in radial magnetic field.

nature of these sensitive detectors, their construction is delicate. Care must be taken to limit the current in them to an amount not significantly exceeding that needed for full scale deflection. An excess current can injure the instrument in one or both of two ways; the wire of the coil or the suspensions may be melted ("burned out"), or the meter may deflect with great violence, injuring the moving parts.

2. *The Table Galvanometer.* The coil of the portable galvanometer is mounted to pivot in jeweled bearings. Current is led into and out of the coil through springs coiled at the bearings. These springs also serve to return the galvanometer to zero when the current ceases. A pointer is attached to the coil, and indicates its position by means of a dial [Fig. G.1(b)]. The zero of this type of galvanometer is usually in the center of the scale so that currents in either direction may be measured. The scale, it should be noted, is not usually calibrated in terms of any standard units, but is divided uniformly so that the scale reading is *proportional* to the deflecting current.

Some table galvanometers are equipped with internal shunts so that only a portion of the total current passes through the movable coil, thus reducing the sensitivity of the meter. A switch selects either the reduced ("low") sensitivity or the full ("high") sensitivity of which the meter is capable. In circuits using galvanometers as null-indicators (that is, to indicate when the current in the galvanometer branch of the circuit has been reduced to zero), the initial readings should always be taken, using the low sensitivity to guard against probable overloads. When the circuit has been adjusted to the desired null condition as nearly as possible, using the low sensitivity, the high sensitivity may be used for the final adjustment.

3. *The Reflecting Galvanometer.* The moving element of this type of galvanometer consists of a rectangular coil wound on a light frame and suspended by a fine wire or ribbon of gold or phosphor bronze that also furnishes the restoring torque to the coil. Moreover, the suspension serves as one current lead to the coil, whereas the other lead consists of a very light wire helix attached to the lower end of the coil. A small mirror is fastened to the top of the coil and turns with it, reflecting an image of a scale into a telescope mounted with the scale before the mirror. This use of reflection corresponds in effect to attaching a weightless pointer to the coil whose length is twice the distance between the mirror and the scale. There are usually slight errors in the position of the zero. In addition, the field is not everywhere truly radial. The resulting errors can usually be minimized by taking the deflection to each side of zero and using the average value.

Adjustment. Complete adjustment of the reflecting galvanometer is beyond the scope of the courses for which this manual is designed. Leveling the instrument and adjusting the suspension should not be attempted. *It is of utmost importance, therefore, not to move the galvanometer in any manner, since this will probably throw it out of level and prevent it from functioning properly.* If one of these adjustments is at fault, the assistance of the instructor should be obtained.

The telescope and scale are usually clamped to a rod that is adjustably attached to the galvanometer. The telescope may have to be adjusted to suit the individual eye. To do this, the eyepiece E should be moved *in* from a position *too far out*, until the cross hairs of the telescope appear in sharp focus. Then, the telescope is aimed directly at the mirror. Unless it is badly out of adjustment, the scale should now appear. It is now possible to slide the second tube within the third tube from a position too far out, until the image of the scale is seen most distinctly. If the scale is too high or too low in the field when the mirror is in the center of the field, raise or lower the scale arm by means of the screw S. The cross hairs should be in the plane of the image, that is, free from parallax (see Note E of this Appendix). If parallax is present, the eyepiece should be readjusted, necessitating a readjustment of the focusing. The scale may be moved with respect to the supporting rod in order to adjust the position of the zero mark.

A maladjustment that does not respond to the above treatment requires the attention of the instructor. The following notes are important in avoiding eyestrain and obtaining a clear image: observations should always be taken with both eyes open; if necessary, cover one eye with a hand. The *scale* should be well illuminated. It should be noted that light from *behind* the observer may be reflected from the glass face of the galvanometer, obscuring the scale image.

Damping. When the coil turns in the magnetic field, it acts as a generator. If the terminals are short-circuited, a current proportional to the angular velocity will be generated, setting up a magnetic field that reacts with the stationary field in such a way as to oppose the motion. Thus instead of allowing the coil to oscillate after each reading, it is possible to bring it to rest at the zero point by use of a damping key, which short-circuits the coil.

4. *The Ballistic Galvanometer.* For many purposes, the reflecting galvanometer as described in the preceding section may be used as a ballistic galvanometer, that is,

to *measure a charge* passed rapidly through it. Accuracy here depends upon the moving coil having a relatively large moment of inertia, so that all the charge has passed through the coil before it has been able to move appreciably. The coil receives an impulsive torque from the passage of the charge, and the resulting swing or throw is proportional to the amount of the charge

$$Q = KD \tag{G.1}$$

where D is the deflection in millimeters, and Q is the charge in coulombs. K is called the ballistic constant or *coulomb sensitivity* (coulombs per millimeter) of the galvanometer.

H. METERS

The basic structure of many meters is the d'Arsonval galvanometer, the most common types having the coil supported in jeweled bearings. A galvanometer may be converted into a voltmeter, an ammeter, or an ohmmeter of any range greater than a certain minimum if two factors are known: the *current sensitivity* (Note G, Sec. 1) and the *resistance of the coil*. In addition, for convenience in use, the number of divisions on the scale of the galvanometer should be taken into account. The computation will be shown in detail in the sections below.

1. *The Voltmeter.* The resistance of the coil of a galvanometer being constant, the deflection will be proportional to the voltage across the coil. Thus the galvanometer in its basic form may be considered to be a voltmeter of a very low range. However, it is generally not a convenient range, since the factor relating the voltage to the deflection is not ordinarily a simple integer.

Let $k =$ current sensitivity,
 $R_g =$ coil resistance,
 $N =$ number of divisions between zero and full scale on galvanometer dial,
 $I_f =$ full-scale current (amount necessary to deflect galvanometer N divisions),
 $E_f =$ full-scale voltage (amount necessary to deflect galvanometer N divisions),
 $R =$ a resistance in series with the galvanometer,
 $R_v =$ total resistance of the voltmeter ($R + R_g$),
 $V =$ voltage at which it is desired that the voltmeter read full scale.

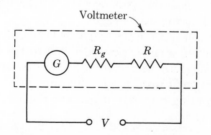

FIGURE H.1   Schematic of voltmeter.

Now the current necessary for full-scale deflection is the amount necessary for a one-division deflection times the number of divisions

$$I_f = kN \tag{H.1}$$

If it is desired that a voltage V cause the voltmeter to deflect to full scale, this voltage must produce a current I_f. Then

$$V = I_f(R_v) = I_f(R + R_g) \tag{H.2}$$

so that the size of the series resistance that it is necessary to add in order to convert the galvanometer into a voltmeter with a range of V volts is

$$R = \frac{V}{I_f} - R_g$$

A voltmeter is supposed to measure the potential difference across any part of an electrical circuit, without changing the circuit. Ideally, this supposition means that the voltmeter resistance R_v should be infinite; e.g., consider the measurement of the voltage drop across a resistance, as shown in Fig. H.2.

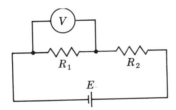

FIGURE H.2   Voltmeter applied to simple d-c circuit.

When the voltmeter is connected across R_1, the net resistance of the combination of R_1 and the voltmeter is less than R_1

$$R_{net} = \frac{R_1 R_v}{R_1 + R_v}$$

The total resistance of the circuit is now reduced by an amount $R_1 - R_{net}$ so that the current is greater, causing a larger IR drop across R_2, and since the applied emf is constant, a smaller drop across R_1 than before the voltmeter was applied. Thus the voltmeter always reads a value lower than the true value. As the ratio of the voltmeter resistance to the circuit resistance increases, however, the error becomes smaller. If the voltmeter resistance is 100 times as large as the circuit resistance, the error is about 1%. For careful work, therefore, the voltmeter resistance must be taken into account. For the highest quality voltmeters, the galvanometer current sensitivity should be very small. (Remember, of course, that the smaller the current sensitivity, the more sensitive is the galvanometer.)

Since the voltmeter is simply a modified galvanometer, the warning at the end of Sec. 1 of Note G applies. *Do not apply more voltage than the full scale amount.*

2. *The Ammeter.* The galvanometer may be used as an ammeter whose range is $I_f = kN$. If larger currents are to be measured, say a maximum current of I, all but the amount I_f must be shunted through a parallel path around the galvanometer (see Fig. H.3). The easiest way to calculate the correct value of S for any particular current range is to consider the following facts: when there is a current in the circuit, point

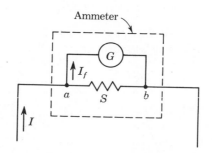

FIGURE H.3 Schematic of ammeter.

a is at some definite potential V_a; point b is at the potential V_b. The *difference* ($V_a - V_b$) must equal E_f when the external current equals I, for by our assumption, the current I in the external circuit is to cause full-scale deflection of the galvanometer. However, the potential difference E_f also exists across the shunt resistance S. The current in this resistance is $I - I_f$. Thus the value of S is given by

$$S = \frac{E_f}{I - I_f} = \frac{I_f}{I - I_f} R_g \qquad \text{(H.3)}$$

An ammeter is supposed to measure the current in any circuit without changing the circuit. Ideally, this means that the ammeter resistance R_a should be zero (consider the measurement of the current in a circuit like the one of Fig. H.4).

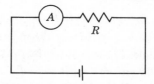

FIGURE H.4 Ammeter connection in simple d-c circuit.

When the ammeter is connected in the circuit, the total resistance of the circuit increases by the amount R_a. Thus the current falls below that when the ammeter is not present. If the resistance of the ammeter is $\frac{1}{100}$ as much as that of the rest of the circuit, the error is 1%. For careful work, therefore, and in cases where the circuits have low resistances, the ammeter resistance must be taken into account. For the highest quality ammeters, the galvanometer should have a low k, which means that R_a will be small.

WARNING: An ammeter must never be placed across a voltage source; it must only be used in *series* with a load that does not draw more than the full scale current.

3. *The Ohmmeter.* The common type of ohmmeter consists of a galvanometer in series with a battery and a resistor, R_0 (see Fig. H.5). It is clear from the sketch that the amount of current in the ohmmeter when connected across an unknown

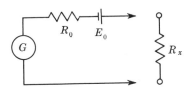

FIGURE H.5 Schematic of ohmmeter.

resistance, R_x, depends upon the values of the resistances R_g, R_0, and R_x. If R_x is an open circuit, that is if $R_x = \infty$, then the current is zero. The rest position of the pointer on the dial may be marked "∞". It is also clear that if R_x is a short circuit, that is if $R_x = 0$, then the current is limited only by $R_g + R_0$. Thus R_0 must be of such a size that the current with $R_x = 0$ is just the full scale current, I_f. Suppose further that $R_x = R_0 + R_g$. The current is now half the amount it was for $R_x = 0$. Thus the point on the dial corresponding to this value of R_x is halfway between the zero and infinity marks. The scale is obviously not linear.

Different scales are obtained by changing both R_0 and E_0.

4. *The Wattmeter.* This instrument is designed to measure the power consumed in an electric circuit element. It is constructed on the electrodynamometer principle and has two windings, a "current" winding and a "voltage" winding (see Fig. H.6). The former winding is placed in series with the circuit element to be measured. Therefore, the current in this winding is the same in phase and magnitude as that in the measured element. The voltage winding is placed "across" the measured element, and since the resistance of this winding is high, the current through it is in the same phase and is proportional to the voltage across the measured element. The deflection of the wattmeter is accomplished in much the same way as the ordinary d'Arsonval movement: the current winding furnishes the stationary field for the movable coil, which is itself the voltage winding; the torque is proportional to the product of the currents in the windings, that is, to the product of E and I in the measured element. Furthermore, the currents in the two windings must be in phase for maximum torque, and, when the phases differ by 90°, a maximum current in one winding corresponds to zero current in the other, giving zero torque. Therefore, it may be shown that the torque will also be proportional to the cosine of the phase angle. The net effective is now seen to be the a-c power equation, $P = EI \cos \theta$, translated into mechanical motion. The factor "$\cos \theta$" is called the *power factor*.

The wattmeter may also be used to measure power in d-c circuits. In this case the power factor $\cos \theta$ is always unity.

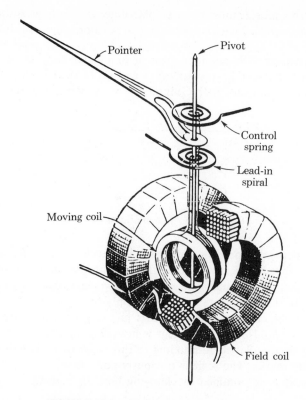

FIGURE H.6 Wattmeter construction.

Note that the wattmeter in its simplest form has four terminals, two current terminals and two voltage terminals. One terminal of each pair is marked "$\pm$". These terminals must both be connected *toward the same end* (high potential or low potential) of the circuit.

I. THE TRANSISTOR

Only a brief description of the action of the transistor is given herewith. Space in this manual is inadequate for a treatment of the development, characteristics, and applications of semiconductor devices and, for this reason, you are encouraged to refer to magazine articles and books on this subject, of which there are many.

Characteristically crystalline materials show an orderly array of atoms. Atoms are held in their positions because of the bonding due to valence electrons. Neighboring atoms share their valence electrons and, when the outer orbit has been "filled" with shared electrons, a pattern has been established that is repeated throughout the crystalline material. Some solids are polycrystalline in structure, others monocrystalline.

Because of the non-uniformity in the polycrystal it is desirable, to consider the single crystal. The semiconductor, as the name implies, lies between the conductor (materials which permit electrons to flow readily) and insulators (materials in which electron flow is practically zero). In the ideal semiconductor crystal the atoms are completely bonded hence there are no free electrons. Conduction electrons may be produced by the application of heat, but a doping process is responsible for some interesting and important properties.

1. *Semiconductor*. In considering the basic structure of the semiconductor one needs to be reminded of the structure of the atom and the bonding that takes place in building molecules and crystals. When the array of homogeneous atoms is such as in germanium, bonding takes place, and the pattern is one that is characterized by its uniformity of spatially positioned atoms. In this substance an atom shares its electrons with its neighbor atoms such that the valence orbit of an atom has been satisfied in the sharing process. The bonding is electrical in nature and accounts for a fairly stable array or arrangement of atoms. In the crystalline structure of the pure element the properties are not particularly interesting; however, when an element of either higher or lower valence is added to the pure element in the formation of the crystal, the properties have changed considerably and the result is a most exciting product.

2. *Doped crystal*. When atoms like arsenic, antimony, and phosphorous are added in minute quantities to germanium or silicon in the formation of the crystal, the contaminant atom locates itself within the lattice arrangement, the same as would the host atom. Each of the doping atoms, however, has one electron more than is necessary for the bonding process with its neighbors. This space electron finds itself free to move about the lattice structure. That being the case electrons can be urged to flow through the crystal when an electric field has been established within the crystal.

In the above case the element which is used to dope the crystal to produce free electrons is called a donor. One would suspect that the opposite process can be achieved, and such atoms as aluminum, gallium, indium, and boron leave the crystal with one less electron per contaminant; hence they are called acceptors. Where the donor atoms provided free electrons, the acceptor atoms produced "holes" or vacancies within the lattice structure. In either case the crystal now is able to "transfer" charge. Consequently, we have the two types, namely: (1) the P-type—the crystal with the acceptor atoms; and (2) the N-type—the semiconductor with donor atoms.

3. *P-N Junction*. In Fig. I.1 an N-type semiconductor has been fused with a P-type to form a P-N junction. When connections are made as shown in Fig. I.2, the direction and magnitude of the current are dependent upon the battery polarities. In (a) of Fig. I.2, I_f is the forward current and is relatively large, whereas in (b), with the opposite polarity, the current is very small. The junction in (a) is forward biased; that in (b) is reverse biased.

Where the junction is biased in the forward direction, the majority carriers constitute the current; whereas in reverse bias the current is determined by the minority

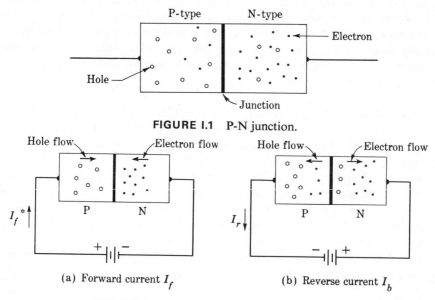

FIGURE I.1   P-N junction.

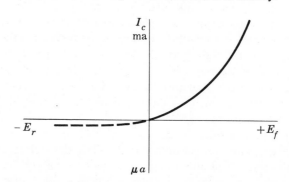

(a) Forward current I_f

(b) Reverse current I_b

FIGURE I.2   P-N junction connected to battery.

FIGURE I.3   Static conduction curve for P-N junction.

carriers. A graph (Fig. I.3) illustrates best the action of the junction under the two conditions.

Note that the forward bias occurs when the polarity of the battery matches the majority carrier of the layer; i.e., the anode of the battery is connected to the P layer, etc.

At the immediate vicinity of the junction (Fig. I.1) some holes and electrons diffuse to form a depletion region. The entire P-N crystal, however, is not depleted of positive and negative ions because of the formation of a junction barrier. As the depletion process takes place (union of electrons and holes), the fixed atoms constitute a space-charge region. The donor region has been depleted of its spare electrons,

*Direction of conventional flow of charge.

hence left positive. On the other hand, the acceptor region has added negative ions to an electrically neutral region, hence it has assumed a negative charge. This depleted region in the immediate vicinity of the junction behaves as a battery in direct opposition to a continuance of the fusion of electrons and holes. Carriers with sufficient energy to overcome the barrier determine the current within the junction.

4. *Diode Action.* The P-N junction shows a large disparity in charge conduction, the current being determined not only by the magnitude of the applied voltage but also by whether the junction is forward or reverse biased. For this reason it serves

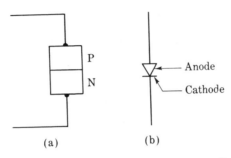

(a)   (b)

FIGURE I.4   Symbolism; P-N junction, diode.

as a good rectifier, though not perfect. The asymmetrical conductivity of the P-N junction accounts for numerous applications such as power-supply rectification, signal rectification, switching, and many others.

5. *Triode Action.* Fig. I.5 shows two ways by which P-N junctions may be fused to form sets of 3 layers. The behavior of each is similar and the major difference in

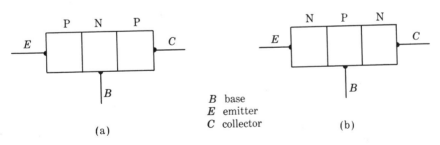

(a)   (b)

FIGURE I.5   Two types of transistors.

the application is the polarity requirements. In order to abbreviate the discussion on the transistor we shall confine our attention to the NPN type, knowing that the PNP requires opposite d-c polarity but acts similarly.

In Fig. I.6 the power sources are applied to an NPN transistor such that the *E-B* junction is forward biased and the *C-B* junction is reverse biased. Represented

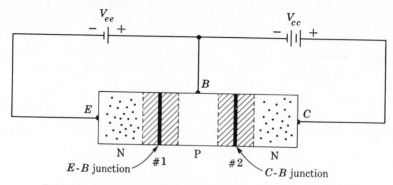

FIGURE I.6   Common-base connection of N-P-N transistors.

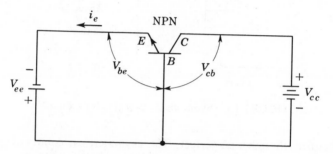

FIGURE I.7   Schematic of common-base connection.

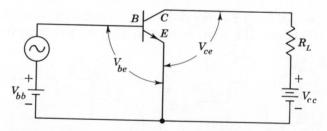

FIGURE I.8   Schematic of common-emitter connection.

on the diagram are the depletion regions where barrier #1 has been reduced because of forward biasing and barrier #2 has been increased due to reverse biasing. The same circuit is schematically shown in Fig. I.7.

If Fig. I.7 is modified to include a variable input signal and a load resistor in series with d-c source #2 (see Fig. I.8), then a small change in V_{be} causes a large change in V_{ce}. This voltage amplification is caused by a gain in the collector current, i.e.,

$$\Delta I_c > \Delta I_b \quad \text{or} \quad \frac{\Delta I_c}{\Delta I_b} > 1$$

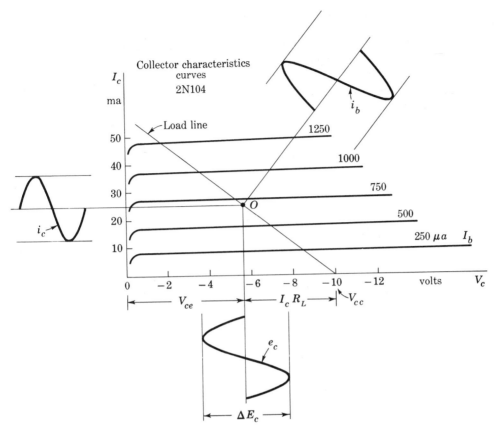

FIGURE I.9 Typical family of collector characteristics curves; load line, operating point, and current and voltage variations.

(See Fig. I.9.) The behavior of the circuit is similar to that in which the vacuum tube was used to amplify signals. There are several similarities and differences that might be appropriately pointed out at this time, namely:

1. The action of a vacuum tube is dependent on a flow of electrons, whereas both the positive holes and the negative electrons constitute the flow of charge in the transistor.

2. The source of the electrons in the vacuum tube is the *heated* cathode, yet in the transistor a heater is not required.

3. In class A operation no grid current flows in the vacuum tube, but an emitter-base current flows in most transistor amplications.

4. The current in the plate circuit of a vacuum-tube amplifier is controlled by the voltage of the grid. In the transistor the emitter-base voltage controls the collector current.

5. The input impedance of a vacuum-tube stage is usually quite high, though it is quite small in the transistor amplifier.

J. THE SPECTROMETER

The spectrometer is essentially an instrument for the measurement of angles of deviation of light rays resulting from reflection, refraction, and diffraction. Its essential features are represented diagrammatically in Fig. J.1. A circular horizontal plate K, the edge of which is graduated in degrees, is supported upon a vertical mounting that also carries a telescope T and a collimator C. The latter consists merely of a tube

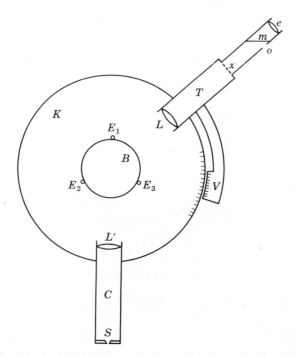

FIGURE J.1 Diagram showing spectrometer components.

carrying a slit S of adjustable width, and so mounted that it may be placed in the principal focal plane of the lens L'. This arrangement makes it possible to regard S as an infinitely distant source of light, for waves (rays) that originate at S become plane waves (parallel rays) after passing through the lens L'.

The telescope T is mounted so it will rotate about the vertical axis of the plate K. The angular position of the telescope with reference to the graduations on the plate is read by means of a vernier scale (or scales) V attached to the telescope. A small piece of plane glass m is inserted in the eyepiece e of the telescope so as to make an angle of 45° with the axis of the telescope. The purpose of this arrangement is to make it possible to illuminate the cross hairs at x by projecting a beam of light into the eyepiece through the circular opening at O. Such an eyepiece is called a Gauss eyepiece.

A second smaller circular plate B, the spectrometer table, is mounted at the center of the large plate K. It may be leveled by means of leveling screws E_1, E_2, E_3. The optical device that is being examined is placed on this table, e.g., prism, diffraction grating, and so on. This table may be rotated about its vertical axis independently of the rotation of the telescope. It may also be adjusted for height.

An examination of the spectrometer will reveal that there is a large number of screws and clamps on it that are to be used in adjusting the instrument. For example, there are leveling screws and clamps on both the telescope and the collimator in addition to focusing devices. There are clamps for the adjustment of the spectrometer table, either as to height or as to rotation. The telescope and verniers attached to it may be clamped in position, except for slow motion afforded by a tangent screw. The same is true for the collimator. Therefore, it is necessary for the student to examine carefully the functions of these various screws and clamps before trying to use the spectrometer (see Fig. J.2).

FIGURE J.2   TK = leveling clamp of telescope; TL = leveling screw of telescope; TC = telescope clamp; TS = tangent screw for slow motion of telescope; TP = rack and pinion for focusing telescope. Corresponding clamps and leveling screws exist for the collimator. S = slit of collimator.

Adjustment of Spectrometer. The complete adjustment of the spectrometer is a long and arduous task for the novice. For highly accurate work, it is necessary to have the spectrometer in complete adjustment. This adjustment is usually done in advanced work in experimental optics, but it is hardly advisable in beginning courses in optics. It will serve our purpose to make an approximate adjustment of the spectrometer only. In making this approximate adjustment, the following steps should be taken in the order given.

1. Focus the eyepiece on the cross hairs of the telescope. This focusing may be

done by directing the telescope toward some bright object, drawing out the *eyepiece tube* of the telescope to its full amount, then slowly pushing it in until the cross hairs are *first* distinctly observed. Focusing in this manner places the cross hairs in the focal plane of the eyepiece. Light rays from the cross hairs that enter the eye are, therefore, parallel rays. This adjustment gives the most comfortable vision for most people.

2. Focus the telescope for parallel rays. In most experiments with the spectrometer it is necessary that rays coming from the collimator be parallel. Therefore, the telescope must be adjusted for parallel rays. This adjustment may be done by focusing the telescope on a distant object. Direct the telescope toward an open window at some distant object. Using the rack and pinion focusing device on the telescope (which moves both eyepiece and cross hairs), rack out the tube as far as it will go, then rack it in slowly until a clear image of the distant object *first* comes into distinct view. Focusing in this manner places the real image of the distant object (produced by the objective lens of the telescope) in the focal plane of the eyepiece, and therefore in coincidence with the cross hairs. Under these conditions, there should be no parallax between the image and the cross hairs (see Appendix II, Note E). If parallax exists, the telescope should be refocused.

3. Alignment. Align the telescope and collimator with the center of the spectrometer table. *In the Gaertner spectrometers this adjustment has already been made by the manufacturer and should not be disturbed.* In other types this adjustment is done by unclamping the telescope or the collimator from its supporting arm, sighting along its barrel, turning it until it points to the center of the table, and then reclamping it.

4. Level the spectrometer. Place a level on the plate K so that it is parallel to a line through two of the spectrometer feet, and level by adjusting the foot screws. Then, place the level at right angles to its first position, and level by adjusting the third foot screw.

5. Level the spectrometer table. Place a level on the table so that it is parallel to a line through two of the leveling screws, and level by adjusting these screws. Then, place the level at right angles to its first position, and level by adjusting the third screw.

6. Level the telescope and collimator. Place the level on top of the barrel of the telescope, and adjust by use of the leveling screw. In a similar manner level the collimator.

7. Adjustment of collimator. Illuminate the slit of the collimator with the light from an ordinary frosted lamp bulb. Have the slit fairly wide ($\frac{1}{2}$ mm) and in a vertical position. Rotate the telescope until it is directly opposite the collimator and in alignment with it. Sight through the telescope for the image of the collimator slit. In general, a blurred image of the slit will be observed. Bring it into coincidence with the cross hairs by rotating the telescope, if necessary. Cause this image to be seen *distinctly, without parallax with respect to the cross hairs* of the telescope, *by sliding the draw tube of the slit. Do not refocus the telescope* in this process; it is already focused for parallel rays. If, after this operation, the cross hairs are not vertically centered on the slit, adjust the level of the collimator until they are. These operations place the slit of the collimator in the focal plane of the collimator lens.

The spectrometer is now in approximate adjustment. These adjustments should not be changed during the course of an experiment.

For accurate work, it is necessary to use a better, though somewhat longer, method of adjustment. Such a method is outlined in the following steps:

1'. Same as 1.

2'. Same as 2.

3'. Adjustment of collimator level and focus. Turn the collimator slit to the horizontal position, and place a lamp with a frosted bulb in front of the slit. Point the telescope at the collimator tube across the center of the spectrometer table. Focus the *collimator slit* (not the telescope) and adjust the collimator level and aperture until the slit appears in the telescope as a narrow, sharp, horizontal line passing exactly through the intersection of the cross hairs.

4'. Adjustment of prism on table. Place the prism on the spectrometer table so that the two optical surfaces are as nearly vertical as possible, and so that each face can be rotated into position for reflection of the light rays from the collimator. Rotate the telescope so that it can receive light from one face of the prism. Adjust the *prism table* (not the telescope) by means of screws E_1, E_2, and E_3 so that the reflected image of the slit is exactly at the height of the cross hairs. Rotate the *prism* to get the reflected image from the other face and repeat. Continue until the prism gives satisfactory reflections from either face, without additional leveling.

5'. Adjustment of telescope. At this point, the spectrometer table should be exactly correct and use can be made of this fact to adjust the telescope, which has only been approximately adjusted in Steps 2' and 3'. Rotate the spectrometer table so that one face of the prism appears approximately perpendicular to the axis of the telescope. Turn on the light at the Gauss eyepiece, illuminating the cross hairs (see Fig. J.1). When the prism face is exactly perpendicular to the axis of the telescope, a *reflection* of the cross hairs will *also* be seen. Rotate the spectrometer table until the cross hairs and reflection are *vertically* aligned. Adjust the telescope level (not the prism) until horizontal alignment is obtained. Refocus the telescope if necessary to to get the brightest image of the cross hairs.

6'. Adjustment of collimator. With the lamp in front of the slit, reflect the image of the slit from one face of the prism into the telescope, as in Step 4'. But now, adjust the level of the *collimator* (not the prism) to bring the image to the right height. If the telescope had to be refocused in Step 5', the collimator will now have to be correspondingly refocused.

As a partial check on the accuracy of the adjustment, either face of the prism should now give a reflected slit image exactly at the right height, without further adjustment, for any position of the telescope for which a reflected image is possible. If this is not the case, repeat Step 3' and all following operations.

Use of the Spectrometer. 1. Reading the verniers. See Appendix II, Note A, Sec. 4 for a discussion of angular measurement with verniers. Most spectrometers are equipped with simple magnifiers for reading the verniers.

2. In order to rotate the telescope, handle *only* the arm to which it is attached

and not the telescope itself. In making adjustments *do not force any moving parts of this instrument.* They should move freely if the proper clamps are loosened.

3. In using a flame as a source of illumination for the collimator, be sure to keep the flame sufficiently far from the slit in order to avoid overheating the slit mechanism.

4. It is frequently advisable to cover the prism table, telescope, and collimator with a black cloth in order to eliminate extraneous light.

K. THE ANALYTICAL BALANCE

1. *General Discussion.* (a) The analytical balance is used to accurately determine the mass of an object. Since it is one of the most delicate (and expensive) pieces of apparatus used in the laboratory, and is easily damaged or put out of adjustment, great care must be exercised in its use. The student should be thoroughly familiar with the functions of its various parts and the precautions to be used in handling them before attempting to use the balance (see Fig. K.1).

(b) The balance consists of a beam, B, pivoting by means of an agate knife-edge, K_1, on an agate surface; two pans, P, hung on the beam at equal distances from K_1 by means of secondary knife-edges, K_2; a long pointer, I, attached to the beam, indicating the position of the beam by means of a short scale, S, behind which a mirror, M, is placed; and a scale, R, attached to the beam, on which a small rider may be placed for fine adjustments in balance. In addition, there is an arm and hook arrangement which lifts and moves the rider without opening the case, and an arrestment which lifts the beam and pans off their knife-edges at all times, except during an actual balancing operation. Two adjustments are provided: C_s, when moved vertically, changes the center of mass of the beam vertically with respect to the knife-edge K_1, and thus changes the sensitivity of the balance; C_z, when moved horizontally, moves the center of mass of the beam in the same direction, allowing adjustment of the unloaded rest position to coincide with the zero point of scale S. *In general, neither of these adjustments is to be attempted by the student.*

(c) An object is "weighed" by placing it on one pan (usually the left) and placing members of a special set of "weights," called analytical weights, in the other pan until balance is obtained. It might be noted here that this process of weighing is essentially equivalent to comparing the mass of the body to the mass of the "weights." It is a valid process regardless of the value of g, the acceleration due to gravity, provided that this acceleration is the same at both pans of the balance.

The set of analytical weights usually consists of a group made of brass, ranging from 1 to 100 gm, and another group made of aluminum or of platinum, ranging from 10 up to 500 mg. The previously mentioned rider provides adjustments from 0.1 up to 10 mg. This last adjustment may be made in one of two different ways, depending on the marking of scale R. If R is divided into 20 divisions, with zero in the center, a 10 mg rider is provided. Moving the rider to the tenth division either way amounts to placing 10 mg on the corresponding pan. Moving it 5 divisions amounts to placing 5 mg on the corresponding pan, and so forth. If the scale R is divided into 10 divisions, with zero to the left, a 5 mg rider is provided. The scales are balanced for the rider at the zero position. Moving the rider 10 divisions amounts to removing

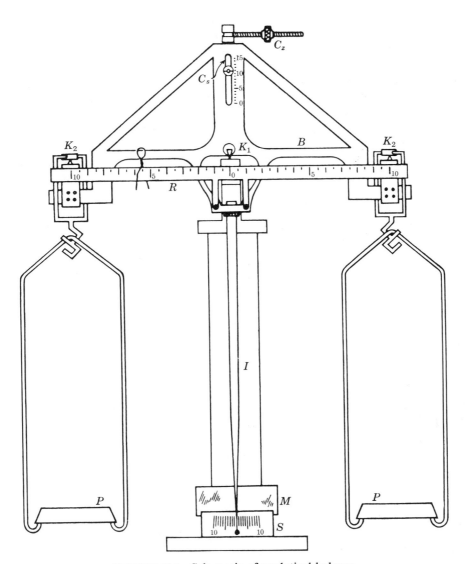

FIGURE K.1 Schematic of analytical balance.

5 mg from the left pan and adding 5 mg to the right pan, a total transaction of 10 mg.

(d) In order to determine the mass of a body, the *unloaded* (normal) *rest point* of the balance must first be found. This point is the position of the pointer I on the scale S when the *unloaded* balance ceases to vibrate; in general, it will not be the center point of the scale. The balance between the object and the analytical weights has been obtained when the pointer comes to rest at this normal rest point (and not at the center point).

In practice, the scales swing for so long a time that it is impractical to wait for them to come to rest. The ultimate rest point of the balance may be found by the *method of swings*, as follows: After the motion has become regular, record three *successive* extremes of the travel of the pointer past the center of scale (two on one side and one on the other). The rest point will lie halfway between the one reading and the average of the other two. For more careful work, record five successive extremes (three on one side and two on the other); the rest point lies halfway between the average of the three and the average of the two. We may show the logic of this procedure by postulating some natural rest point, say left 3, and by assuming that the balance loses equal amounts of amplitude each swing because of friction and air-damping, say $\frac{1}{2}$ division. Then, with an initial swing to the left of say 5 divisions beyond the rest point, the succeeding swings will go 4.5 divisions to the right of the rest point, 4.0 to the left of it, 3.5 to the right, and 3.0 to the left, giving readings of L8, R1.5, L7, R0.5, and L6. The averages are L7 and R1.0; the mean of these is L3, the originally postulated rest point. It is to be noted that this procedure, when followed with the pans empty, gives the *unloaded rest point*. When followed with the object and weights in place, it gives the *loaded rest point*. When enough weights have been added so that these two rest points coincide, the scales are balanced and the two masses are equal.

2. *Precautions.* (a) The arrestment of the balance must always be engaged, except when actually measuring swings. It must be engaged when changing weights, even when shifting the rider. The knife edges are sharp and brittle, and any shock may ruin them permanently. The arrestment should be engaged and disengaged *gently*.

(b) The balance is enclosed in glass to avoid the effects of air currents. When the scales are balanced to the point where swings may be recorded, the front window must be closed. After the window is closed, wait a few seconds for any air currents to die out before releasing the arrestment.

(c) In reading the extremes of the swings, use the mirror behind the scale S to avoid parallax (see Note E). The reading is correct when the pointer and its image in the mirror coincide, i.e., when the pointer exactly covers its image.

(d) The analytical weights are precisely calibrated. They are to be used only with the analytical balance. They must be *handled with the forceps* and never with the fingers, because oil and moisture from the hands will change their weights. This method of handling applies equally to the largest and the smallest weights. Each weight has a particular place in its box. When finished with a weight, *return it to its proper place in the box*. In order to avoid error in totaling the group of weights used in weighing a load, record each one individually in a column as they are returned to their box, record the contribution of the rider, and add later.

(e) Since there are two sizes of rider available, as noted in Sec. 1(c) above, be sure that the balance to be used is equipped with the proper one. This condition may be checked by placing the rider in one pan and a 10 mg weight in the other, and seeing if the scales balance. If they are far out of balance, the rider is of the 5 mg size.

3. *Weighing Procedure.* (a) With the case closed, insert the arrestment knob, and *gently* lower the beam onto its knife-edges. If it swings freely, accurately find the

unloaded rest point, as indicated above, using swings of approximately 8 or 10 divisions total amplitude. The size of the swing can be controlled by the care with which the beam is lowered onto its knife-edges. If the balance does not swing freely, it is out of adjustment, and the instructor should be called. The student should not attempt to adjust it himself. Record the unloaded rest point and engage the arrestment.

(b) Open the case and place the object to be "weighed" on the left-hand pan. On the right-hand pan place a single large weight of approximately the same mass as the object. *Gently* lower the arrestment. It will immediately become clear whether the mass of this weight is larger or smaller than that of the object, since the beam will definitely lean toward one side or the other as soon as the arrestment permits. Reengage the arrestment, and replace the weight by one nearer the mass of the object. Always use as the largest weight, the first one that fails to overbalance the object.

Reengage the arrestment and add the next smaller weight to the pan. Gently release the arrestment to determine whether this addition is too large or too small. If too small, leave the weight on the pan and add the next smaller weight, after reengaging the arrestment.

Continue in this fashion until (with the rider in its zero position) an added 10 mg weight in the right-hand pan overbalances the object being weighed. Remove the 10 mg weight and finish the balancing operation with the rider, using the method of swings and taking only three successive extremes until a good balance is obtained, i.e., until the loaded rest point very nearly equals the unloaded rest point. For ordinary work, the mass is determined to the nearest milligram, since in order for the tenths of a milligram to be significant, special corrections for the buoyant force of the air, and so forth, are usually necessary.

(c) For work that demands an 0.1 mg accuracy, the *sensitivity* of the balance should be obtained. Determine the unloaded rest point as above, using the method of swings with five successive extremes. Repeat. Then, by means of the rider, add the equivalent of 1 mg to the right-hand pan and determine the new rest point. *The shift in the rest point for the addition of 1 mg is known as the sensitivity of the balance.* Knowing the sensitivity, it is no longer necessary to bring the loaded rest point exactly into coincidence with the unloaded rest point when determining the mass of a body. Having determined the distance of the loaded from the unloaded rest point, the number of milligrams which *would be* necessary to reduce this distance to zero may easily be calculated.

(d) When finding the *difference* between two very nearly equal weights, the effect of small errors in the analytical weights may be radically reduced by using the *same* large weights to balance both loads. In this way only errors in the small weights making up the difference will enter into the final computations. A still better way to proceed is to balance the heavier object, first by placing sufficient weights in the right pan. Then, *leaving* these weights in the *right* pan of the balance, replace the heavier object by the lighter object in the left pan. Finally, add small weights to the *left* pan until balance is again restored. The total of the weights *added* to the *left* pan will equal the *difference* in the weights of the two objects. In this manner any error due to inequality of the arms of the balance will be eliminated.

L. THE CATHODE-RAY OSCILLOSCOPE

1. *General Description.* Essentially, the oscilloscope consists of a cathode-ray tube, sweep circuits, a synchronization circuit, high and low voltage supplies, and horizontal and vertical amplifiers.

2. *The Cathode-Ray Tube.* The cathode-ray tube is an evacuated chamber, thereby allowing a beam of electrons to travel the length of the tube without colliding with residual gas molecules before striking the fluorescent screen. It consists mainly of an electron emitter (cathode), a device to control intensity (grid), a focusing element (anode A_1), an accelerating electrode (anode A_2), deflection plates—vertical and horizontal, and a screen which fluoresces when electrons impinge on it. All of the components above are illustrated in the schematic Fig. L.1. Where magnetic deflection of the electron beam is used, the deflection plates are not needed.

3. *Electron Emission.* The means of obtaining electrons in this tube is no different from that employed in the common radio tube, i.e., the heated cathode serves as the electron source. The filament (heater) is electrically insulated from, yet is thermally connected to, the electron emitter (cathode). The cathode, which is coated with a material such as MgO, furnishes electrons profusely at temperatures well above room temperature.

4. *Beam Intensity.* As in the radio tube (triode), the grid (wire screen or similar electrode) is in close proximity to the cathode and usually has a negative charge such that any off-axis electron will be urged to change its course so as to join the axial electrons. The function or action of the two anodes is similar to that of a positive lens in an optical system.

Each of the elements has an axial opening in a conductor so that those electrons which are near the axis of the tube can emerge from anode A_2 with very nearly the same velocity. However, not all electrons emerge from the second anode with precisely the same energy; this, for the most part, is due to the difference in energies of the electrons as they are emitted from the cathode.

5. *Beam Deflection.* When the tube is to be operated by electrostatic deflection, two sets of plates are mounted between the second anode and the screen. As illustrated in Fig. L.1, one pair of plates, mounted in vertical planes, would cause an electron to move horizontally when an electric field exists between the plates. Because these plates control horizontal displacements of the electron, they are referred to as the horizontal deflection plates. Similarly, horizontally mounted plates have vertical electrical fields and, when actuated, cause the electron to move either up or down. Actuation of both sets of plates causes a resultant displacement of the electron which is determined by the vector sum of the vertical and horizontal motions.

For several reasons it is necessary to have control over the position of the spot on the screen. In a tube in which the beam is not perfectly aligned, the horizontal and vertical position knobs make it possible to center the spot on the screen. The posi-

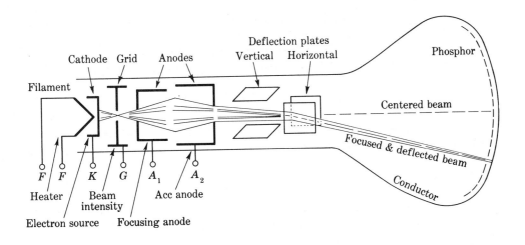

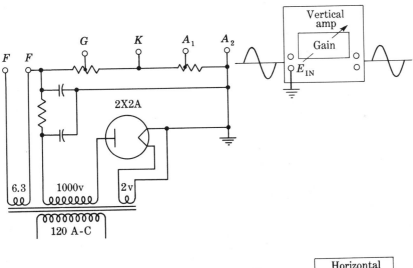

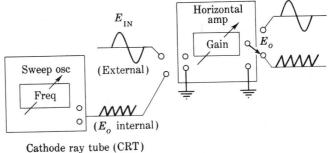

Cathode ray tube (CRT)

FIGURE L.1 The cathode ray tube and the electrical connections in an oscilloscope.

tioning of the spot is accomplished by applying d-c potentials to each of the two sets of deflecting plates in such magnitude and polarity that the electron beam strikes the screen at a predetermined point. In general, the starting point of the spot on the screen may be shifted at will when H and V position knobs are provided.

6. *Screen.* Once the axial electron has emerged through the opening of anode A_2, it is entirely on its own; i.e., it is in free flight, since it is not submerged in an electric field except as fields are set up between deflecting plates. For all practical purposes we can ignore the effect due to gravity. The electron would strike the center of the screen if it is travelling along the axis of the tube. When it does strike the screen of the tube, it collides with materials that are optically activated by the electron impact.

Light is emitted when atoms of the phosphor coating, which are in higher energy states as the result of electron bombardment, return to their normal states. There is a graduation, however, in the persistence of phosphors. Some have very long persistence which makes them desirable for the study of a single trace that is not repeated. Others have very short persistence time. Such phosphors have "short memory," so that individual events may be separated one from the other. Phosphors having medium persistence are usually found on the general-purpose oscilloscopes.

Aquadag or some similar conducting material is used to coat the inside of the neck of the tube, thereby allowing the electrons to "leak off" to the anode A_2 and eventually through the power supply to the cathode. In Fig. L.1, electrode A_2 may be connected to the chassis as indicated; in which case, the cathode is very much negative with respect to the chassis potential.

7. *Sweep Circuit Oscillator* (*Linear Time Base*). If an alternating signal voltage is applied to one set of deflector plates only, the image on the screen as seen by the eye is merely a line. All wave forms or varying signals with the same maximum values would produce a straight line of the same length. This happens because the persistence of the eye and the persistence of the phosphor make it impossible to see the spot as it travels across the screen. A sweep circuit is connected to one set of deflector plates, usually the vertical set, which produces a horizontal displacement of the electron beam and thus spreads out the deflections resulting from the incoming signal to the other set of deflecting plates. The action of a horizontal sweep circuit of the oscilloscope can be simulated by assembling a simple relaxation oscillator (Fig. L.2). The neon lamp is inserted into the circuit to allow discharging of the capacitor after a certain voltage across the lamp has been reached. Below the breakdown voltage, the lamp is nonconducting.

If the neon lamp in the circuit of Fig. L.2 is nonconducting and the resistance between the input terminals of the oscilloscope is very high, the voltage of the capacitor will build up from zero at the beginning of a cycle (closing of the switch) to the full battery voltage. The equation representing the voltage is

$$V_C = V_0(1 - e^{-t/RC})$$

where V_C is the voltage across the capacitor, V_0 is the emf of the battery, R is the resis-

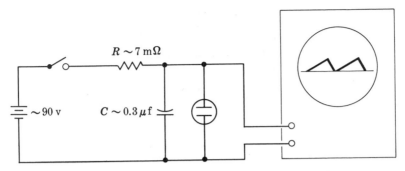

FIGURE L.2 Relaxation oscillator circuit. Saw tooth wave.

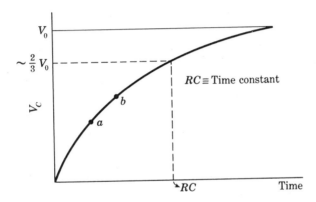

FIGURE L.3 Voltage versus time of charging of a capacitor.

tance in series with the battery, and C is the capacitance. The curve for the voltage V_C plotted against time is shown in Fig. L.3. The quantity RC, called the *time constant*, is that time required for the voltage V_C to build up to about 63% of its final value V_0.[1]

The time required for a complete cycle of buildup of charges and then discharge of the capacitor, that is the frequency of oscillation of the relaxation oscillator, can be adjusted by changing the value of RC. The frequency also depends on the value of the applied voltage and that at which the neon lamp will "break down," i.e., become conducting, thereby causing the capacitor to discharge.

The graph (Fig. L.3), at first glance, seems unsuitable for sweep voltages where the designer requires a linear relationship of voltage to time. However certain portions of the graph do approximate a straight line. The circuitry can be modified so that one operates between chosen points on the graph such as a and b.

With the vertical input of the oscilloscope connected across the capacitor of the simple oscillator (Fig. L.2), the instantaneous voltages cause the spot to move vertically. If the spot moves vertically due to the signal from the relaxation oscillator applied to the vertical amplifier and moves horizontally due to the built-in sweep of

[1] $\left(1 - \dfrac{1}{e}\right) \times 100\%$ or $\left(1 - \dfrac{1}{2.7}\right) \times 100\%$ or 63%.

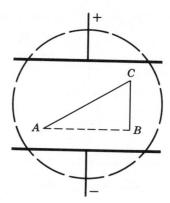

FIGURE L.4 The saw tooth pattern on the scope screen.

the scope, the resultant path is AC in Fig. L.4. When the voltage across the capacitor reaches the striking potential of the neon bulb, the capacitor discharges very quickly, and the path of the electron beam on the screen is nearly vertical (CB) as the deflecting voltage drops to zero. The excursion for one cycle is AC plus CB. The sweep oscillator of the scope moves the spot from B back to A at which point and time it is ready to repeat the cycle. The pattern is similar to the edge of a saw, hence its name.

Usually the saw-toothed generator built into a scope is more complicated than that illustrated in the figure. This may be necessary in order to (1) improve the linear response, (2) provide greater flexibility in the sweep frequency, (3) maintain greater circuit stability, and (4) start the sweep cycle only when the input signal to the scope has attained certain values.

8. *Synchronization.* Synchronization signifies that events occur simultaneously. Usually the two events to be synchronized are (1) the input signal to the scope, and (2) the sweep signal of the scope. When these signals are synchronized, the events are made to appear at the same time. In other words, the beginning of a pulse of a wave train is caused to appear at the beginning of the sweep trace. When this occurs, the signal, if it is periodic, will appear to be stationary on the screen. Should the "synch" control be a little out of adjustment, the wave pattern will move across the screen either left to right or right to left. Such a pattern requires a slight adjustment of the fine "synch" knob to "freeze" the pattern on the screen. With more complicated circuits, it is possible to have the sweep field between the vertical plates (produced by the internal circuits) start only when the input signal has attained a predetermined magnitude or even a predetermined rate of change or slope. It is also possible to incorporate a time delay after the beginning of the input signal before the sweep field is applied. For some purposes, a sweep signal from an external source is desired, and terminals are provided to allow direct connection to the vertical plates or to the scope amplifier connected to these plates.

9. *Power Supplies.* Transformers, rectifiers, filters, etc., in a built-in "power supply" furnish the necessary (a) filament voltages for the vacuum tubes, (b) low

voltage (approximately 250 v) for the circuitry tubes, and (c) high voltage (1–15 kv) for the cathode-ray tube.

10. *Amplifiers.* Horizontal and vertical amplifiers are needed to amplify small signals, i.e., to increase the deflection in the vertical and horizontal directions without distortion of the incoming signals. Amplifiers sensitive to weak signals and responsive and useful in the low and high frequency ranges have been designed.

M₁. THE FP400 DIODE

This tube has a pure tungsten filament located axially in a cylindrical zirconium-coated nickel anode. It is built with sufficient precision so that it may be used in a variety of emission and space-charge experiments. The anode has a small hole centrally located so that a short portion of the filament may be viewed through it for optical-pyrometer and traveling-microscope measurements. The following tables show the characteristics of the filament and anode.

Filament:

Voltage ....4.0 v; max: 4.75 v
Current ....2.25 amp
Lead resistance ....0.08 ohm (mostly in positive lead)
Length ....1.25 in. Effective length, 1.0 in. (corrected for end effect)
Diameter (cold)..............0.005 in.

Anode:

Voltage ....125 v max
Current ....25 ma max
Power dissipation...................15 w max
Length ....1.5 in.
Diameter (inside)..................0.620 in.

The base connections are shown in Fig. M.1. The view is from the bottom of the tube.

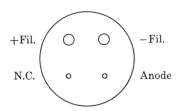

+Fil.   −Fil.

N.C.   Anode

FIGURE M.1 FP400; base connections.

M₂. THE GRD7 DIODE (FERRANTI)

This tube is a directly heated diode with a pure tungsten filament and with cylindrical co-axial electrodes (anode plus guard rings). It may be used in a variety of emission and space-charge experiments. The anode has a small hole centrally located so that the heated filament (center) may be observed for temperature determinations.

Ratings (*Continuous Operations*)

Max Filament Voltage ....6.3 v
Max Anode Voltage ....300 v
Max Anode Dissipation ....2 w

Characteristics

Anode Length ....14.5 mm
Anode Internal Diameter ....6.5 mm
Filament Diameter (cold) ....0.125 mm
Effective Filament Length..................14.5 mm

The base connections (bottom view) are shown in Fig. M.2.

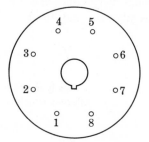

FIGURE M.2   GRD7; base connections.

Pin 1 ⎫[2] Filament
Pin 2 ⎭

Pin 3 Anode

Pin 4 No connection

Pin 5 Guard Rings

Pin 6 No connection

Pin 7 ⎫[2] Filament
Pin 8 ⎭

N. LEEDS AND NORTHRUP STUDENT POTENTIOMETER

This potentiometer has been designed for use in the general physics laboratory. It is direct reading, simple in form, and quite reliable for making potential measurements of moderate precision. The instrument is provided with three ranges: 0–1.6 v, 0–0.16 v, and 0–0.016 v. The high range is useful for measuring the emf of cells, meter calibration, and so forth, whereas the low range is especially useful for temperature measurements with thermocouples.

The internal circuit of the potentiometer is shown in Fig. N.1, along with the external connections that must be made in using the instrument. The long slide wire in the simple potentiometer (Exp. 37) is replaced in this instrument by a set of fifteen

[2]It is advisable that one filament lead shall be connected to both Pins 1 and 2, and the other lead to both Pins 7 and 8.

50 ohm coils (shown at *A*) and a 50 ohm circular slidewire (shown at *B*). The auxiliary resistances, *C*, *D*, *F*, and *F'*, in the instrument serve the purpose of keeping the effective resistance of the potentiometer constant, no matter what range is being used. Separate end-terminals, *H*, *H'*, *L*, *L'*, for the circular slidewire extend the usefulness of the instrument.

The working current in the potentiometer chain of resistances (*A* + *B*) for the normal range (0–1.6 v) is 0.002 amp. With this current, the potential drop across each of the 50 ohm coils, as well as across the 50 ohm slidewire, is just 0.1 v. The slidewire scale has 200 divisions so that each division represents a potential drop of 0.0005 v. Therefore, any emf or potential difference in the normal range may be determined to within 0.0005 v.

For the lower ranges, 0–0.16 v and 0–0.016 v, the working currents in the potentiometer chain are reduced to 0.0002 amp and 0.00002 amp respectively by proper rearrangements of the auxiliary resistances *C*, *D*, *F*, and *F'*. This rearrangement does not affect the overall resistance of the instrument. Thus, it continues to draw the

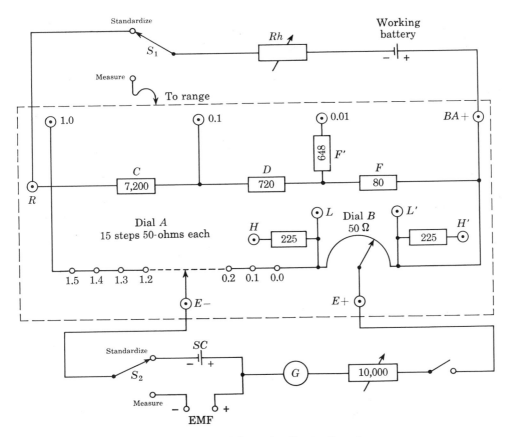

FIGURE N.1 Schematic of potentiometer.

same current (0.0022 amp) from the external battery regardless of the range being used. Since this current is quite small, the working battery may be two dry cells in series. The emf of the working battery must, of course, be larger than 1.6 v.

In order to standardize the current in the potentiometer (for any one of the three ranges), the switches S_1 and S_2 are set to *standardize*, as shown in Fig. N-1. The potentiometer dials A and B are set to the certified emf value of the standard cell (SC). Then the rheostat (Rh) in the battery circuit is adjusted to make the galvanometer (G) deflection zero. The galvanometer is protected with a 10,000 ohm resistor P, which is used in the normal manner in getting a balance point. This standardization procedure ensures a current of just 0.002 amp in the potentiometer chain $A + B$.

After the current has been standardized, switches S_1 and S_2 are set to *measure*. The measure terminal on S_1 is connected to the appropriate *range* terminal on the potentiometer; i.e., 1, 0.1, or 0.01. The measure terminal on S_2 is connected to the negative pole of the emf to be measured. Then, dials A and B are adjusted (*not Rh*) to obtain the new balance point. The value of the emf in volts is the reading of $A + B$ times the multiplier 1, 0.1, or 0.01, corresponding to the range used.

In Fig. N.1, the switches S_1 and S_2, as shown, are completely independent of one another. In actual practice, both must be manipulated together; i.e., both must be set on standardize or on measure for correct connections. To insure this manipulation, these two switches should be replaced by a single double-pole double-throw switch, half of which replaces S_1 and the other half replaces S_2.

Before using this potentiometer, the student should be sure that he thoroughly understands the potentiometer principles involved and the procedures for measuring an unknown emf. This knowledge may best be gained by a careful study of Fig. N.1 in conjunction with the preceding discussion, or by study of the Leeds and Northrup *Directions Manual*, which comes with the instrument.

Tables

Table A Natural Trigonometric Functions

sin

	.0	.1	.2	.3	.4	.5	.6	.7	.8	.9		
0°	.0000	.0017	.0035	.0052	.0070	.0087	.0105	.0122	.0140	.0157	.0175	89°
1°	.0175	.0192	.0209	.0227	.0244	.0262	.0279	.0297	.0314	.0332	.0349	88°
2°	.0349	.0366	.0384	.0401	.0419	.0436	.0454	.0471	.0488	.0506	.0523	87°
3°	.0523	.0541	.0558	.0576	.0593	.0610	.0628	.0645	.0663	.0680	.0698	86°
4°	.0698	.0715	.0732	.0750	.0767	.0785	.0802	.0819	.0837	.0854	.0872	85°
5°	.0872	.0889	.0906	.0924	.0941	.0958	.0976	.0993	.1011	.1028	.1045	84°
6°	.1045	.1063	.1080	.1097	.1115	.1132	.1149	.1167	.1184	.1201	.1219	83°
7°	.1219	.1236	.1253	.1271	.1288	.1305	.1323	.1340	.1357	.1374	.1392	82°
8°	.1392	.1409	.1426	.1444	.1461	.1478	.1495	.1513	.1530	.1547	.1564	81°
9°	.1564	.1582	.1599	.1616	.1633	.1650	.1668	.1685	.1702	.1719	.1736	80°
10°	.1736	.1754	.1771	.1788	.1805	.1822	.1840	.1857	.1874	.1891	.1908	79°
11°	.1908	.1925	.1942	.1959	.1977	.1994	.2011	.2028	.2045	.2062	.2079	78°
12°	.2079	.2096	.2113	.2130	.2147	.2164	.2181	.2198	.2215	.2233	.2250	77°
13°	.2250	.2267	.2284	.2300	.2317	.2334	.2351	.2368	.2385	.2402	.2419	76°
14°	.2419	.2436	.2453	.2470	.2487	.2504	.2521	.2538	.2554	.2571	.2588	75°
15°	.2588	.2605	.2622	.2639	.2656	.2672	.2689	.2706	.2723	.2740	.2756	74°
16°	.2756	.2773	.2790	.2807	.2823	.2840	.2857	.2874	.2890	.2907	.2924	73°
17°	.2924	.2940	.2957	.2974	.2990	.3007	.3024	.3040	.3057	.3074	.3090	72°
18°	.3090	.3107	.3123	.3140	.3156	.3173	.3190	.3206	.3223	.3239	.3256	71°
19°	.3256	.3272	.3289	.3305	.3322	.3338	.3355	.3371	.3387	.3404	.3420	70°
20°	.3420	.3437	.3453	.3469	.3486	.3502	.3518	.3535	.3551	.3567	.3584	69°
21°	.3584	.3600	.3616	.3633	.3649	.3665	.3681	.3697	.3714	.3730	.3746	68°
22°	.3746	.3762	.3778	.3795	.3811	.3827	.3843	.3859	.3875	.3891	.3907	67°
23°	.3907	.3923	.3939	.3955	.3971	.3987	.4003	.4019	.4035	.4051	.4067	66°
24°	.4067	.4083	.4099	.4115	.4131	.4147	.4163	.4179	.4195	.4210	.4226	65°
25°	.4226	.4242	.4258	.4274	.4289	.4305	.4321	.4337	.4352	.4368	.4384	64°
26°	.4384	.4399	.4415	.4431	.4446	.4462	.4478	.4493	.4509	.4524	.4540	63°
27°	.4540	.4555	.4571	.4586	.4602	.4617	.4633	.4648	.4664	.4679	.4695	62°
28°	.4695	.4710	.4726	.4741	.4756	.4772	.4787	.4802	.4818	.4833	.4848	61°
29°	.4848	.4863	.4879	.4894	.4909	.4924	.4939	.4955	.4970	.4985	.5000	60°
30°	.5000	.5015	.5030	.5045	.5060	.5075	.5090	.5105	.5120	.5135	.5150	59°
31°	.5150	.5165	.5180	.5195	.5210	.5225	.5240	.5255	.5270	.5284	.5299	58°
32°	.5299	.5314	.5329	.5344	.5358	.5373	.5388	.5402	.5417	.5432	.5446	57°
33°	.5446	.5461	.5476	.5490	.5505	.5519	.5534	.5548	.5563	.5577	.5592	56°
34°	.5592	.5606	.5621	.5635	.5650	.5664	.5678	.5693	.5707	.5721	.5736	55°
35°	.5736	.5750	.5764	.5779	.5793	.5807	.5821	.5835	.5850	.5864	.5878	54°
36°	.5878	.5892	.5906	.5920	.5934	.5948	.5962	.5976	.5990	.6004	.6018	53°
37°	.6018	.6032	.6046	.6060	.6074	.6088	.6101	.6115	.6129	.6143	.6157	52°
38°	.6157	.6170	.6184	.6198	.6211	.6225	.6239	.6252	.6266	.6280	.6293	51°
39°	.6293	.6307	.6320	.6334	.6347	.6361	.6374	.6388	.6401	.6414	.6428	50°
40°	.6428	.6441	.6455	.6468	.6481	.6494	.6508	.6521	.6534	.6547	.6561	49°
41°	.6561	.6574	.6587	.6600	.6613	.6626	.6639	.6652	.6665	.6678	.6691	48°
42°	.6691	.6704	.6717	.6730	.6743	.6756	.6769	.6782	.6794	.6807	.6820	47°
43°	.6820	.6833	.6845	.6858	.6871	.6884	.6896	.6909	.6921	.6934	.6947	46°
44°	.6947	.6959	.6972	.6984	.6997	.7009	.7022	.7034	.7046	.7059	.7071	45°
	.9	.8	.7	.6	.5	.4	.3	.2	.1	.0		

cos

	.0	.1	.2	.3	.4	.5	.6	.7	.8	.9		
45°	.7071	.7083	.7096	.7108	.7120	.7133	.7145	.7157	.7169	.7181	.7193	44°
46°	.7193	.7206	.7218	.7230	.7242	.7254	.7266	.7278	.7290	.7302	.7314	43°
47°	.7314	.7325	.7337	.7349	.7361	.7373	.7385	.7396	.7408	.7420	.7431	42°
48°	.7431	.7443	.7455	.7466	.7478	.7490	.7501	.7513	.7524	.7536	.7547	41°
49°	.7547	.7559	.7570	.7581	.7593	.7604	.7615	.7627	.7638	.7649	.7660	40°
50°	.7660	.7672	.7683	.7694	.7705	.7716	.7727	.7738	.7749	.7760	.7771	39°
51°	.7771	.7782	.7793	.7804	.7815	.7826	.7837	.7848	.7859	.7869	.7880	38°
52°	.7880	.7891	.7902	.7912	.7923	.7934	.7944	.7955	.7965	.7976	.7986	37°
53°	.7986	.7997	.8007	.8018	.8028	.8039	.8049	.8059	.8070	.8080	.8090	36°
54°	.8090	.8100	.8111	.8121	.8131	.8141	.8151	.8161	.8171	.8181	.8192	35°
55°	.8192	.8202	.8211	.8221	.8231	.8241	.8251	.8261	.8271	.8281	.8290	34°
56°	.8290	.8300	.8310	.8320	.8329	.8339	.8348	.8358	.8368	.8377	.8387	33°
57°	.8387	.8396	.8406	.8415	.8425	.8434	.8443	.8453	.8462	.8471	.8480	32°
58°	.8480	.8490	.8499	.8508	.8517	.8526	.8536	.8545	.8554	.8563	.8572	31°
59°	.8572	.8581	.8590	.8599	.8607	.8616	.8625	.8634	.8643	.8652	.8660	30°
60°	.8660	.8669	.8678	.8686	.8695	.8704	.8712	.8721	.8729	.8738	.8746	29°
61°	.8746	.8755	.8763	.8771	.8780	.8788	.8796	.8805	.8813	.8821	.8829	28°
62°	.8829	.8838	.8846	.8854	.8862	.8870	.8878	.8886	.8894	.8902	.8910	27°
63°	.8910	.8918	.8926	.8934	.8942	.8949	.8957	.8965	.8973	.8980	.8988	26°
64°	.8988	.8996	.9003	.9011	.9018	.9026	.9033	.9041	.9048	.9056	.9063	25°
65°	.9063	.9070	.9078	.9085	.9092	.9100	.9107	.9114	.9121	.9128	.9135	24°
66°	.9135	.9143	.9150	.9157	.9164	.9171	.9178	.9184	.9191	.9198	.9205	23°
67°	.9205	.9212	.9219	.9225	.9232	.9239	.9245	.9252	.9259	.9265	.9272	22°
68°	.9272	.9278	.9285	.9291	.9298	.9304	.9311	.9317	.9323	.9330	.9336	21°
69°	.9336	.9342	.9348	.9354	.9361	.9367	.9373	.9379	.9385	.9391	.9397	20°
70°	.9397	.9403	.9409	.9415	.9421	.9426	.9432	.9438	.9444	.9449	.9455	19°
71°	.9455	.9461	.9466	.9472	.9478	.9483	.9489	.9494	.9500	.9505	.9511	18°
72°	.9511	.9516	.9521	.9527	.9532	.9537	.9542	.9548	.9553	.9558	.9563	17°
73°	.9563	.9568	.9573	.9578	.9583	.9588	.9593	.9598	.9603	.9608	.9613	16°
74°	.9613	.9617	.9622	.9627	.9632	.9636	.9641	.9646	.9650	.9655	.9659	15°
75°	.9659	.9664	.9668	.9673	.9677	.9681	.9686	.9690	.9694	.9699	.9703	14°
76°	.9703	.9707	.9711	.9715	.9720	.9724	.9728	.9732	.9736	.9740	.9744	13°
77°	.9744	.9748	.9751	.9755	.9759	.9763	.9767	.9770	.9774	.9778	.9781	12°
78°	.9781	.9785	.9789	.9792	.9796	.9799	.9803	.9806	.9810	.9813	.9816	11°
79°	.9816	.9820	.9823	.9826	.9829	.9833	.9836	.9839	.9842	.9845	.9848	10°
80°	.9848	.9851	.9854	.9857	.9860	.9863	.9866	.9869	.9871	.9874	.9877	9°
81°	.9877	.9880	.9882	.9885	.9888	.9890	.9893	.9895	.9898	.9900	.9903	8°
82°	.9903	.9905	.9907	.9910	.9912	.9914	.9917	.9919	.9921	.9923	.9925	7°
83°	.9925	.9928	.9930	.9932	.9934	.9936	.9938	.9940	.9942	.9943	.9945	6°
84°	.9945	.9947	.9949	.9951	.9952	.9954	.9956	.9957	.9959	.9960	.9962	5°
85°	.9962	.9963	.9965	.9966	.9968	.9969	.9971	.9972	.9973	.9974	.9976	4°
86°	.9976	.9977	.9978	.9979	.9980	.9981	.9982	.9983	.9984	.9985	.9986	3°
87°	.9986	.9987	.9988	.9989	.9990	.9990	.9991	.9992	.9993	.9993	.9994	2°
88°	.9994	.9995	.9995	.9996	.9996	.9997	.9997	.9997	.9998	.9998	.9998	1°
89°	.9998	.9999	.9999	.9999	.9999	1.000	1.000	1.000	1.000	1.000	1.000	0°
		.9	.8	.7	.6	.5	.4	.3	.2	.1	.0	

cos

tan

	.0	.1	.2	.3	.4	.5	.6	.7	.8	.9	
0°	.0000	.0017	.0035	.0052	.0070	.0087	.0105	.0122	.0140	.0157	.0175	89°
1°	.0175	.0192	.0209	.0227	.0244	.0262	.0279	.0297	.0314	.0332	.0349	88°
2°	.0349	.0367	.0384	.0402	.0419	.0437	.0454	.0472	.0489	.0507	.0524	87°
3°	.0524	.0542	.0559	.0577	.0594	.0612	.0629	.0647	.0664	.0682	.0699	86°
4°	.0699	.0717	.0734	.0752	.0769	.0787	.0805	.0822	.0840	.0857	.0875	85°
5°	.0875	.0892	.0910	.0928	.0945	.0963	.0981	.0998	.1016	.1033	.1051	84°
6°	.1051	.1069	.1086	.1104	.1122	.1139	.1157	.1175	.1192	.1210	.1228	83°
7°	.1228	.1246	.1263	.1281	.1299	.1317	.1334	.1352	.1370	.1388	.1405	82°
8°	.1405	.1423	.1441	.1459	.1477	.1495	.1512	.1530	.1548	.1566	.1584	81°
9°	.1584	.1602	.1620	.1638	.1655	.1673	.1691	.1709	.1727	.1745	.1763	80°
10°	.1763	.1781	.1799	.1817	.1835	.1853	.1871	.1890	.1908	.1926	.1944	79°
11°	.1944	.1962	.1980	.1998	.2016	.2035	.2053	.2071	.2089	.2107	.2126	78°
12°	.2126	.2144	.2162	.2180	.2199	.2217	.2235	.2254	.2272	.2290	.2309	77°
13°	.2309	.2327	.2345	.2364	.2382	.2401	.2419	.2438	.2456	.2475	.2493	76°
14°	.2493	.2512	.2530	.2549	.2568	.2586	.2605	.2623	.2642	.2661	.2679	75°
15°	.2679	.2698	.2717	.2736	.2754	.2773	.2792	.2811	.2830	.2849	.2867	74°
16°	.2867	.2886	.2905	.2924	.2943	.2962	.2981	.3000	.3019	.3038	.3057	73°
17°	.3057	.3076	.3096	.3115	.3134	.3153	.3172	.3191	.3211	.3230	.3249	72°
18°	.3249	.3269	.3288	.3307	.3327	.3346	.3365	.3385	.3404	.3424	.3443	71°
19°	.3443	.3463	.3482	.3502	.3522	.3541	.3561	.3581	.3600	.3620	.3640	70°
20°	.3640	.3659	.3679	.3699	.3719	.3739	.3759	.3779	.3799	.3819	.3839	69°
21°	.3839	.3859	.3879	.3899	.3919	.3939	.3959	.3979	.4000	.4020	.4040	68°
22°	.4040	.4061	.4081	.4101	.4122	.4142	.4163	.4183	.4204	.4224	.4245	67°
23°	.4245	.4265	.4286	.4307	.4327	.4348	.4369	.4390	.4411	.4431	.4452	66°
24°	.4452	.4473	.4494	.4515	.4536	.4557	.4578	.4599	.4621	.4642	.4663	65°
25°	.4663	.4684	.4706	.4727	.4748	.4770	.4791	.4813	.4834	.4856	.4877	64°
26°	.4877	.4899	.4921	.4942	.4964	.4986	.5008	.5029	.5051	.5073	.5095	63°
27°	.5095	.5117	.5139	.5161	.5184	.5206	.5228	.5250	.5272	.5295	.5317	62°
28°	.5317	.5340	.5362	.5384	.5407	.5430	.5452	.5475	.5498	.5520	.5543	61°
29°	.5543	.5566	.5589	.5612	.5635	.5658	.5681	.5704	.5727	.5750	.5774	60°
30°	.5774	.5797	.5820	.5844	.5867	.5890	.5914	.5938	.5961	.5985	.6009	59°
31°	.6009	.6032	.6056	.6080	.6104	.6128	.6152	.6176	.6200	.6224	.6249	58°
32°	.6249	.6273	.6297	.6322	.6346	.6371	.6395	.6420	.6445	.6469	.6494	57°
33°	.6494	.6519	.6544	.6569	.6594	.6619	.6644	.6669	.6694	.6720	.6745	56°
34°	.6745	.6771	.6796	.6822	.6847	.6873	.6899	.6924	.6950	.6976	.7002	55°
35°	.7002	.7028	.7054	.7080	.7107	.7133	.7159	.7186	.7212	.7239	.7265	54°
36°	.7265	.7292	.7319	.7346	.7373	.7400	.7427	.7454	.7481	.7508	.7536	53°
37°	.7536	.7563	.7590	.7618	.7646	.7673	.7701	.7729	.7757	.7785	.7813	52°
38°	.7813	.7841	.7869	.7898	.7926	.7954	.7983	.8012	.8040	.8069	.8098	51°
39°	.8098	.8127	.8156	.8185	.8214	.8243	.8273	.8302	.8332	.8361	.8391	50°
40°	.8391	.8421	.8451	.8481	.8511	.8541	.8571	.8601	.8632	.8662	.8693	49°
41°	.8693	.8724	.8754	.8785	.8816	.8847	.8878	.8910	.8941	.8972	.9004	48°
42°	.9004	.9036	.9067	.9099	.9131	.9163	.9195	.9228	.9260	.9293	.9325	47°
43°	.9325	.9358	.9391	.9424	.9457	.9490	.9523	.9556	.9590	.9623	.9657	46°
44°	.9657	.9691	.9725	.9759	.9793	.9827	.9861	.9896	.9930	.9965	1.000	45°
	.9	.8	.7	.6	.5	.4	.3	.2	.1	.0	

cot

	.0	.1	.2	.3	.4	.5	.6	.7	.8	.9		
45°	1.000	1.003	1.007	1.011	1.014	1.018	1.021	1.025	1.028	1.032	1.036	44°
46°	1.036	1.039	1.043	1.046	1.050	1.054	1.057	1.061	1.065	1.069	1.072	43°
47°	1.072	1.076	1.080	1.084	1.087	1.091	1.095	1.099	1.103	1.107	1.111	42°
48°	1.111	1.115	1.118	1.122	1.126	1.130	1.134	1.138	1.142	1.146	1.150	41°
49°	1.150	1.154	1.159	1.163	1.167	1.171	1.175	1.179	1.183	1.188	1.192	40°
50°	1.192	1.196	1.200	1.205	1.209	1.213	1.217	1.222	1.226	1.230	1.235	39°
51°	1.235	1.239	1.244	1.248	1.253	1.257	1.262	1.266	1.271	1.275	1.280	38°
52°	1.280	1.285	1.289	1.294	1.299	1.303	1.308	1.313	1.317	1.322	1.327	37°
53°	1.327	1.332	1.337	1.342	1.347	1.351	1.356	1.361	1.366	1.371	1.376	36°
54°	1.376	1.381	1.387	1.392	1.397	1.402	1.407	1.412	1.418	1.423	1.428	35°
55°	1.428	1.433	1.439	1.444	1.450	1.455	1.460	1.466	1.471	1.477	1.483	34°
56°	1.483	1.488	1.494	1.499	1.505	1.511	1.517	1.522	1.528	1.534	1.540	33°
57°	1.540	1.546	1.552	1.558	1.564	1.570	1.576	1.582	1.588	1.594	1.600	32°
58°	1.600	1.607	1.613	1.619	1.625	1.632	1.638	1.645	1.651	1.658	1.664	31°
59°	1.664	1.671	1.678	1.684	1.691	1.698	1.704	1.711	1.718	1.725	1.732	30°
60°	1.732	1.739	1.746	1.753	1.760	1.767	1.775	1.782	1.789	1.797	1.804	29°
61°	1.804	1.811	1.819	1.827	1.834	1.842	1.849	1.857	1.865	1.873	1.881	28°
62°	1.881	1.889	1.897	1.905	1.913	1.921	1.929	1.937	1.946	1.954	1.963	27°
63°	1.963	1.971	1.980	1.988	1.997	2.006	2.014	2.023	2.032	2.041	2.050	26°
64°	2.050	2.059	2.069	2.078	2.087	2.097	2.106	2.116	2.125	2.135	2.145	25°
65°	2.145	2.154	2.164	2.174	2.184	2.194	2.204	2.215	2.225	2.236	2.246	24°
66°	2.246	2.257	2.267	2.278	2.289	2.300	2.311	2.322	2.333	2.344	2.356	23°
67°	2.356	2.367	2.379	2.391	2.402	2.414	2.426	2.438	2.450	2.463	2.475	22°
68°	2.475	2.488	2.500	2.513	2.526	2.539	2.552	2.565	2.578	2.592	2.605	21°
69°	2.605	2.619	2.633	2.646	2.660	2.675	2.689	2.703	2.718	2.733	2.747	20°
70°	2.747	2.762	2.778	2.793	2.808	2.824	2.840	2.856	2.872	2.888	2.904	19°
71°	2.904	2.921	2.937	2.954	2.971	2.989	3.006	3.024	3.042	3.060	3.078	18°
72°	3.078	3.096	3.115	3.133	3.152	3.172	3.191	3.211	3.230	3.251	3.271	17°
73°	3.271	3.291	3.312	3.333	3.354	3.376	3.398	3.420	3.442	3.465	3.487	16°
74°	3.487	3.511	3.534	3.558	3.582	3.606	3.630	3.655	3.681	3.706	3.732	15°
75°	3.732	3.758	3.785	3.812	3.839	3.867	3.895	3.923	3.952	3.981	4.011	14°
76°	4.011	4.041	4.071	4.102	4.134	4.165	4.198	4.230	4.264	4.297	4.331	13°
77°	4.331	4.366	4.402	4.437	4.474	4.511	4.548	4.586	4.625	4.665	4.705	12°
78°	4.705	4.745	4.787	4.829	4.872	4.915	4.959	5.005	5.050	5.097	5.145	11°
79°	5.145	5.193	5.242	5.292	5.343	5.396	5.449	5.503	5.558	5.614	5.671	10°
80°	5.671	5.730	5.789	5.850	5.912	5.976	6.041	6.107	6.174	6.243	6.314	9°
81°	6.314	6.386	6.460	6.535	6.612	6.691	6.772	6.855	6.940	7.026	7.115	8°
82°	7.115	7.207	7.300	7.396	7.495	7.596	7.700	7.806	7.916	8.028	8.144	7°
83°	8.144	8.264	8.386	8.513	8.643	8.777	8.915	9.058	9.205	9.357	9.514	6°
84°	9.514	9.677	9.845	10.02	10.20	10.39	10.58	10.78	10.99	11.20	11.43	5°
85°	11.43	11.66	11.91	12.16	12.43	12.71	13.00	13.30	13.62	13.95	14.30	4°
86°	14.30	14.67	15.06	15.46	15.89	16.35	16.83	17.34	17.89	18.46	19.08	3°
87°	19.08	19.74	20.45	21.20	22.02	22.90	23.86	24.90	26.03	27.27	28.64	2°
88°	28.64	30.14	31.82	33.69	35.80	38.19	40.92	44.07	47.74	52.08	57.29	1°
89°	57.29	63.66	71.62	81.85	95.49	114.6	143.2	191.0	286.5	573.0	∞	0°
	.9	.8	.7	.6	.5	.4	.3	.2	.1	.0		

Table B   Table of Logarithms to Base 10

Note: $\log_e N = \log_e 10 \log_{10} N = 2.3026 \log_{10} N$, $\log_{10} e^x = x \log_{10} e = 0.43429x$

N	0	1	2	3	4	5	6	7	8	9	P.P. 1	2	3	4	5
10	0000	0043	0086	0128	0170	0212	0253	0294	0334	0374	4	8	12	17	21
11	0414	0453	0492	0531	0569	0607	0645	0682	0719	0755	4	8	11	15	19
12	0792	0828	0864	0899	0934	0969	1004	1038	1072	1106	3	7	10	14	17
13	1139	1173	1206	1239	1271	1303	1335	1367	1399	1430	3	6	10	13	16
14	1461	1492	1523	1553	1584	1614	1644	1673	1703	1732	3	6	9	12	15
15	1761	1790	1818	1847	1875	1903	1931	1959	1987	2014	3	6	8	11	14
16	2041	2068	2095	2122	2148	2175	2201	2227	2253	2279	3	5	8	11	13
17	2304	2330	2355	2380	2405	2430	2455	2480	2504	2529	2	5	7	10	12
18	2553	2577	2601	2625	2648	2672	2695	2718	2742	2765	2	5	7	9	12
19	2788	2810	2833	2856	2878	2900	2923	2945	2967	2989	2	4	7	9	11
20	3010	3032	3054	3075	3096	3118	3139	3160	3181	3201	2	4	6	8	11
21	3222	3243	3263	3284	3304	3324	3345	3365	3385	3404	2	4	6	8	10
22	3424	3444	3464	3483	3502	3522	3541	3560	3579	3598	2	4	6	8	10
23	3617	3636	3655	3674	3692	3711	3729	3747	3766	3784	2	4	5	7	9
24	3802	3820	3838	3856	3874	3892	3909	3927	3945	3962	2	4	5	7	9
25	3979	3997	4014	4031	4048	4065	4082	4099	4116	4133	2	3	5	7	9
26	4150	4166	4183	4200	4216	4232	4249	4265	4281	4298	2	3	5	7	8
27	4314	4330	4346	4362	4378	4393	4409	4425	4440	4456	2	3	5	6	8
28	4472	4487	4502	4518	4533	4548	4564	4579	4594	4609	2	3	5	6	8
29	4624	4639	4654	4669	4683	4698	4713	4728	4742	4757	1	3	4	6	7
30	4771	4786	4800	4814	4829	4843	4857	4871	4886	4900	1	3	4	6	7
31	4914	4928	4942	4955	4969	4983	4997	5011	5024	5038	1	3	4	6	7
32	5051	5065	5079	5092	5105	5119	5132	5145	5159	5172	1	3	4	5	7
33	5185	5198	5211	5224	5237	5250	5263	5276	5289	5302	1	3	4	5	6
34	5315	5328	5340	5353	5366	5378	5391	5403	5416	5428	1	3	4	5	6
35	5441	5453	5465	5478	5490	5502	5514	5527	5539	5551	1	2	4	5	6
36	5563	5575	5587	5599	5611	5623	5635	5647	5658	5670	1	2	4	5	6
37	5682	5694	5705	5717	5729	5740	5752	5763	5775	5786	1	2	3	5	6
38	5798	5809	5821	5832	5843	5855	5866	5877	5888	5899	1	2	3	5	6
39	5911	5922	5933	5944	5955	5966	5977	5988	5999	6010	1	2	3	4	6
40	6021	6031	6042	6053	6064	6075	6085	6096	6107	6117	1	2	3	4	5
41	6128	6138	6149	6160	6170	6180	6191	6201	6212	6222	1	2	3	4	5
42	6232	6243	6253	6263	6274	6284	6294	6304	6314	6325	1	2	3	4	5
43	6335	6345	6355	6365	6375	6385	6395	6405	6415	6425	1	2	3	4	5
44	6435	6444	6454	6464	6474	6484	6493	6503	6513	6522	1	2	3	4	5
45	6532	6542	6551	6561	6571	6580	6590	6599	6609	6618	1	2	3	4	5
46	6628	6637	6646	6656	6665	6675	6684	6693	6702	6712	1	2	3	4	5
47	6721	6730	6739	6749	6758	6767	6776	6785	6794	6803	1	2	3	4	5
48	6812	6821	6830	6839	6848	6857	6866	6875	6884	6893	1	2	3	4	4
49	6902	6911	6920	6928	6937	6946	6955	6964	6972	6981	1	2	3	4	4
50	6990	6998	7007	7016	7024	7033	7042	7050	7059	7067	1	2	3	3	4
51	7076	7084	7093	7101	7110	7118	7126	7135	7143	7152	1	2	3	3	4
52	7160	7168	7177	7185	7193	7202	7210	7218	7226	7235	1	2	2	3	4
53	7243	7251	7259	7267	7275	7284	7292	7300	7308	7316	1	2	2	3	4
54	7324	7332	7340	7348	7356	7364	7372	7380	7388	7396	1	2	2	3	4

N	0	1	2	3	4	5	6	7	8	9	P.P. 1	2	3	4	5
55	7404	7412	7419	7427	7435	7443	7451	7459	7466	7474	1	2	2	3	4
56	7482	7490	7497	7505	7513	7520	7528	7536	7543	7551	1	2	2	3	4
57	7559	7566	7574	7582	7589	7597	7604	7612	7619	7627	1	2	2	3	4
58	7634	7642	7649	7657	7664	7672	7679	7686	7694	7701	1	1	2	3	4
59	7709	7716	7723	7731	7738	7745	7752	7760	7767	7774	1	1	2	3	4
60	7782	7789	7796	7803	7810	7818	7825	7832	7839	7846	1	1	2	3	4
61	7853	7860	7868	7875	7882	7889	7896	7903	7910	7917	1	1	2	3	4
62	7924	7931	7938	7945	7952	7959	7966	7973	7980	7987	1	1	2	3	3
63	7993	8000	8007	8014	8021	8028	8035	8041	8048	8055	1	1	2	3	3
64	8062	8069	8075	8082	8089	8096	8102	8109	8116	8122	1	1	2	3	3
65	8129	8136	8142	8149	8156	8162	8169	8176	8182	8189	1	1	2	3	3
66	8195	8202	8209	8215	8222	8228	8235	8241	8248	8254	1	1	2	3	3
67	8261	8267	8274	8280	8287	8293	8299	8306	8312	8319	1	1	2	3	3
68	8325	8331	8338	8344	8351	8357	8363	8370	8376	8382	1	1	2	3	3
69	8388	8395	8401	8407	8414	8420	8426	8432	8439	8445	1	1	2	3	3
70	8451	8457	8463	8470	8476	8482	8488	8494	8500	8506	1	1	2	2	3
71	8513	8519	8525	8531	8537	8543	8549	8555	8561	8567	1	1	2	2	3
72	8573	8579	8585	8591	8597	8603	8609	8615	8621	8627	1	1	2	2	3
73	8633	8639	8645	8651	8657	8663	8669	8675	8681	8686	1	1	2	2	3
74	8692	8698	8704	8710	8716	8722	8727	8733	8739	8745	1	1	2	2	3
75	8751	8756	8762	8768	8774	8779	8785	8791	8797	8802	1	1	2	2	3
76	8808	8814	8820	8825	8831	8837	8842	8848	8854	8859	1	1	2	2	3
77	8865	8871	8876	8882	8887	8893	8899	8904	8910	8915	1	1	2	2	3
78	8921	8927	8932	8938	8943	8949	8954	8960	8965	8971	1	1	2	2	3
79	8976	8982	8987	8993	8998	9004	9009	9015	9020	9025	1	1	2	2	3
80	9031	9036	9042	9047	9053	9058	9063	9069	9074	9079	1	1	2	2	3
81	9085	9090	9096	9101	9106	9112	9117	9122	9128	9133	1	1	2	2	3
82	9138	9143	9149	9154	9159	9165	9170	9175	9180	9186	1	1	2	2	3
83	9191	9196	9201	9206	9212	9217	9222	9227	9232	9238	1	1	2	2	3
84	9243	9248	9253	9258	9263	9269	9274	9279	9284	9289	1	1	2	2	3
85	9294	9299	9304	9309	9315	9320	9325	9330	9335	9340	1	1	2	2	3
86	9345	9350	9355	9360	9365	9370	9375	9380	9385	9390	1	1	2	2	3
87	9395	9400	9405	9410	9415	9420	9425	9430	9435	9440	0	1	1	2	2
88	9445	9450	9455	9460	9465	9469	9474	9479	9484	9489	0	1	1	2	2
89	9494	9499	9504	9509	9513	9518	9523	9528	9533	9538	0	1	1	2	2
90	9542	9547	9552	9557	9562	9566	9571	9576	9581	9586	0	1	1	2	2
91	9590	9595	9600	9605	9609	9614	9619	9624	9628	9633	0	1	1	2	2
92	9638	9643	9647	9652	9657	9661	9666	9671	9675	9680	0	1	1	2	2
93	9685	9689	9694	9699	9703	9708	9713	9717	9722	9727	0	1	1	2	2
94	9731	9736	9741	9745	9750	9754	9759	9763	9768	9773	0	1	1	2	2
95	9777	9782	9786	9791	9795	9800	9805	9809	9814	9818	0	1	1	2	2
96	9823	9827	9832	9836	9841	9845	9850	9854	9859	9863	0	1	1	2	2
97	9868	9872	9877	9881	9886	9890	9894	9899	9903	9908	0	1	1	2	2
98	9912	9917	9921	9926	9930	9934	9939	9943	9948	9952	0	1	1	2	2
99	9956	9961	9965	9969	9974	9978	9983	9987	9991	9996	0	1	1	2	2

Table C Densities

(Grams per cubic centimeter)

Gases (0°C, 76 cm Hg)

Air (dry)	0.001293	Hydrogen	0.00008988
Carbon dioxide	0.001965	Oxygen	0.001429
Helium	0.0001784		

Liquids (20°C)

Alcohol, ethyl	0.789	Mercury	13.546
Carbon tetrachloride	1.60	Water	0.998
Ether, ethyl	0.715		

Solids

Aluminum	2.70	Silver	10.5
Brass (70% Cu; 30% Zn)	8.44	Tungsten	18.8
Copper	8.87	Wood:	
Glass (common)	2.4 to 2.6	maple	0.6 to 0.9
Gold	19.3	oak	0.6 to 0.9
Iron	7.87	pine	0.4 to 0.7
Platinum	21.5	cork	0.2 to 0.3

Table D Saturated Aqueous Vapor Pressure

Temperature, °C	Vapor pressure, mm Hg	Temperature, °C	Vapor pressure, mm Hg
−10	2.0	16	13.6
− 9	2.1	17	14.5
− 8	2.3	18	15.5
− 7	2.6	19	16.5
− 6	2.8	20	17.6
− 5	3.0	21	18.7
− 4	3.3	22	19.8
− 3	3.6	23	21.1
− 2	3.9	24	22.4
− 1	4.2	25	23.8
0	4.6	26	25.2
1	4.9	27	26.8
2	5.3	28	28.4
3	5.7	29	30.1
4	6.1	30	31.9
5	6.5	31	33.7
6	7.0	32	35.7
7	7.5	33	37.8
8	8.0	34	40.0
9	8.6	35	42.2
10	9.2	36	44.6
11	9.8	37	47.1
12	10.5	38	49.8
13	11.2	39	52.5
14	12.0	40	55.4
15	12.8		

Table D (Continued)

Temperature, C°	Vapor pressure, mm Hg
97.0	682.0
97.2	687.0
97.4	692.0
97.6	697.1
97.8	702.2
98.0	707.3
98.2	712.4
98.4	717.5
98.6	722.7
98.8	728.0
99.0	733.3
99.2	738.5
99.4	743.8
99.6	749.1
99.8	754.5
100.0	760.0
100.2	765.5
100.4	771.0
100.6	776.5
100.8	782.0

Table E   Coefficients of Linear Expansion; Specific Heats

Substance	Coefficient	Specific heat
Aluminum	$22.2 \times 10^{-6}/°C$	0.210
Brass (70% Cu; 30% Zn).........	18.8	0.089
Copper	16.2	0.092
Iron, steel	11.7	0.104
Glass	8.0	0.19
Lead	29.4	0.031
Pyrex	3.3	0.20

Table F   Barometer Correction

(Brass scale correct at 0°C. The correction is to be subtracted from the observed height.)

Temperature °C	Correction in mm Hg if observed height is		
	720 mm	740 mm	760 mm
15	1.8	1.8	1.9
16	1.9	1.9	2.0
17	2.0	2.0	2.1
18	2.1	2.2	2.2
19	2.2	2.3	2.3
20	2.3	2.4	2.5
21	2.5	2.5	2.6
22	2.6	2.6	2.7
23	2.7	2.8	2.8
24	2.8	2.9	3.0
25	2.9	3.0	3.1
26	3.0	3.1	3.2
27	3.2	3.2	3.3
28	3.3	3.4	3.5
29	3.4	3.5	3.6
30	3.5	3.6	3.7
31	3.6	3.7	3.8
32	3.7	3.8	4.0
33	3.9	4.0	4.1
34	4.0	4.1	4.2
35	4.1	4.2	4.3

Table G   Indexes of Refraction

Substance	Sodium D line (5890 A)	Mercury green line (5461 A)
Water (20°C)	1.3330	1.3345
Glass, crown	1.5170	1.5191
Glass, flint	1.6499	1.6546
Alcohol, amyl	1.41	
Alcohol, ethyl	1.367	
Carbon tetrachloride	1.464	
Lucite	1.51	
Turpentine	1.47	

Table H   Principal Spectrum Lines of Certain Elements

Wavelength shown in angstrom units. Letters indicate following colors: red, orange, yellow, green, blue, violet.

Mercury arc: 4047 v
4078 v
4358 v
4916 bg
5461 g
5770 y
5791 y
6152 r
6234 r

Lithium flame: 6104 o
6708 r

Potassium flame: 4044 v
4047 v
5802 y
7668 r
7702 r

Sodium flame: 5890 y (D_2 line)
5896 y

Table J   Coefficients of Kinetic Friction

Surfaces	Coefficient
Metals on hardwood, dry	0.5 to 0.6
Metals on hardwood, wet	0.24 to 0.26
Metals on metals, dry	0.15 to 0.20
Metals on metals, wet	0.3
Wood on wood, dry	0.25 to 0.5
Steel on agate, dry	0.20
Steel on agate, oiled	0.107
Smooth surfaces, oiled	0.05 to 0.08
Rolling friction	0.004 to 0.006

Table K₁ Thermal Electromotive Force for Copper-Constantan Thermocouple, Standard

(One junction at 0°C)

Junc temp, °C	Emf, mv	Diff, mv	Junc temp, °C	Emf, mv	Diff, mv
−200	5.54		0	0.00	
−190	5.38	0.16	10	0.39	0.39
−180	5.20	0.18	20	0.79	0.40
−170	5.02	0.18	30	1.19	0.40
−160	4.82	0.20	40	1.61	0.42
−150	4.60	0.22	50	2.03	0.42
−140	4.38	0.22	60	2.47	0.44
−130	4.14	0.24	70	2.91	0.44
−120	3.89	0.25	80	3.36	0.45
−110	3.62	0.27	90	3.81	0.45
−100	3.35	0.27	100	4.28	0.47
− 90	3.06	0.29	110	4.75	0.47
− 80	2.77	0.29	120	5.23	0.48
− 70	2.46	0.31	130	5.71	0.48
− 60	2.14	0.32	140	6.20	0.49
− 50	1.81	0.33	150	6.70	0.50
− 40	1.47	0.34	160	7.21	0.51
− 30	1.11	0.36	170	7.72	0.51
− 20	0.75	0.36	180	8.23	0.51
− 10	0.38	0.37	190	8.76	0.53
− 0	0.00	0.38	200	9.29	0.53
			210	9.82	0.53
			220	10.36	0.54
			230	10.91	0.55
			240	11.46	0.55
			250	12.01	0.55
			260	12.57	0.56
			270	13.14	0.57
			280	13.71	0.57
			290	14.28	0.57
			300	14.86	0.58

Table K$_2$ Thermal Electromotive Force for Iron-Constantan Thermocouple, Standard

(One junction at 0°C)

Junc temp, °C	Emf, mv	Diff, mv	Junc temp, °C	Emf, mv	Diff, mv
0	.00		200	10.99	
10	.52	.52	210	11.55	.56
20	1.05	.53	220	12.11	.56
30	1.58	.53	230	12.67	.56
40	2.12	.54	240	13.23	.56
50	2.66	.54	250	13.79	.56
60	3.20	.54	260	14.35	.56
70	3.75	.55	270	14.90	.55
80	4.30	.55	280	15.45	.55
90	4.85	.55	290	16.00	.55
100	5.40	.55	300	16.55	.55
110	5.95	.55	310	17.11	.56
120	6.51	.56	320	17.66	.55
130	7.07	.56	330	18.21	.55
140	7.63	.56	340	18.76	.55
150	8.19	.56	350	19.31	.55
160	8.75	.56	360	19.86	.55
170	9.31	.56	370	20.41	.55
180	9.87	.56	380	20.96	.55
190	10.43	.56	390	21.51	.55
200	10.99	.56	400	22.06	.55

Table L Miscellaneous Constants and Numbers

Name	Symbol	Value
Alpha particle, rest mass	m_α	6.598×10^{-24} gm
Angstrom unit	A	10^{-8} cm
1 atomic mass unit	u	1.66×10^{-24} gm
		931 Mev
Avogadro's number	N_0	6.023×10^{23}/mol
Boltzman's constant	k	1.3805×10^{-16} erg/deg
1 millicurie	mc	3.71×10^7 dis/sec
e (base of natural logs)	e	2.71828
$\log_{10} e$	..	0.43429
Earth's magnetic field, horizontal component (Minneapolis)	H	0.17 oersted
Earth's magnetic field, vertical component (Minneapolis)	V	0.59 oersted
Electron, charge	e	1.6008×10^{-19} coulomb
Electron, rest mass	m_0	9.1066×10^{-24} gm
Electron, charge/mass ratio	e/m	1.759×10^7 abs emu/gm
Electron, rest energy	$M_0 C^2$	.5107 Mev
Electron volt; one ev is equivalent to	ev	1.59×10^{-12} erg
		1.16×10^4 deg abs
		1.07×10^{-9} mass unit
		1.77×10^{-33} gm
Faraday	F	96,489 coulombs/equivalent
Gas constant	R	8.314×10^7 erg/mol deg
Gravitational constant	G	6.67×10^{-11} nt-m²/kg²
Horsepower	hp	550 ft-lb/sec
Inch	in.	2.540 cm
Latent heat of fusion of H_2O	L_f	79.6 cal/gm
Latent heat of vaporization of H_2O	L_v	539 cal/gm
Mechanical equivalent of heat	J	4.185 joules/cal
Pi	π	3.14159
$\log_{10} \pi$	..	0.49715
Planck's constant	h	6.624×10^{-27} erg sec
Pound	lb	453.59 gm
Pressure coefficient, perfect gas	α_p	0.00367/°C
Specific heat ratio, c_p/c_v (air)	γ	1.402
Standard conditions	P_0, T_0	760 mm Hg, 0°C
Surface tension of H_2O (at 20°C)	T	73 dynes/cm
Triple point of water	T_{tr}	273.16°K
Velocity of light (vacuum)	c	2.9989×10^{10} cm/sec
Velocity of sound (dry air at 0°C)	v_0	3.3136×10^4 cm/sec

Table M   Elastic Constants

(dynes per square centimeter)

Substance	Elasticity (Young's modulus)	Rigidity (shear modulus)	Volume elasticity (bulk modulus)
Aluminum	6.8 to 7.1 $\times 10^{11}$	2.4 to 2.6 $\times 10^{11}$	7.5 $\times 10^{11}$
Brass	9 to 11	3.5 to 4.1	11
Copper	11 to 13	4.1 to 4.7	13
Glass	4 to 6	1.6 to 2.4	4
Mercury	....	....	0.26
Steel	20.0 to 20.5	7.3 to 8.3	16
Water	....	....	0.20

Table N   Work Functions

(Adapted from Smithsonian tables.)

Work Function (Electron Volts)

Element	Thermionic	Photoelectric
Cesium	1.81	1.91
Copper	4.38	4.46
Iron	4.48	4.63
Nickel	4.61	
Platinum	5.32	
Tantalum	4.19	4.05
Tungsten	4.52	4.3–4.5
Zirconium	4.13	
Carbon	4.34	4.82
Barium	2.10	2.48–2.51

Table P   Acceleration due to Gravity

Latitude	g (sea level)	Altitude	Subtract
20°	978.64 cm/sec²	500 ft	0.05 cm/sec²
25	978.96	1000	0.09
30	979.33	1500	0.14
35	979.74	2000	0.18
40	980.17	2500	0.23
45	980.62	3000	0.23
50	981.07	3500	0.32
55	981.51	4000	0.37
		4500	0.41
		5000	0.46

The acceleration due to gravity at various latitudes and altitudes above sea level may be obtained by interpolation from the table below, or by use of the following approximate formula:

$$g = 978.04 + 5.17 \sin^2 \lambda - 0.000092A, \text{ (cm/sec}^2)$$

where λ is the latitude in degrees and A is the altitude above sea level in feet.

Table Q   Transmittance Curves: Wratten Filters

Wave length	Percent transmittance			
	No. 22	No. 24	No. 25	No. 29
400 mμ	—	—	—	—
10	—	—	—	—
20	—	—	—	—
30	—	—	—	—
40	—	—	—	—
50	—	—	—	—
60	—	—	—	—
70	—	—	—	—
80	—	—	—	—
90	—	—	—	—
500	—	—	—	—
10	—	—	—	—
20	—	—	—	—
30	—	—	—	—
40	—	—	—	—
50	0.25	—	—	—
60	19.0	—	—	—
70	60.0	—	—	—
80	81.0	4.55	—	—
90	87.0	37.3	12.6	—
600	88.5	72.3	50.0	—
10	89.0	82.9	75.0	10.5
20	89.5	86.4	82.6	45.0
30	89.8	87.8	85.5	73.5
40	90.0	88.5	86.7	84.2
50	90.1	89.0	87.6	87.8
60	90.2	89.3	88.2	89.2
70	90.3	89.7	88.5	89.8
80	90.4	89.9	89.0	90.3
90	90.5	90.2	89.3	90.4

Stability: BAC

Stability: AAA

Stability: AAB

Stability: AAA

Table R Temperature of a Tungsten Filament*

The absolute temperature of a long uniform tungsten filament
is related to the current in the filament, i_f, and to the diameter
of the filament, d, as shown in the following table (end effects
neglected).

$T°K$	$\frac{i_f}{d^{3/2}}$ amp/cm$^{3/2}$	Differences
500	47.6	
600	75.2	27.6
700	108	33.2
800	148	40
900	193	45
1000	244	51
1100	301	57
1200	363	62
1300	431	68
1400	504	73
1500	581	77
1600	662	81
1700	747	85
1800	836	89
1900	927	91
2000	1022	95
2100	1119	97
2200	1217	98
2300	1319	102
2400	1422	103
2500	1526	104
2600	1632	106
2700	1741	109
2800	1849	108
2900	1961	112
3000	2072	111
3100	2187	115
3200	2301	114
3300	2418	117
3400	2537	119
3500	2657	120
3600	2777	120

*Adapted from *The Characteristics of Tungsten Filaments
as Functions of Temperature*, by Howard A. Jones and Irving
Langmuir. *General Electric Review* **30**, pp. 310–319. June, 1927.

Table S Normal Distribution Functions

$$\phi(z) = \frac{1}{\sqrt{2\pi}} e^{-z^2/2}, \qquad z = \frac{X - \mu}{\sigma^2} \cong \frac{X - \bar{X}}{s^2}$$

$$\psi(z) = \int_{-\infty}^{z} \phi(z')\, dz'$$

z	$\phi(z)$	$\psi(z_+)$	$\psi(z_-)$	$\psi(z_+) - \psi(z_-)$
0.0	0.399	0.500	0.500	0.000
0.5	0.352	0.691	0.308	0.383
1.0	0.242	0.841	0.158	0.683
1.5	0.130	0.933	0.067	0.866
2.0	0.054	0.977	0.023	0.954
2.5	0.018	0.994	0.006	0.988
3.0	0.004	0.999	0.002	0.997
∞	0.000	1.000	0.000	1.000

Index

Index